# Lecture Notes in Geoinformation and Cartography

Publications of the International Cartographic Association (ICA)

More information about this series at http://www.springer.com/series/10036

Mirela Altić • Imre Josef Demhardt •
Soetkin Vervust
Editors

# Dissemination of Cartographic Knowledge

## 6th International Symposium of the ICA Commission on the History of Cartography, 2016

Springer

*Editors*
Mirela Altić
Institute of Social Sciences
Zagreb, Croatia

Imre Josef Demhardt
Department of History
University of Texas at Arlington
Arlington, Texas
USA

Soetkin Vervust
Department of Geography
Ghent University
Ghent, Belgium

ISSN 1863-2246 ISSN 1863-2351 (electronic)
Lecture Notes in Geoinformation and Cartography
ISSN 2195-1705 ISSN 2195-1713 (electronic)
Publications of the International Cartographic Association (ICA)
ISBN 978-3-319-61514-1 ISBN 978-3-319-61515-8 (eBook)
DOI 10.1007/978-3-319-61515-8

Library of Congress Control Number: 2017949529



Printed on acid-free paper

This Springer imprint is published by Springer Nature
The registered company is Springer International Publishing AG
The registered company address is: Gewerbestrasse 11, 6330 Cham, Switzerland

# Preface

This volume comprises most of the research papers presented at the 6th International Symposium of the ICA Commission on the History of Cartography which took place in Dubrovnik, Croatia, on 13–15 October 2016. As such, it is already the fourth volume in a series of proceedings which has been made possible through the partnership between the International Cartographic Association (ICA) and Springer International Publishing.

The history of cartography covers a vast field of knowledge and includes all maps and map-like graphics made by humankind since prehistoric times. Map compilation and map use today are, however, seldom dependent on maps which were produced before early modern times. With this in mind, the ICA Commission decided to concentrate on the history of cartography since the Enlightenment—without neglecting research on earlier periods relevant to the understanding of modern cartographic developments.

The 6th International Symposium of the ICA Commission on the History of Cartography had as its general theme "The Dissemination of Cartographic Knowledge: Production—Trade—Consumption—Preservation." Under the leadership of the ICA Commission on the History of Cartography, the Symposium was jointly organized with the ICA Commission on Map Production and Geoinformation Management and the ICA Commission on Use, User, and Usability Issues, and with the collaboration of the Institute of Social Sciences "Ivo Pilar" (Zagreb, Croatia). The various paper and poster sessions were held at the Centre for Advanced Academic Studies of the University of Zagreb which is situated in the heart of the historic city of Dubrovnik.

The papers published in this volume discuss the historical dimension of the dissemination of cartographic knowledge and the effectiveness thereof in diverse cartographic cultures and their related user groups around the globe. This includes the technological and conceptual aspects of cartographic production (maps, charts, globes, atlases, educational tools, etc.), the usability of these techniques and the resulting products, as well as the conditions of the map trade as a changing network of private enterprises and official institutions, and the role of diverse audiences in the creation, circulation, consumption, and ultimate preservation of knowledge. Although the Symposium was primarily focused on the eighteenth to twentieth centuries, it was also open to contributions on the sharing of cartographic knowledge during earlier centuries, as they often formed the background for later important developments.

We would like to acknowledge our gratitude to the University of Zagreb and especially to its Centre for Advanced Academic Studies in Dubrovnik and their Head of Office, Mrs. Vlasta Brunsko, for her guidance and logistic support.

Zagreb, Croatia — Mirela Altić
Arlington, TX, USA — Imre Josef Demhardt
Ghent, Belgium — Soetkin Vervust

# Contents

# Editors and Contributors

## About the Editors

**Mirela Altić** is a chief research fellow at the Institute of Social Sciences in Zagreb, Croatia. At the Department of History, University of Zagreb, Dr. Altic holds the rank of full professor and lectures on the history of cartography and historical geography. Besides her specialization in South Eastern and Central European map history, over the last few years she has also published extensively on the Jesuit cartography of the Americas and conducted research in European and American archives and libraries. She is the author of twelve books and numerous scholarly papers and a contributor to *The History of Cartography* encyclopedia project. Since 2015, she serves as Vice-Chair of the ICA Commission on the History of Cartography.

**Imre Josef Demhardt** is interested in post-enlightenment cartography, colonialism, and regional studies with a focus on Central Europe, Sub-Saharan Africa, and North America. Besides numerous articles and several books on these subjects, he is involved as coeditor of Vol. 5 (Nineteenth Century) in the encyclopedia project on the *History of Cartography*. He holds the Garrett Chair in the History of Cartography at the University of Texas at Arlington and currently serves as Chair of the ICA Commission on the History of Cartography.

**Soetkin Vervust** is a postdoctoral research fellow at the Free University of Brussels (Belgium) and Newcastle University (UK). Her research interests lie in eighteenth- and nineteenth-century military cartography, the use of digital techniques for the study of old maps, and their applicability to historical geography and landscape archaeology. Since 2015, she serves as Executive Secretary of the ICA Commission on the History of Cartography.

## List of Contributors

**Mirela Altić** Institute of Social Sciences, Zagreb, Croatia

**Merve Senem Arkan** Eötvös Loránd University, Budapest, Hungary

**Edward Boyle** Center for Asia-Pacific Future Studies, Kyushu University, Fukuoka, Japan

**Andreas Christoph** Friedrich Schiller University Jena, Jena, Germany

**Robert Clancy** University of Newcastle, Callaghan, NSW, Australia

**Martin Davis** Canterbury Christ Church University, Canterbury, UK

**Enali De Biaggi** Université Jean Moulin, Lyon, France

**Karen De Coene** Department of Geography, Ghent University, Ghent, Belgium

**Philippe De Maeyer** Department of Geography, Ghent University, Ghent, Belgium

**Bart De Wit** Department of Geography, Ghent University, Ghent, Belgium

**Imre Josef Demhardt** University of Texas at Arlington, Arlington, TX, USA

**Isabelle Devos** Department of History, Ghent University, Ghent, Belgium

**Natalia Mikhailovna Erman** S.I. Vavilov Institute for the History of Science and Technology of the Russian Academy of Sciences, Moscow, Russia

**Jelena M. Glišović** National Library of Serbia, Belgrade, Serbia

**Francis Herbert** RGS-IBG, London, UK

**David Howell** Bodleian Libraries, University of Oxford, Oxford, UK

**Peter Kang** National Donghwa University, Hualien City, Taiwan

**Alexander J. Kent** Canterbury Christ Church University, Canterbury, UK

**Rink Kruk** National Geographic Institute, Brussels, Belgium

**Lamberto Lesti** University of Cassino, Cassino, Italy

**Eric Losang** Leibniz-Institut für Länderkunde, Leipzig, Germany

**Vinicije B. Lupis** Institute of Social Sciences, Dubrovnik, Croatia

**Nick Millea** Bodleian Libraries, University of Oxford, Oxford, UK

**John Moore** University of Glasgow Library, Glasgow, UK

**Viacheslav A. Nizovtsev** Lomonosov Moscow State University, Moscow, Russia

**Kristien Ooms** Department of Geography, Ghent University, Ghent, Belgium

**Ferjan Ormeling** University of Amsterdam, Amsterdam, The Netherlands

**Fernanda Padovesi Fonseca** Departamento de Geografia, Universidade de São Paulo, São Paulo, Brasil

**Cosimo Palagiano** Sapienza University of Rome, Rome, Italy

**Elien Ranson** Department of Geography, Ghent University, Ghent, Belgium

**Martijn Storms** Leiden University Libraries, Leiden, The Netherlands

**Tillmann Tegeler** Institute for East and Southeast European Studies, Regensburg, Germany

**Alice Tonkinson** State Library of New South Wales, Sydney, NSW, Australia

**Zsolt Győző Török** Eötvös Loránd University, Budapest, Hungary

**Michiel Van den Berghe** Department of Geography, Ghent University, Ghent, Belgium

**Sven Vrielinck** Department of History, Ghent University, Ghent, Belgium

**Melanie Wasmuth** Basel University, Basel, Switzerland

Allard Pierson Museum Amsterdam, Amsterdam, The Netherlands

**Torsten Wiedemann** Department of History, Ghent University, Ghent, Belgium

**Anne Winter** Department of History, Vrije Universiteit Brussel, Brussel, Belgium

# Part I
# Mapping Urban Spaces

# A View of Dubrovnik as a Spectacle: Presentations of the Dubrovnik Earthquake of 1667 in European Commercial Cartography

**Vinicije B. Lupis**

**Abstract** In spite of Dubrovnik's exceptional maritime and political power, European knowledge of the appearance and topography of this city in the early Modern Age was quite poor. This was reflected particularly in the very inadequate and incomplete cartographic representations of Dubrovnik. A lack of city maps or *vedute* depicting a realistic view of Dubrovnik prevented European cartographers from including adequate presentations of the city in their publications. Right up until the mid-seventeenth century, practically the only cartographic portrayal available to the European public consisted of the schematic vedute of Dubrovnik published in the Venetian *isolarii*. After Dubrovnik was hit by a catastrophic earthquake in 1667, European cartographers expressed great interest in passing on the sensation of Dubrovnik's big earthquake. Thus, a significant number of cartographic representations appeared for the first time shortly after 1667, mainly *vedute* showing the city at the moment of the earthquake and immediately after it. In this paper, we give an overview of the cartographic portrayals of Dubrovnik made on the occasion of the big earthquake of 1667 and analyze their accuracy. We offer answers as to the possible sources the cartographers used in their representation of the city and their evaluation of the consequences of the earthquake. In addition, we look in detail at the way the information was transferred from one *veduta* to another, i.e., how these *vedute* influenced the dissemination of knowledge about the real appearance of the city and the true impact of the earthquake.

## 1 About Cartography in Old Dubrovnik

As a maritime state, the interest in cartography and the need for it in Dubrovnik grew from its very beginnings. After gaining independence from the Croatian-Hungarian kingdom in 1358, Dubrovnik developed into one of the most powerful merchant navies in the world. Maritime dominance required cartographic knowledge to aid navigation. On 23 August 1546 a map was recorded in the assets of

V.B. Lupis (✉)
Institute for Social Sciences, Dubrovnik, Croatia
e-mail: Vinicije.Lupis@pilar.hr

 
M. Altić et al. (eds.), *Dissemination of Cartographic Knowledge*, Lecture Notes in Geoinformation and Cartography, DOI 10.1007/978-3-319-61515-8_1

Nicholas Benedict de Fornaris as "uno Mappa mundi—map of the world". It was already observed long ago that during the earlier years of the Dubrovnik state very few maps presenting this part of the Adriatic coast were published. The reason for this scarcity lies not so much in the imperfections of measuring instruments, but in the mistrust towards foreign countries and their cartographers, who were perceived as spies. Dubrovnik only allowed its own subjects or foreigners to create maps of the area at times when Turkey was weak and when they could not reprimand it, meaning that in the seventeenth century special maps were made which included the Dubrovnik Republic.

In 1664 Nicolas Sanson (1600–1671) published a map of the Dubrovnik Republic, which he titled "The map of the Dalmatian coast that belonged to Venice, Dubrovnik Republic and Turkey" (*Coste de Dalmacie ou sont…*). Somewhat later, Vincenzo Maria Coronelli (1650–1718), in his atlases *Mari, golfi, isole* (Venice, 1688) and *Corso geografico* (Venice, 1695) published a map entitled *Stato di Ragusi, Bocca del Fiume Narenta, Isole di Lesina e Curzola nella Dalmazia possedute Dalla Serenissima Republica di Venetia…*(Kozličić and Lozić 1994: 15–16).

The maps Dubrovnik produced from 1699 to 1737 not only served the Republic, but were also used as the means to solicit services from Austria, as long as it was winning. On 9 March 1700 the Dubrovnik Senate rejected a proposal to pay a fee to Miho Alegreti, the Secretary of State, for making maps, and finally a fee was only paid in April of the same year. When it seemed that Austria would expel Turkey from southeastern Europe, Dubrovnik had its eye on nearby parts of Turkish territory, and sent its diplomat Peter Bianchi, at his own request, a map of the Dubrovnik Republic from Ponta Oštro to Ponta Lovište. A map made by priest Miho Pešić—Dubrovnik diplomat, writer and painter- dates to a slightly later period. This manuscript map, which is now kept in the State Archives in Dubrovnik, shows the territorial division of the Republic and was made between 1718 and 1746.

In eighteenth century Dubrovnik, numerous paintings of its urban features were made. However, there was no geographical school that could develop a modern map of the state. It is true that Dubrovnik cartographers living outside their home country contributed significantly to the development of its cartography, such as Vicko Volčić, who in 1596 founded his own nautical and cartographical school in Livorno or Kristo Antić who worked in Livorno at the nautical school in the late eighteenth and early nineteenth centuries (Luetić 1984: 91–96). It was not until 1804, after a whole series of political troubles, that an Austrian cartographic expedition led by officer Catinelli produced the first topographic map of the Dubrovnik Republic and established a trigonometric network throughout the state that could be linked to the system in the Bay of Kotor and in Dalmatia. This topographic map may have been the one that the Dubrovnik Senate refused to give to the French consul in 1806, when he asked for it for the benefit of the French army that was moving south (Muljačić 2006: 93–104).

## 2 Early Representations of the City of Dubrovnik

For centuries, Dubrovnik was an intermediate station on the journey to and from the Levant. In the narratives dealing with the Dubrovnik history there are a number of graphics, produced in the fifteenth and sixteenth centuries, which bear no resemblance to Dubrovnik. Milan Rešetar is one of the first who in 1924 published a number of these prints titled "Pictures of old Dubrovnik". Among the first visual representations of Dubrovnik is a copper engraving from the book *Supplementum chronicarum* by Jacopo Filippo from Bergamo, printed in Venice in 1490. It is considered to be an ideogram of the town, with the portrayal of the Great Onofrio's fountain deemed to be its only credible feature. All of the graphics were created by printing houses in their efforts to enrich editions of a book. The same graphic but this time titled "Veduta di Ancona" by Filippo Jacopo di Jesi (1540), was printed in the *Guida di Ancona* by Giorgio Manganese and Valeria Pazzi in 1991 and therefore used to depict both towns.

Another idealized representation of Dubrovnik is found in the colored drawing of the German pilgrim Bernard von Breydenbach printed in 1485 in the works *Peregrinatio* and *Terra Sancta*. A close variant of this view of Dubrovnik, which was both the fruit of imagination as well as based on information about the actual situation of the protected harbor and the western fort Lovrijenac, was copied by the German pilgrim Konrad von Grünenberg in the work: *Pilgerreise von Konstantz nach Jerusalem*, in 1486.

From the late sixteenth century there are two schematic graphical portrayals of Dubrovnik, which are basically very similar. The first is the view of Dubrovnik from the book of Giuseppe Rosaccio—*Viaggio da Venetia a Constantinopoli* from around 1580, the second the representation of Dubrovnik by Francesco Valesi in his *Teatro delle più illustri et famose città di tutto il mondo* from 1571. A similar representation with breakwater Kaše is found on the map by Simone Pinargenti from 1573 in the book *Isole che son to Venetia nella Dalmatia*, and there is a view of Herceg Novi and Kotor (Polič Bobić 2000: 20). Finally, the most beautiful and most credible presentation of the appearance of the city can be found in the plans of the city and its port from the archives of Turin. Considering some of the details from the Turin drawings, one notices the fortress of St. Ivan does not yet have its present form (which it had from 1557). The tower of Mul is still there with a semicircular bastion and there is no Revelin fortress (finished in 1551, after construction started in 1538). It is also noticeable that in the Square of Arms by the Revelin fortress there is no parapet, which was built on the basis of a resolution of the Senate of 14 May 1594. The pier of Ponta is still made from wood, while the construction of the stone pier began in 1566. Together these clues testify that the display originates before the first half of the sixteenth century, and that Dubrovnik maps were not necessarily produced in 1607 or 1608, but rather earlier, and not at the same time, because it is due to its exceptional strategic position that Dubrovnik was involved in the wartime events of the First and Second Holy League. Duke Charles Emanuel I of Savoy (1598–1630) was engaged in the war against the Turks during which

Dubrovnik became a military base. Emanuel Filiberto began to collect a selection of maps, and those of Dubrovnik should be linked to the formation of his collections (Ničetić 2005: 157–160).

At the same time the printed maps show the schematic representation of Dubrovnik and its port. Giovanni Francesco Camocio printed in Venice in 1571 and 1574 his *Isole famose, porti, fortezze, e terre maritimme sottoposte alla Ser (enissi)ma Sig(no)ria di Venetia, ad altri Principi Christiani, et al. Sig(n)or Turco, novame(n)te poste in luce in Venetia alla libraria del segno di S(an) Marco.)*, with a schematized representation of Dubrovnik—*Citta di Ragvsi* (Kozličić 1995: 120, 129, 130). A similar schematized portrayal of the city appears on the map made by Simone Pinargenti*: Isole che sono da Venetia nella Dalmatia e per tutto l'Archipelago fino a Constantinopoli*, Venice 1573 (Kozličić 1995: 136, 145). In 1579 an isolario by Giacomo Franco appeared. His presentation of Dubrovnik is based on Camocio's work, which he supplemented with the new vistas of Dubrovnik (Kozličić 1995: 146, 149).

The isolario of Angielo Degli Oddi, *Viaggio de la provincie di mare della Signoria di Venetia* (Venice, 1584), documented the existence of three ports in the Dubrovnik Republic (Old harbor, Porto di Santa Croce and Ragusa Vechia) as well as the appearance of the city of Dubrovnik itself. For the first time Dubrovnik is presented together with Lokrum and some other non-existent island, Lovrijenac fortress, and a somewhat credible direction of the port and the three main bell towers. The other contents of the graphics are figments of the imagination (Kozličić 1995: 160–161, 166).

A schematized representation of Dubrovnik also appears in *Tabula Hydrographica* by Willem Barents, printed in Amsterdam in 1595 (Kozličić 1995: 192–193). Giuseppe Rosaccio in his *Viagio da Venetia a Constantinopoli…*, printed in Venice in 1606, repeats the schematized portrayal of Dubrovnik previously published by Giacomo Franco (Kozličić 1995: 210). Through his work *Description Villes, Havres et Isles du Golfe de Venice du cote'Oriental…* (Paris, 1686), Jacques Peeters belongs to the top of a similar cartographic material in terms of its artistic value, but in terms of cartography it does not present a breakthrough. Peeters's schematic representation of Dubrovnik is repeated by Vincenzo Maria Coronelli in his *Isolario dell' atlante veneto* (Venice, 1696).

## 3 Drawings and Paintings of the City: Modern Portrayals of Pre-earthquake Dubrovnik

In the visual arts there are many preserved portrayals of the pre- and post-earthquake city, the most valuable of which is the one on a silver relief from the collegiate church of St. Blaise (first half of the fifteenth century) and in a polyptych of Dubrovnik painted by Nikola Božidarević made at the beginning of the sixteenth century in the Dubrovnik Dominican Monastery (Gjukić-Bender 1999/2000). In

Dubrovnik a number of painted panoramic views before the big earthquake of 1667 and portrayals of the city (*ritratti di città*) from the seventeenth century have also been preserved. This genre encompasses a *veduta* with the family crest Božidarević/Bosadri in the Rector's Palace, panoramic views in the altarpiece of Our Lady in Glory with St. Blaise and St. Francis in the Dominican Monastery and, finally, a panoramic view from the Saraka family (property of the Society of Friends of Dubrovnik Antiquities), which offers without doubt the most detailed and realistic portrayal of pre-earthquake Dubrovnik (Majer Jurišić and Šurina 2016: 18, 19). The Dominican altarpiece of Madonna in Glory with St. Blaise and St. Francis by painter Antonio de Bellis, painted between 1657 and 1658, is considered to have been prepared based on a bird's eye view by Didier Barr (Majer Jurišić and Šurina 2016: 18). The detailed panoramic view belonging to the Saraka family demonstrates that in the seventeenth century the visualization of Dubrovnik became clear, in contrast to the unreliable representations of the city in maps.

## 4 Historical Narratives of the Dubrovnik Earthquake

The disastrous Dubrovnik earthquake of 1667 aroused the curiosity of historians. Much more than for any other episode in the history of Dubrovnik, a multitude of testimonies of survivors and witnesses as well as descriptions of the disaster have been preserved. Most of them have been printed in the last 300 years. The list of the most relevant references on the earthquake is given by Radovan Samardžić in his book *Borba Dubrovnika za opsztanak posle velikog zemljotresa 1667* (Beograd, 1960).

An anonymous essay on the earthquake had the greatest commercial success. It was entitled: *A true relation of the terrible earthquake which happened at Ragusa and several other cities in Dalmatia and Albania: the sixth of April 1667 as we have it in a particular account from Venice* (London, 1669). This essay was printed in Italian in Venice in 1667, 2 years later in Paris in French, and then in London in English. The Netherlands were notified of the earthquake by the envoy in Sublime Porte Van Dam, whose report was printed in The Hague in 1672, and translated in German in 1718. Among the scholars of the time who showed a special interest in the Dubrovnik earthquake was a Dutch classical philologist and historian at the University of Utrecht, Johann Georg Greve (1632–1703), and the famous Danish scientist, anatomist and founder of modern geology Niels Steensen (1638–1686), who in 1670 asked for detailed descriptions of the Dubrovnik earthquake (Krasić 2013: 317). The Venetian historian Battista Nani only 12 years after the earthquake described it in his *Historia della Republica Veneta* (Venice, 1679), which demonstrates the first modern perception of the events to the wider European public. The most interesting description of the earthquake may be the one by Dubrovnik Dominican friar Serafin Crijević in his work *Sacra metropolis Ragusina* (1744: I: 161–183). Interest in seismic phenomena was shown by a cartographer and district construction engineer in Dubrovnik, Lorenzo Vitelleschi, who in 1827 described the phenomenon of earthquakes in the Dubrovnik area in the 1820s (Lupis 2002: 37). Vicko Adamović from Dubrovnik published a thesis on earthquakes in the city of Dubrovnik, in which he systematically studied earthquakes from Dubrovnik's past. Mijo Kišpatić in the book *Potresi u Hrvatskoj* deals in detail with the natural cataclysm of 1667.

## 5 Graphic Representations of Dubrovnik in the Earthquake

In parallel with the written news of the catastrophic earthquake there were also printed graphic sheets showing the event. Several of them have been preserved in Dubrovnik's cultural institutions. As already mentioned, Jacques Peeters' *Description Villes, Havres et Isles du Golfe de Venice du cote'Oriental* (Paris, 1686) presents Dubrovnik in the most realistic way when compared to other maps of the time. His perspective is from the southeast, so the island of Lokrum is in the forefront. Next is the entrance to the harbor, but without the Kaše breakwater. To the southern and western outside defensive wall he added a channel and wide mass of land which in reality are non-existent. The attention is drawn to a picture of a small Dubrovnik galleon. Apart from the names of Ragusa and galleon, the graphic bears no other similarity with Dubrovnik, and the depiction of a mosque with a minaret in the city makes it more unrealistic.

A *veduta* of Dubrovnik depicting the earthquake by Matthäus Merian Jr. (1621–1687), the son of Swiss graphic master Matthäus Merian (1593–1650), appears around 1690. It is divided into two imaginary images of the city (see Fig. 1). The top half of the sheet shows the coastal side of the city that is laid in ruins by the earthquake. The *veduta* in the lower zone was produced by Matthäus Merian Sr. in 1649. The upper section showing the disaster is actually merely a modification of the lower panorama of Dubrovnik, to which a representation of its demolition by the earthquake of 1667 has been added. This can also be read in the German descriptions above each representation: *Eigentliche Vorstellung des erschrecklichen Erdbebens so den 6(t) April 1667. In und ausser der Stadt Ragusa in Dalmatia enstanden* (a real representation of the horrific earthquake that happened on 6 April 1667 in and outside the city of Ragusa in Dalmatia)—*Abriss der Schönen Stad Ragusa in Dalmatia, wie solche vor begonnenem Erdfall an zusehen gewesen* (A depiction of the beautiful city of Ragusa in Dalmatia, as it looked prior to the earthquake) (Budiša 1991: 32–36; Kisić 1988: 7; Krasić 2013: 323; Ničetić 2003: 75–89). It should be noted that the imaginary *veduta* is based on the representation of Dubrovnik which was published as an inset on Willem Barentzs map *Tabula Hydrographica* (Amsterdam, 1595) and a graphic of Dubrovnik published in the isolario of Simone Pinargenti (Venice, 1573) (Budiša 1991: 32–37). Thus, graphic

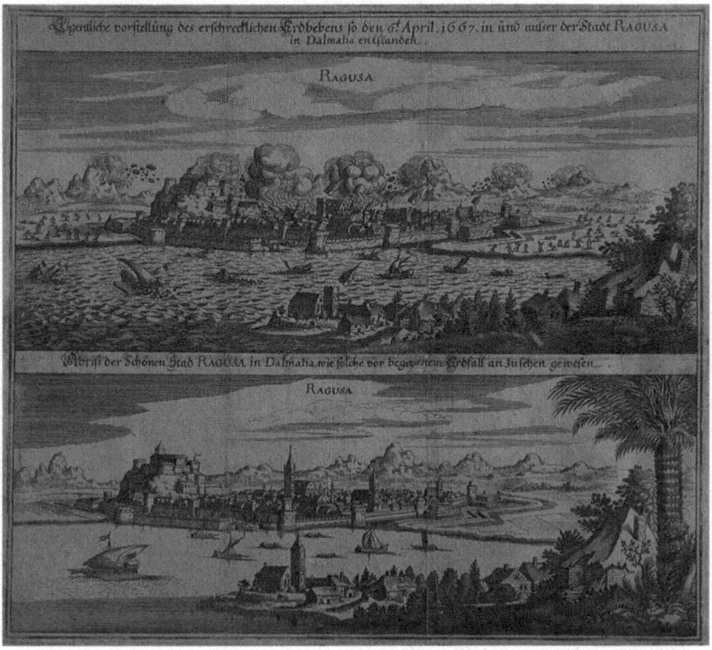


**Fig. 1** Engraving depicting Dubrovnik before the earthquake and during the earthquake of 1667. Matthäus Merian Jr., Frankfurt, c. 1690 (Courtesy of Dubrovnik Museums, KPM GR – 512)

artists Merian Sr. and Jr. used templates which were almost a century old as the basis for their own prints of the Dubrovnik earthquake.

The graphic artist and publisher Matthäus Merian Jr. printed another graphic sheet of Dubrovnik and the Bay of Kotor depicting the earthquake of 1667 (see Fig. 2). This time around he used his father's copperplate from 1649 which he supplemented with the data from the book *Archontologia Cosmica*, written by German chronicler Johann Philip Abelin (pseudonym J. L. Gottfried). Above the display of Dubrovnik being destroyed there is an inscription in German: *Erschröcklicher Untergang und Verbrennung der Statt Ragvsa* (Horrific ruination and fire of the city of Dubrovnik) (Budiša 1991: 32–37; Kisić 1988: 7; Krasić 2013: 322–323; Ničetić 2003: 75–89).

Among the contemporary graphic sheets that show the destruction of Dubrovnik, there is also a late seventeenth century drawing of the city before and during the earthquake, printed by Johann Kristoff Lochner Jr. (Hol et al. 2009: 12). Because several Dutchmen perished in the earthquake, it is not surprising that a number of prints portraying the disaster were created by Jan Luyken (1649–1712), a poet, illustrator, and engraver. The graphic sheet "The Earthquake in Dubrovnik" from 1698 by Jan Luyken and Pieter van der Aa shows the moment ancient buildings are

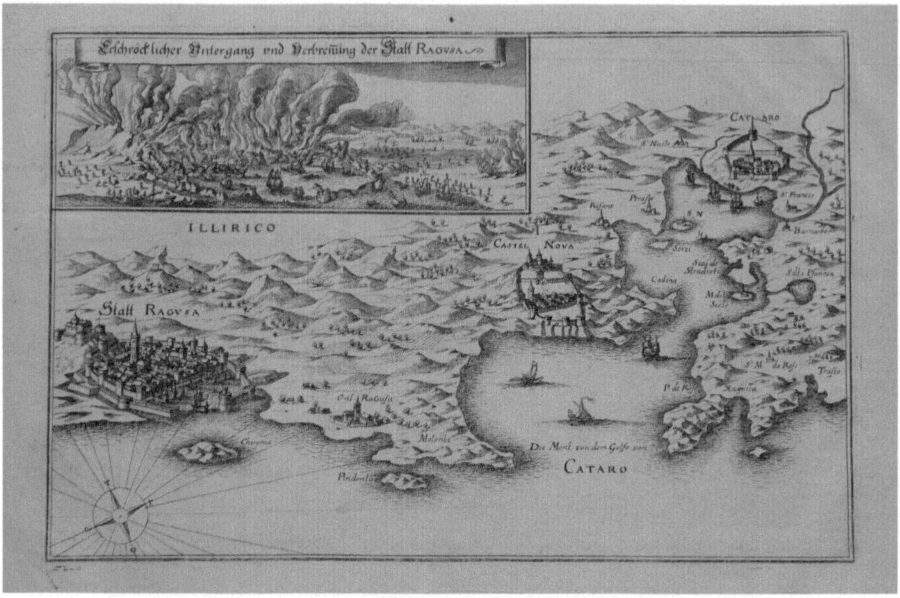


**Fig. 2** Engraving depicting the great earthquake of 1667. Matthäus Merian Jr., Frankfurt, c. 1677 (Courtesy of Dubrovnik Museum, KPM GR – 460)

demolished and people are injured, albeit without any similarity to Dubrovnik in that period. Obviously the artists tried to provoke sympathy. They also created similar iconography for San Severino in Sicily and for Indonesia. Furthermore, Jan Luyken made a simpler version of his representation of the Dubrovnik earthquake, supposedly depicting a fire in Trondheim in Norway.

When comparing drawings and paintings to the maps and prints depicting the earthquake, there is a clear discrepancy between the realistic portrayal in paintings and drawings as opposed to the unrealistic one in the graphics. Fortunately, some drawings of Dubrovnik's urban tissue have also been preserved in the post-earthquake letters by abbot Stjepan Gradić (1613–1683). He is one of the most outstanding Dubrovnik mathematicians, physicists and astronomers of the seventeenth century, and also a director of the Vatican Library. Some of his experimental studies are cited by Jacob Bernoulli, and his tractate about navigation inspired Gottfried Wilhelm Leibniz to discuss the problem of steering ships using helms. Gradić's book *Dissertationes physico-mathematicae quatuor* was published in Amsterdam in 1680. He died in Rome, and according to his last wish was buried in the Croatian church of St. Jerome. The name of Stjepan Gradić is connected in a very special way to the consequences of the great earthquake. Dubrovnik, as a city and a state, had been brought to its knees. From that time on Gradić's life changed completely. He subordinated everything to the need to come to the aid of his native city as fast as he could, becoming the central personality of Dubrovnik's politics and diplomacy, via whom from then on all communications with the world were conducted. Rome thus became the real capital of the Dubrovnik Republic (Körbler 1915: 1–99; Krasić 1987: 1–441; Krasić 2013: 222–331). Gradić's link to Amsterdam publishers, along with the perishing of the Dutch, points to the possible emergence of new prints.

In his letters to the Senate of the Dubrovnik Republic, Gradić advocates the expansion of the city outside the walls to Pile and suggests building forts in Ilijina glavica, drawing a sketch of the new fortifications. He devises the plan in the document *Discorso sopra la fortificazione della Città* with the collaboration of the engineer Paolo Falconieri. The next sketch by Gradić—*Esempio del dissegno*—shows the new fortifications that are supposed to protect the city up to Boninova in the west and the nearby slopes of mount Srđ (Krasić 2013: 222–231). Gradić's sketches are a telling example of where respected European printing houses could get the basis for their often imaginary representations which also included some realistic details. The earthquake in Dubrovnik in 1667 thus not only left its mark on the city itself, which at that time played an extremely important role in the global maritime industry, but also on its cartography.

## 6 Conclusion

Dubrovnik holds a special place in world history and has always been an important maritime center, but despite this fact it never developed its own cartographic school to supply the Dubrovnik seafarers with necessary maps. This fact is all the more interesting since several people from Dubrovnik enjoyed a reputation of being excellent cartographers.

Graphical representations of Dubrovnik from the previous centuries are not realistic. A significant milestone in Dubrovnik's cartography was a catastrophic earthquake in 1667, which is depicted on numerous maps and *vedute* trying to show the extent of the damage. In this sense the phenomenon of the earthquake is an interesting cartographic experience in the wider scientific community of the time, but also still important today, since it serves to promote national interests.

## References

Adamović V (1883) O trešnjama grada Dubrovnika [About the earthquakes of Dubrovnik]. Tiskarnica Joza Flori, Dubrovnik

Budiša D (1991) Dubrovnik i dubrovački motivi na knjižnim grafikama u 15. i 16. stoljeću [Dubrovnik and scenes from Dubrovnik in book illustrations of the 15th and 16th centuries]. In: Fisković I (ed) Likovna kultura Dubrovnika 15. i 16. Stoljeća. Znanstvena djela Muzejsko-Galerijskog Centra 2, Zagreb

Crijević S (1744) Sacra Metropolis Ragusina, sive Ragusinae provinciae pontificum series variis ecclesiarum monumentis atque historicis, chronologicis, criticis commentariis illustrate. Ragusii

Gjukić-Bender V (1999/2000) Prikazi Dubrovnika u slikarstvu [Presentation of Dubrovnik in Paintings]. Prilozi povijesti umjetnosti u Dalmaciji 38:215–244

Hol RC, Kooten M, Van Dam J (2009) The Ragusa-disaster of 1667: A Dutch eyewitness account of the Earthquake and Tsunami in Ragusa in April 1667. Nationaal Archief, Den Haag

Kisić A (1988) Obala Hrvatske na pomorskim i geografskim kartama od 16. do 19. stoljeća [Croatian coastline on maps and sea charts from 16th to 19th century]. Dubrovački muzeji, Dubrovnik

Kišpatić, M (1892) Potresi u Hrvatskoj [Earthqukes in Croatia]. Rad Jugoslavenske akademije znanosti i umjetnosti, Zagreb

Körbler Đ (1915) Život Stjepana Gradića i njegov rad za Dubrovnik [Life of Stjepan Gradić and his work for Dubrovnik]. In: Körbler Đ (ed) Pisma opata Stjepana Gradića Dubrovčanima senatu Republike Dubrovačke od godine 1677. do 1683. Jugoslavenska akademija znanosti i umjetnosti, Zagreb

Kozličić M (1995) Kartografski spomenici hrvatskog Jadrana. Izbor karata, planova i veduta do kraja 17. stoljeća [The cartographic sources on the Croatian Adriatic. The selection of maps, plans and views by the end of the seventeenth century]. AGM, Zagreb

Kozličić M, Lozić V (1994) Starije zemljopisne karte u Sveučilišnoj knjižnici u Splitu [Older geographical maps in national library of split]. Sveučilišna knjižnica Split, Split

Krasić S (1987) Stjepan Gradić (1613.–1683.): život i djelo [Stjepan Gradić (1613–1683): Life and Work]. Jugoslavenska akademija znanosti i umjetosti, Zagreb

Krasić S (2013) Obnovitelj našega grada i slobode [Restorer of our City and our Freedom]. In: Vilać P (ed) Stjepan Gradić otac domovine. Dubrovački muzeji, Dubrovnik

Luetić J (1984) Pomorci i jedrenjaci Republike Dubrovačke [Sailors and Sailboats of the Republic of Dubrovnik]. Nakladni zavod Matice hrvatske, Zagreb

Lupis VB (2002) Arhitekt Lorenzo Vitelleschi i njegovo vrijeme [Architect Lorenzo Vitelleschi and his Time]. In: Lupis Vinicije B (ed) Lorenzo Vitelleschi, Povijesne i statističke bilješke u Dubrovačkom okrugu, prikupio okružni inženjer Lorenzo Vitelleschi, Dubrovnik, 1827, Matica hrvatska ogranak u Dubrovniku, Dubrovnik

Majer Jurišić K, Šurina E (2016) Velika fontana u Dubrovniku kroz stoljeća, povijesnoumjetnička i konzervatorska studija [Great fountain in Dubrovnik over the centuries, art history and conservation study. Ministarstvo kulture Uprava za zaštitu kulturne baštine, Zagreb

Muljačić Ž (2006) Prva moderna geografska karta dubrovačkog područja [The first modern geographic map of Dubrovnik area]. In: Hekman J (ed) Iz dubrovačke prošlosti, Matica hrvatska, Zagreb

Nani B (1679) Historia della Republica veneta. Parte seconda. Combi & La Nou, Venice

Ničetić A (2003) Grafike koje ipak ne prikazuju Dubrovnik 15. i 16. stoljeća [Unfaithful representation of Dubrovnik on fifteenth and sixteenth century graphics]. Anali HAZU 41:75–89

Ničetić A (2005) Nove spoznaje o postanku Dubrovnika, o njegovu brodarstvu i plovidbi svetoga Pavla [New findings on the origin of Dubrovnik, on the shipping and navigation of St. Paul]. Sveučilište u Dubrovniku, Dubrovnik

Polič Bobić M (2000) Među križem i polumjesecom, dubrovačke dojave španjolskom dvoru o Turcima u XVI. Stoljeću [Between the cross and the crescent: Ragusan reports to the Spanish court regarding the Ottomans in the sixteenth century]. Naklada Ljevak, Zagreb

Rešetar M (1924) Slike starog Dubrovnika Narodna starina 2(8, III):176–189

Samardžić R (1960) Borba Dubrovnika za opstanak posle velikog zemljotresa 1667 g: arhivska građa 1667–1670 [The fight for the survival of Dubrovnik after the big earthquake in 1667 g: archival sources 1667–1670. Srpska akademija nauka i umetnosti, Beograd

**Vinicije B. Lupis** graduated in 1992 in the double-major program in history and archeology from Zadar University. In 1995 he obtained a second degree in the field of art history, also from Zadar University. He obtained his PhD in 2004. His work in the field began in 1992 as a conservator-archaeologist at the Regional Office in Split and since 1993 as a conservator-art historians in the State Directorate for the Protection of Cultural Monuments in Dubrovnik. From 1995 onwards he was specialist archivist and head of the Department of nineteenth and twentieth century documents. Throughout his interdisciplinary work, he connects several humanistic fields: history, art history, archaeology and archival science. Since 2007 he works at the Institute of Social Sciences Ivo Pilar, and since 2008 he is head of the Regional Center of Dubrovnik Institute. He has published hundreds of scientific papers and several books on the topics of religious heritage, history and art history of the Dubrovnik region and the Bay of Kotor.

# Mapping the Metropolis: Analysing Map Production in Lyon and São Paulo in the First Half of the Twentieth Century

**Enali De Biaggi and Fernanda Padovesi Fonseca**

**Abstract** A project of cooperation between the universities of São Paulo and Lyon has produced exchanges on the subjects of urbanization dynamics and spatial representations. The approach of two cities that differ largely in what involves periods and intensity of growth has been an interesting way to compare how mapping procedures developed in the first half of the twentieth century. In a moment of important urban sprawl, both cities have incorporated a range of new techniques, such as aerial photography or new geodetic networks. This was done in an attempt to cover the overflow of the city beyond its original borders.

Mapping activities were the basis for a series of new urban plans, even if they were not conducted in the same way or by the same kind of institution in each city. In Lyon the role of local private surveyors was particularly important in the new series proposed since 1910: 1/2000 mappings were used, building on a 1/500 cartography started in the 1860s. In São Paulo, the role of foreign companies, as the SARA mapping in the late 1920s, created a new reference map in the scales of 1/5000 and 1/1000 contributing to the idea of a modern city. In both cases, new projects for road maps set the stage for a city where daily mobility involved larger distances based on new means of transportation. This caused the need for a change in the scale and the way to portray a city that would be decisive for its further development.

## 1 Introduction

The idea of comparing how cities have been mapped is not new and many books have demonstrated the development of urban representations as a means of understanding cities themselves (Black 2015 and Knox 2014). In many cases, the idea of

E. De Biaggi (✉)
Université Jean Moulin Lyon 3, Lyon, France
e-mail: enali.debiaggi@univ-lyon3.fr

F. Padovesi Fonseca
Departamento de Geografia, Universidade de São Paulo, São Paulo, Brazil
e-mail: ferpado@gmail.com

 
M. Altić et al. (eds.), *Dissemination of Cartographic Knowledge*, Lecture Notes in Geoinformation and Cartography, DOI 10.1007/978-3-319-61515-8_2

comparing maps from different continents is justified as a means of exploring particular kinds of cities and encourages the discussion of urban development in general. This paper proposes to focus on a comparison of just two cities: Lyon in France and São Paulo in Brazil. One parameter is the same—the number of inhabitants living in each city. We will then focus on the period when both cities seemed to have had about the same population—around the decade of the 1920s, at the end of the nineteenth until the beginning of the twentieth century. We hope this will help us better understand how mapping techniques evolved during that period in two different parts of the world, that were subjected to very different dynamics, that have two very different cultures and in general, how these two cities became what they are today. Following, the mapping techniques and procedures used in a particular moment of urban growth, we will try to establish how different the two cities really are. At the same time, we will consider the possibility of employing a methodological approach capable of dealing with a diversity of situations[1]. Mapping projects have to be compared to other data produced in city administration, as we try to understand in what way there were underlying traditions that survived the eventful years between 1850 and 1950, and what can be seen as new not only in the mapping techniques, but in the way the city is portrayed. First we will analyze each city separately before comparing the two.

## 2 Following São Paulo's Cartography

São Paulo in the beginning of the second half of the nineteenth century still looked like a colonial city—as seen by some views of the city produced at that time[2]. The first official census from 1872 indicates 35,385 inhabitants, when not only the city but also the land to the west of the city was subject to important changes. The expansion of coffee plantations in the countryside at the moment when slavery was finally abolished (1888) attracted new immigrants to help occupy the new land, and many of them stayed in the city, which explains why there were 65,000 people in 1890 and almost 200,000 at the turn of the century. Looking at the maps produced during this the period, we go from picturesque maps produced in the 1870s that used scales such as 1:3000, to much bigger maps that used the scale 1:20000 in the

[1]This work has been part of a larger project between the universities of Lyon in France and São Paulo in Brazil—the USP-COFECUB project Uc Sh 152-14 ("Dynamiques d'urbanisation et représentations spatiales: l'approche géohistorique des territoires sous SIG"). Many people involved in the project have contributed to the discussions leading to this paper, and some of their names will appear as sources in the present article.

[2]The website made by the city of São Paulo to display historical data can be the first resource with official sources: http://smdu.prefeitura.sp.gov.br/historico_demografico/. For the first period, many of the photographs come from the *Álbum Comparativo da Cidade de São Paulo, 1862–1887*, from Militão Augusto de Azevedo, existing in the *Acervo IMS*. See also Fernandes et al. (2012).

1900s. Indeed, comparing the scope of the first scattered production of maps with the succession of new representations that appear in the later part of the nineteenth century, we can understand how the picture had changed. It grew from a few streets forming the first nucleolus around the Jesuitical settlement on top of a hill, close to the Tamanduatei River, to a city that incorporated several surrounding villages and would become the biggest urban area of the country during the following century.

## 2.1 Traditional Mapping in São Paulo

Most colonial maps picturing São Paulo are centered on the hill formed between 2 rivers, the Anhangabau and Tamanduatei, the latter being larger. The Tamanduatei is often shown on the top of the page (thus orienting the map to the east), or at least positioned where it could be immediately seen as a reference. The meanders and plains of the two rivers will be the object of intense urbanization with the use of canalization and drainage from 1896 on. This created space for new housing compounds in the adjacent areas, as can be seen in the map made by Jules Martin in 1890 (Fig. 1). On this map, the "old image" of the city is still seen in the upper right corner of the map, a plan not yet north-oriented as future maps will progressively indicate and incorporating new waterways that will soon become the new landmarks of the city, such as the *Tietê* river in the north.

New avenues and viaducts were built, streets were enlarged as new buildings, squares and parks were planned by the city government largely inspired by European projects from the late nineteenth century[3]. Maps were produced in order to plan new water and sewage systems for the city[4], new railways and stations were being constructed by English companies and horse-drawn trams would soon become electric tramways as part of the "The São Paulo Tramway Light and Power Com", an Anglo-Canadian firm founded in 1901.

Kuvasney (2016) mentions how the city administration was completely overwhelmed by the incredible growth and multiple demands for road and network services and how building permits were subject to an attempt to control the city according to official maps. The 1897 general map made by Gomes Cardim in the scale of 1:20,000 shows indeed a much larger area of the city (Fig. 2). Sometimes road networks were not even named, in an attempt to portrait the projects being developed by private companies that did not wait for the official assignment or recognition of new areas. Compared to the previous map from 1890, the 1897 map

[3]See the work of Beatriz Piccolotto Siqueira Bueno (2016) that retraces the city's thorough transformation from the nineteenth to the beginning of the twentieth century.

[4]Like the 1881 maps produced by Henry P. Joyner (Cia. Cantareira de Águas e Esgotos) *Planta da Cidade de São Paulo*, Scale: 1:5000, which better depict buildings and the real occupation of new city areas, cf. at http://smdu.prefeitura.sp.gov.br/historico_demografico/img/mapas/1881.jpg

**Fig. 1** Planta da Capital do Estado de São Paulo e seus Arrabaldes, from Jules Martin (1890). Approximate scale: 1:6000 [Courtesy of São Paulo Historical Archives (AHSP)]

shows a much broader surface area (see the extent indicator in Fig. 2) but it also changed the scope of what was shown in the cartouche appearing in the bottom left corner of the map. Instead of retracing the old image of downtown São Paulo, the author introduces an overview map, setting the stage for an even larger image of the city.

A new map produced in 1905, was officially named as a reference by the city council and its departments, shows, in fact, a smaller expansion area (Fig. 3). The

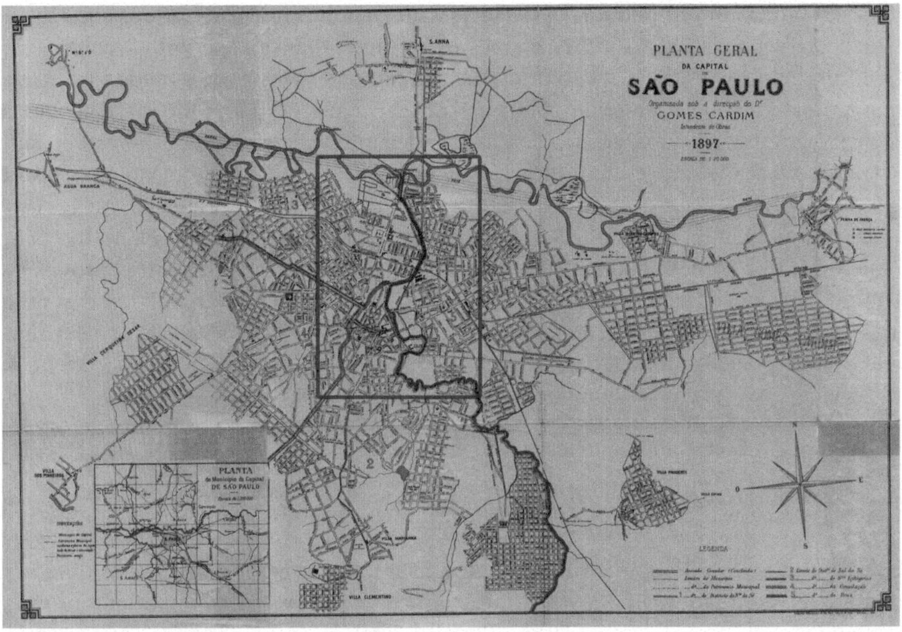


**Fig. 2** "Planta Geral da Capital de São Paulo", from Gomes Cardim (1897). Scale: 1:20,000—indicating the scope of the previous map from Jules Martin in blue [Courtesy of São Paulo Historical Archives (AHSP)]

**Fig. 3** "Planta Geral da Cidade de São Paulo", from Alexandre Mariano Cococi e Luís Fructoso e Costa (1905). Scale: 1:20,000 [Courtesy of the Brazilian Digital National Library (BNDigital)]

map key covers the 11 districts of the city, with its police and fire stations and all the streets are represented with their names in an index that opened with "remarkable buildings and institutions" presenting the official city. This map included an insert that indicated the triangulation points used in the construction of the map by one of the city engineers. This corroborated the idea of what should be a correct and scientifically based representation of São Paulo. Indeed, new attempts to rectify and create new maps according to a scientific approach can be seen later in 1914 when the Geological and Geographic Commission (CGG), formed in São Paulo in 1886, issued a series of new plans of the city, and in 1916, with yet another "provisory edition" produced by the cadastral division of the municipality. In both cases a new grid is applied to the map, announcing a much bigger cartographic project that would take place in the 1920s.

## 2.2 New Mapping Procedures: The SARA Campaign

By the 1920s, São Paulo had more than half a million inhabitants, more than the double its population at the beginning of the century. This important growth demanded new means of controlling the city, even if economic crises appeared in the horizon. In 1927 the local government made a bid for a new detailed mapping of the entire city, and the SARA (*Società Anonima Rilevamenti Aerofotogrammetrici*) mapping campaign began.

Italian technology was adopted by the city administration in a pioneer project to map an important part of the urban sprawl of the moment (Fig. 4). Between 1929 and 1933, a topographic survey was carried out using the Nistri aerial photogrammetry method, a pioneering work that made São Paulo one of the first cities in the world to have a register of articulated maps of great precision in a detailed scale (1: 1000 and 1: 5000) based on photogrammetry. 132 sheets printed by the Geographic Institute of Agostini in Novara (Italy) were delivered to the city hall, 20 photo-cards and the collection of aerial photographs (vertical and oblique), which were the source of the survey (Dutenkefer 2015: 156–159). The first contracts mention 82 sheets to be delivered in the 1:5000 scale, and mention only the possibility of a 1:1000 delivery of some areas. In fact, the topographic chart in the 1: 1000 scale decided on in later agreements would allow the elaboration of joint plans in the central zone (high value zone, as indicated by Machado 2012: 22); while those of 1:5000 were considered sufficient for preliminary work on water supply, road paving, etc. Finally, the 82 sheets produced in 1:5000 became 69 sheets, 59 being considered "simple" while those depicting rural areas were assembled in double or even triple sheets (Mendes 2014: 115).

More than 4 years were necessary to conduct the work of this survey for the Municipality of São Paulo, although the contract scheduled a period of 2 years for

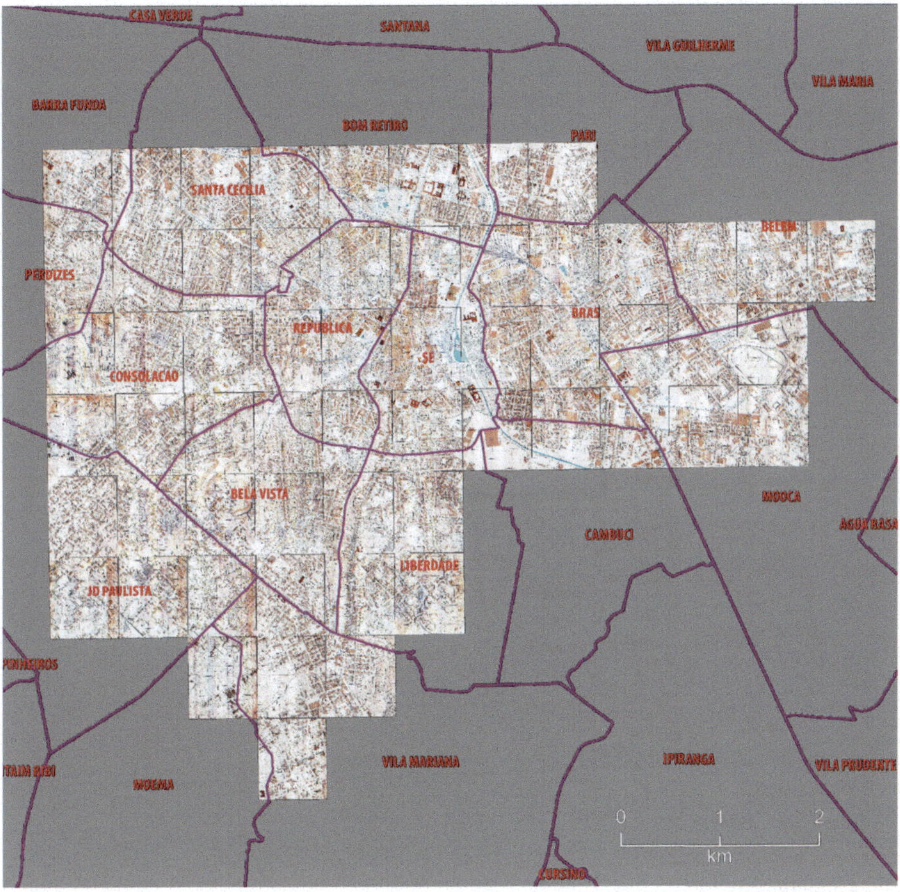


**Fig. 4** The SARA mapping in two scales from São Paulo Historical Archives (AHSP). Above, Index sheet SARA 1930 58 maps in the scale 1:1000. Below, Index sheet SARA 1930–1967 maps in the scale 1:5000, the blue line shows the area of the first mosaïque)

the delivery of all the maps. Still, several things had not been anticipated, such as the heavy rains of 1929, the Revolution of 1930, unforeseen events in the organization of the company. The SARA mapping remains a unique collection, depicting all the new projects that had been launched at the beginning of the twentieth century, a new cathedral, new streets and new housing compounds. It would remain a basic cartographic source for almost 20 years, until a similar project, this time using local companies, would be conducted in the beginning of the 1950s (Mendes 2014: 116–119).

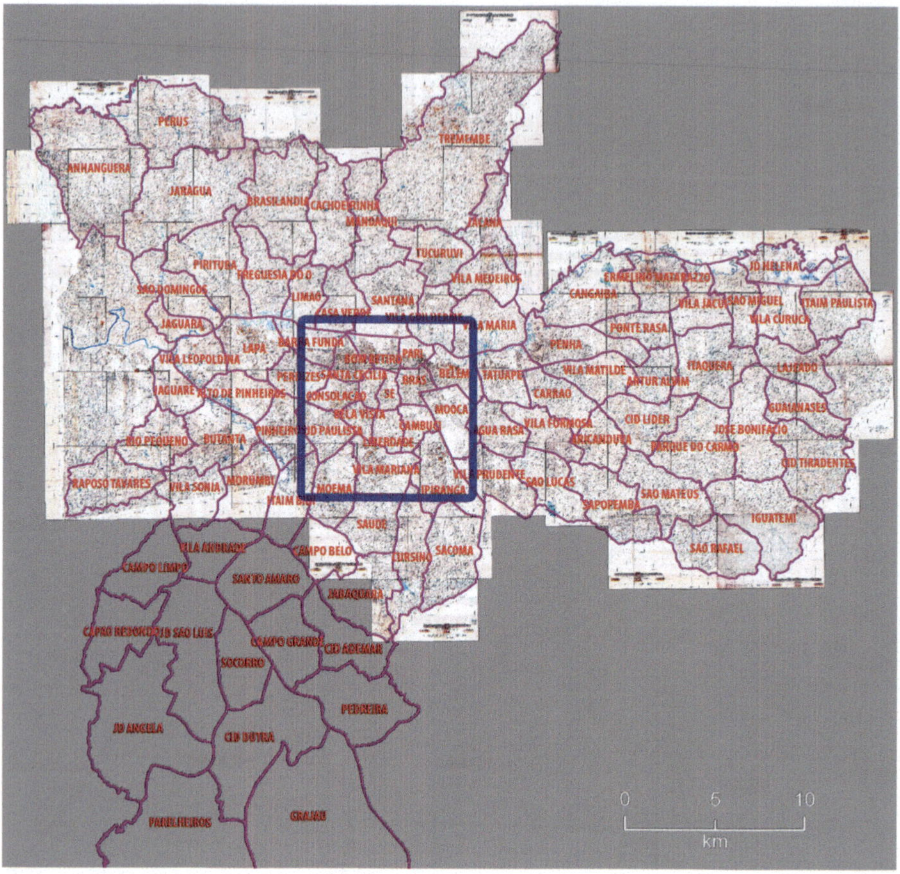


**Fig. 4** (continued)

## 3 Mapping in Lyon in the Nineteenth Century

In the nineteenth century, Lyon was already a mature industrial city, built on the hills and plains of two rivers, the Saône running from the north, and the *Rhône*, coming from the Alps. Traditional views of the city emphasize the river Sâone and the Fourvière hill on its left bank, with an astonishing collection of high and densely packed buildings dating from the seventeenth to eighteenth centuries[5].

Recovering from the French Revolution, a first cadastral map was completed around 1831–1832, according to the law dating from September 1807, referred to as

[5]For an interpretation of the diverse views of Lyon see the exhibition catalogue that portrays Lyon in the nineteenth century "L'esprit d'un siècle—Lyon 1800–1914" and Petermann (2016). For a detailed and long-term analysis of the city's buildings, see the various work of B. Gauthiez, especially Gauthiez (2006, 2012, 2014) where there is mention to the precocity of high-rise buildings and also alignment plans in Lyon.

the Napoleonic cadaster. This was the first complete and precise cartographic survey of the whole city (Gauthiez 2008: 57). Lyon was the setting of numerous rebellions, led mainly by silk industry workers (*canuts*) in the 1830s. The city was under total siege on more than one occasion. After many politically turbulent years, it was during the period of the Second Empire (1852–1870) that important changes took place in the organization of the city. This was set up and implemented by a new administration, which was aided by an ambitious and more detailed mapping program. The city would be the model for many other cities in France with the concerns of a provincial city administration executing a project decided on by the central government in Paris. A representative was sent by Paris to lead the local government (Gauthiez 2012: 324–325). The first one, Claude Vaïsse, was at the same time prefect of the department, police delegate and mayor, which provided him with extensive power not only in Lyon but also in neighboring cities, some of which were attached to Lyon in the process. Many services would thus extend beyond the original city limits, while the center, itself, underwent a complete transformation with new "boulevards" and massive reconstruction. The important growth of the city throughout the century would be depicted in different maps, either partial or global and as in the case of São Paulo, we can see how the image of the city changes along with them.

### 3.1 Traditional Mapping in Lyon

In the beginning of the nineteenth century, maps of Lyon still contain the same scope that had been assumed during the previous century. The maps were normally west oriented, with the Fourvière hill on the top of the page, like the old bird-eye views portraying the city. Maps from Lyon rarely show what would be considered as the surroundings of the city on the eastern side of the Rhône[6]. For the most part the region "under" the Rhône river was used to place an index of places or buildings, as much as to position the map cartouches (Fig. 5). Many of the maps produced at that time used the word "new" in their title and as the metrical scale becomes the norm (gradually going from toises or braces to more round scales of 1:7000), engravings lead the way to lithography.

The Dignoscyo-Riembilienki family is considered one of the most representative of the mapping made in Lyon during the nineteenth century[7]. Indeed, Laurent Dignoscyo started with the usual placing of the city (Fig. 5), but his work would

[6]Exceptions can be found in some of the development plans made in the eighteenth century, like those of Morand, the *Projet d'un plan général de la ville de Lyon et de son agrandissement, en forme circulaire, dans les terrains des Brotteaux*, made in 1764 (see AML 3S0115), or Moithey's *Plan historique de la ville de Lyon et de ses quartiers, assujetti à ses accroissements, embellissements et projets* made in 1786 (AML 3S0694), but that doesn't occur frequently.

[7]See the texts assembled in the *Forma urbis* exhibition, from P. Sarocchi (1999) and Vanario (1999) for more specific information.

**Fig. 5** Nouveau plan géométrique de Lyon, seconde ville du royaume de France, avec ses projets d'agrandissement et d'embellissement. Laurent Dignoscyo, 1821 [Courtesy of Lyon Municipal Archives (A.M.L.)]

soon be valued as it included a larger portion situated in the dynamic eastern part of the city. His 1835 map *Plan de Lyon, de ses environs et des forts* (Fig. 6) with a scale 1:20,000 still was west-oriented, highlighting the same public buildings as in 1821, but it was enlarged to present a larger view in what was going soon to become the new part of Lyon. L. Dignoscyo would be on the service of one of the most important agencies of Lyon at the time, the HCL (*Hospices Civils de Lyon*), which possessed a major part of the land situated in the left banks of the Rhône. The mainly rural areas in the east that Dignoscyo would be inspecting were surrounded by a defensive belt with forts and ramparts. Even though these rural areas were often affected by the flooding of the Rhône, they were rapidly occupied[8], partially by the population coming from the many buildings destroyed in the central areas of Lyon. Already in 1847, a new map made by Dignoscyo and his son in law, Eugene Rembielinski, shows the rapid growth of the Brotteaux and Guillotière areas, that would welcome new railway and industrial plants[9]. E. Rembilienski's detailed maps of Lyon and the Villeurbanne canton to the east, gave thorough statistics

[8]According to Sarocchi (1999), L. Dignoscyo had been responsible for more than 200 detailed reports ("*dossiers de masse*" in the language of the HCL)—which detailed with many maps and texts, the terms under which each parcel was to be administered.

[9]From Polish origin, E. Rembielinski was a political refugee in France since 1832. Cartographer and engraver by training, his career was divided between Paris and Lyon and his son, Jules Rembielinski would first collaborate with his father and afterwards with his uncle, Claude Dignoscyo, see Sarocchi (1999).

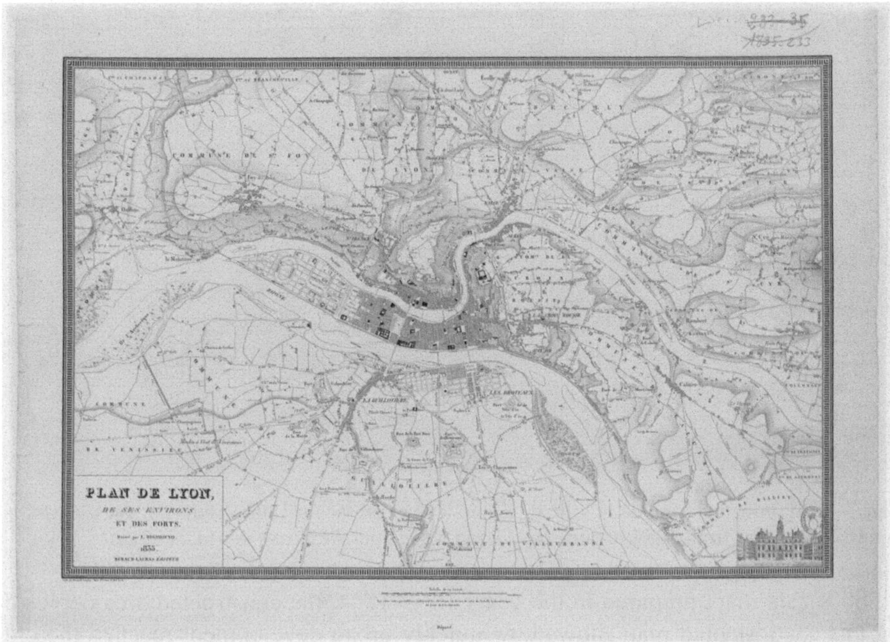




**Fig. 6** Plan de Lyon, de ses environs et des forts/L. Dignoscyo; Lith. de Béraud-Lauras, 1835 [Courtesy of the French National Library (BnF)]

and would also be used in the new projects to tame the Rhône river in 1855 (see A.M.L. 3S0279 and 3S0279a), in an area that will soon obtain more and more attention[10].

Indeed, in the 1850s, a new railway linked Lyon not only to Paris and the South of France, but also followed the ancient ramparts in the east to create the Lyon-Geneva line. They constituted the new borderline of most maps made at that time, and started to occupy a special place on them. E. Rembielinski would be one of the first to orient the mapping of Lyon to the east in the 1860s, with his *Carte du Rhône et de la Saône—Plan general de la ville de Lyon* (A.M.L. 2S1322), showing the Lyon-Geneva line on the top of his plan. Claude Dignoscyo, will join his father Laurent Dignoscyo and his brother in law and produce a similar map in 1863, that would be used as a base map for most of the official city maps from the 1860s, under the administration of chief-engineer G. Bonnet[11]. In the 1885 edition of a huge new

[10]These two maps can be seen nowadays on the website of the municipal archives from Lyon: http://www.fondsenligne.archives-lyon.fr/. One interesting feature is that they are both north-oriented, in the scale of 1:40,000. The first map is dedicated to the new Prefect, C. Vaïsse, imposed on Lyon by the Parisian government—Rembielinski's address is also located in Paris.

[11]See for example in the municipal archives of Lyon (A.M.L.) the maps 2S0149 and 2S0149a from Claude Dignoscyo, that are certainly the source of the many maps used and designed by G. Bonnet,

**Fig. 7** Plan topographique de la ville de Lyon, levé et dressé en 1863. Noms des rues. Surfaces bâties, 3rd edition from 1884 (Courtesy of Lyon Municipal Archives, A.M.L. 1S015)

1:5000 scale map, prepared in the years 1863–1864, the constructed area depicted on this east-oriented map shows how quickly urban development reached the old fortresses and the railway line, leaving few free spaces unoccupied on the left bank of the Rhone River (Fig. 7). Alongside the new city government, a totally new mapping effort would be needed to follow the intense urban sprawl of the second half of the nineteenth century.

## *3.2 New Mapping During the Second Empire*

C. Dignoscyo had prepared a map of triangulation points in Lyon in 1856, but we can only assume how inefficient such a support would be as the city underwent such important expansion. A very ambitious large-scale mapping project is launched in the 1860s that contributed to the affirmation of newly installed municipal services. In addition to the finalization and homogenization of a new general triangulation of the city, initiated by Fouque between 1861 and 1869 and continued by Grisard from 1868, a series of surveyors are called to work under contract in what was known as the "great topographic and cadastral survey of the city of Lyon" (Chinal 1999), a project that would continue for 110 years (1860–1970) (Fig. 8).

A complete new geodetic network was decided on, counting on private surveyors already responsible for projects in the traditional part of the city and using

chief-engineer of the municipality in his *Plan officiel des rues et places publiques de la ville de Lyon* (2S0167, 2S0167a, 2S0170, 2S0172) and so on.

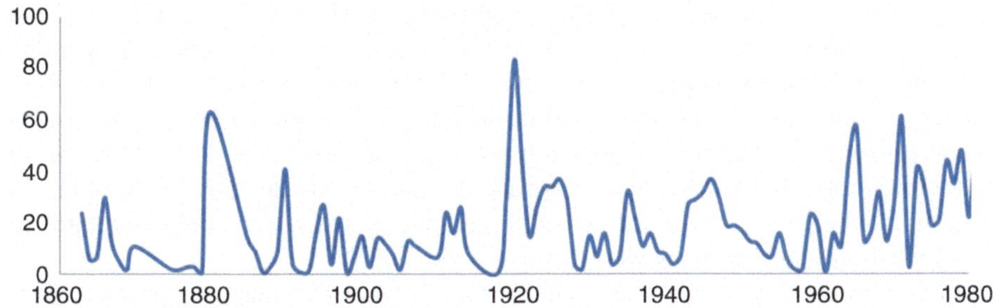


**Fig. 8** Number of charts edited per year at the 1:500 scale mapping from Lyon—series A.M.S. 4S. (Orig. De Biaggi 2016: 78)

the other surveyors already implicated in work in the newly acquired communes[12]. First completed in the scale of 1: 500, the cadastral plans were elaborated in parallel to the work of a common triangulation, starting with the areas where the number of geodesic points already designated would be sufficient. Initially, it was expected to create 342 sheets or sets of sheets representing the entire city of Lyon at that time. Each would measure 1 m wide and 0.6 m long and cover an area of 15 hectares. These maps, made under the direction of the Municipal Department, favored the representation of "urban objects", such as sidewalks, fountains, chandeliers, kiosks, benches, but also showed the planned development projects that gradually changed streets and buildings. Thus one can see the arrival and disappearance of the tram tracks, as well as new construction projects or alignments, sometimes not even carried out, but visible in overhead tracing on the available plans. The buildings appear with their street numbers on the public streets and, depending on the period, the names of their owners. Currently available on the website of the Lyon Municipal Archives (A.M.L.)[13], it is easy to see the evolution of the different editions produced over time, reaching a total of 604 charts, for the total coverage of the city (2479 maps in the 4S series of the Lyon Municipal Archives). The first editions that

---

[12]A first analysis of the authors and dates of publication shows the existence of different "waves" of publication and a very logical distribution of the areas of the city: a great part of the old city (center and west of Lyon, with the district of Vaise) is entrusted first to Balthazard Vignat, which produced 111 sheets between 1863 and 1876, at the time of the launch of the project; the left bank of the Rhône River is shared between E. Simon (10 sheets created in 1869), G. Mermet (45 sheets in 1878–1879 and 1905–1912), later associated with A. Vernay to produce 98 maps between 1885 and 1896. If G. Mermet guarantees continuity in the passage from one century to another, at a time when the rhythm of publication seems to diminish, in the twentieth century other names are associated with this systematic survey, especially Joanny Guillermain, which reaches 88 plans made between 1913 and 1926 and guarantees a significant recovery, visible shortly after World War I. The other surveyors involved, for example, Georges Berthier (responsible for 23 sheets between 1965 and 1969), Perraud (4 in 1966), Jean-Claude Sorro (a plan in 1970, and Roger Arnaud in 1970) complete the collection in the second half of the twentieth century, emphasizing mainly during the recovery of production that occurs after World War II.

[13]See http://www.archives-lyon.fr/archives/sections/fr/archives_en_ligne/le_territoire/cartes_et_plans

appeared in 1863 are still handwritten and one can still find the signature of the person in charge. This becomes less and less frequent as the updates occur at different times for each sheet, on average of 4–5 issues per sheet until 1980. This collection is an important resource for those who want to study the transformations of the urban area of Lyon, even though the pace of production is far from being uniform or regular for the whole territory. Some periods are obviously richer in editions, as the period of the 1880s, 1920s and the 1960s, which could be related to times when new urban plans were being developed.

In fact, in the end of the nineteenth century another two mapping projects were launched, in order to encompass the whole of the new city. From 1887 the 1:500 charts would serve as the basis for a new city mapping project at 1:2000 and afterwards in the beginning of the twentieth century, a 1:1000 series, both by way of the reduction of the existing 1:500. The first 1: 2000 maps appear after 1910 and concern mainly the sectors of the commune subject to new limits: to the west, the border between Lyon and the city of Tassin-la-Demi-lune and, to the east, the boundaries with Villeurbanne. Most of the 32 sheets in 1:2000, obtained from the pantograph reduction of 16 original sections made in 1:500, were edited together at the end of World War I in 1920 (28-sheet edition). This new series, produced first in color in the early 1920s, quickly resumes the appearance of topographic maps in black and white, made in 1: 500, indicating the names and numbers of streets, but only naming the main public and religious institutions. In the same way as the 1:500 series, the 1:2000 mapping will be re-launched several times (between 3 and sometimes up to 12 times, in the case of sheet 14 (Fig. 9)[14]. Through the rhythm of the edition, one can imagine that they were especially valued in the second half of the twentieth century, when more than half of the existing maps were re-edited, which demonstrates their importance for municipal management. The mapping attempt in 1:1000, on the contrary, did not result in general city coverage, effectively being performed for the part located in the center and east in the city (map AML 2S0543). This scale does not seem to have been considered adequate to translate the planning concerns of the territory at the moment, which went far beyond the city of Lyon.

## 4 Comparing Mapping Processes

As it can undoubtedly be seen in both cases, major changes in the cities can be observed by major changes in the way the city is mapped. The two cities, although far away from each other and under different political regimes, in periods of

[14]The establishment of plans at the 1:1000 scale, also by 1:500 reduction techniques, was valued at 14 francs per hectare (2 francs less than the 1: 500 revision) and should be carried out for 4 years. But after several delays and time extensions, the surveyor strove to finish 103 sheets by December 31, 1939!

**Fig. 9** Sheet number 14 from 1920s in the 1:2000 topographic series in Lyon completed by J. Guillermain, based on surveys from 1863, 1902 and 1926 (Courtesy of Lyon Municipal Archives, AM69123_5S014)

particular intense urban sprawl changed the way traditional mapping showed the city, favoring new orientation and new landmarks as references in the process. But as the city grows, new detailed mapping projects appear, even if they sometimes leave the newest regions apart, or to be treated on special occasions.

## *4.1 Mapping the Metropolis: Agents in the New Mapping Processes—Scales and Techniques*

Paradoxically, the growth of the area to be governed and managed seems to require a more detailed look at the city. In São Paulo and in Lyon not only was a completely new geodetic system created, but also large scale mapping was attempted (1:500 in Lyon and 1:1000 in São Paulo). One could consider that such an effort would be quite long to complete, the decision to go to smaller scales (1:2000 or 1:1000 in Lyon—1:5000 in São Paulo), could express a more realistic approach to a city that overflows its original borders. In an interesting twist of situations, in Lyon the bigger scale is produced first and it will remain the basis for a long-term mapping, even if it is subject to interruptions, as in the case of the two world wars. In São Paulo it is the contrary: after deciding on a 1:5000 mapping, the idea of producing bigger scale charts, even if only to a reduced central area comes in a major cartographic effort completed in only 4 and a half years.

In the new scales used, new techniques will try to answer a higher demand on maps, but they do not appear at the same pace in both cities. Having to cope with a rapid growth in a less rich availability of cartographic material, São Paulo is one of the first cities to order a full photogrammetric cover of its area in 1929, the SARA mapping of the city, which would be the basis of the large-scale maps produced from 1929 to 1933. As the detailed mapping from Lyon was launched in the 1860s, even if it was still carried on in the 1920s, it would count on the same methods used when the process began, a topographic survey led by local private surveyors. Lyon has some partial aerial photos being taken from the 1920s, but the first complete aerial coverage of the city was made almost 10 years later, showing nevertheless, a long term commitment to mapping activities inexistent in the extremely dynamic São Paulo, that will wait a long time for new projects.

Thus large scale mapping is a long-term project in Lyon, and in being so, once it is started, it is submitted to less change. In São Paulo large-scale mapping appears in a concentrated effort, capable of using the most innovative techniques available, even if they are developed abroad. In fact, the role of foreign companies mapping São Paulo (as the SARA mapping in the late 1920s) created a new reference map at the scale of 1/5000 and 1/1000 while building the idea of a modern city, even if this will be the only reference until the 1950s. Mendes (2014: 118) comments on how foreign companies will try to sell their products throughout the 1930s, as the country tries to develop the same technical support locally.

In Lyon, the private surveyors under contract that assured the mapping started in the 1860s were mostly locally based. They shared the same cabinet and sometimes were connected to some projects in the areas they were mapping. In long term contracts, the large-scale mapping of the city will take decades, and once started, would follow the same standards, even if new techniques are available.

## 4.2 *New Plans for the City, New Images of the City*

While both cities will finally end up with a conventional "north oriented" map as a basis for the local mapping system, many other supports were successively seen in São Paulo and in Lyon. In the latter, we go from a west-oriented map, to an east-oriented one, before the north-oriented maps become the rule in the beginning of the twentieth century[15]. In this city, where floods were important in the second half of the nineteenth century, the construction of docks for the two major rivers, the *Rhône* and the *Saône*, also helped changing the image of the city, fixing their courses and no longer framing it in the western area ("*presqu'île*"), with two parallel waterways—the *Saône* on the top and the *Rhône* below. In a broader context, a new form of the confluence of these two rivers appears as maps are north-oriented: the general maps of the city change and bring out the form of "Y" made by both rivers in Lyon when extended to the north and east. Even if some

[15]For a detailed report on this subject, see De Biaggi (2016).

particular maps of Lyon are still produced facing west in some special instances, such as during international exhibitions[16], the modern city that was planned in the twentieth century will inevitably include the more dynamic and industrial new extension zones annexed in the second half of the nineteenth century.

The rivers constitute also important landmarks in São Paulo, although they are successively changed from those present in the central areas, to the ones lying to the north-east and west of the city (from the *Tamanduatei* to the *Tietê* river). There are not so many changes in the orientation as in Lyon, but official city maps representing São Paulo do feature rivers mostly meandering, without indicating the floodplains or wetlands. All possible topographical constraints are almost nonexistent with an absence of obstacles to the city growth. The Tiete River, the main and largest river of the city, is shown with the plan of the future channel, which fulfills the function of regulating the riverbed to a pre-defined area, unchanging and stable, at least on the maps. The general plans of the city of São Paulo from 1897, 1901 and 1905 show the channel, preparing citizens for the future destiny of this river, which will separate them from its banks and consequently of leisure activities that were related to it.

The new mapping activities are the basis of a series of new urban plans (*Plan Chalumeau* in Lyon, *Plano de Avenidas* in São Paulo), even if they were not conducted in the same way or by the same kind of institution on each side. In both cases, these plans include mainly road maps, setting the stage for a city where daily mobility involves larger distances to travel based on new means of transportation. They imply a change in the scale and the way to show and live in the city that will be decisive for their further development in a new metropolitan context. Indeed, streets are the first sign of new buildings, even if they are not always drawn on the map. Still, the quality of the new city being build, like all the other services often forgotten on the maps, are of course subjects that will also have to be assessed in another period of urban sprawl, in the second half of the twentieth century, but by then, both cities show such differences, that comparing them will require other methods.

## References

Black J (2015) Metropolis: mapping the city. Conway, London

Bueno BPS (2016) Arqueologia da paisagem urbana: lógicas, ritmos e atores na construção do centro histórico de São Paulo (1809–1942). Revista do Instituto de Estudos Brasileiros 64:99–130

Chinal M (1999) Le grand plan parcellaire ou de détails de la ville de Lyon, 1860–1970: la contribution des géomètres privés. In: Forma urbis, les plans généraux de Lyon, XVIe–XXe siècles. Archives municipales, Lyon (Les Dossiers des Archives Municipales n. 10), pp 121–134

[16]For example, see the Reithofer *Lyon picturesque et monumental,* map produced during the 1894 international exhibition held in Lyon (AML 2S0333).

Delfante C, Pelletier J (2006) Plans de Lyon—portraits d'une ville: 1350–2015. Stéphane Bachès, Lyon

Dutenkefer E (2015) Mapa topográfico do Sara Brasil (1930) sobreposto às imagens atuais de satélite do Google Earth. In GEOUSP—Espaço e Tempo, São Paulo 19(1):156–159

De Biaggi E (2016) Dinâmicas de representações espaciais e dados cartográficos em Lyon no início do século XX: uma primeira abordagem. Revista do Instituto de Estudos Brasileiros, Brasil 64:75–98

Fernandes R Jr, Barbuy H, Frehse F (2012) Militão augusto de Azevedo. Cosac Naify, São Paulo

Gauthiez B (2006) Histoire de la cartographie des villes au XVIe–XVIIIe siècles—apparition et évolution des modes de représenttion. Città e Storia I 2:359–376

Gauthiez B (2008) Lyon en 1824–32: un plan de la ville sous forme vecteur d'après le cadastre ancien. Géocarrefour 83 (1): 57–67

Gauthiez B (2012) La transformation de Lyon et Paris au Second Empire: le projet du ministre de l'Intérieur De Persigny, les exécutants Haussmann et Vaïsse. In Fondazioni urbane, città nuove europee dal medioevo al Novecento, dir. Edizioni Kappa, Aldo Casamento, Rome, pp 323–344

Gauthiez B (2014) Le temps dans l'espace des villes, ou l'espace des villes comme accumulation des produits matériels de phénomènes sociaux. Géocarrefour 89(4):283–296

Knox P (2014) Atlas of cities. Princeton University Press/Ivy Press, Princeton

Kuvasney E (2016) Os mapas como "operadores espaciais" na construção da cidade de São Paulo do início do século XX. Revista do Instituto de Estudos Brasileiros 64:167–182

Machado A (2012) O levantamento aerofotogramétrico da cidade de São Paulo. In: Custódio V (ed) Fundamentos teórico-metodológicos do ensino e da pesquisa em geografia: textos selecionados das primeiras publicações da Associação dos Geógrafos Brasileiros (AGB)—Geografia (1935–1936) e Boletim da AGB (1941–1944). São Paulo

Mendes R (2014) S.A.R.A. Brasil: restituindo o Mapa Topográfico do Município de São Paulo. Informativo Arquivo Histórico De São Paulo 10(37)

Oliva JT et al (2016) Cartografia digital geo-histórica: mobilidade urbana de São Paulo de 1877 a 1930. Revista do Instituto de Estudos Brasileiros 64:31–166

Pelletier J, Delfante C (2004) Atlas historique du Grand Lyon. Xavier Lejeune-Libris, Seysinnet-Pariset

Petermann D (2016) A imagem de Lyon nos guias de viagem (1900–1950): uma abordagem geo-histórica das representações urbanas. Revista do Instituto de Estudos Brasileiros 65:120–144

Sarocchi P (1999) Une dynastie d'ingénieurs-géographes lyonnais au XIXe siècle, les Dignoscyo-Rembielinski. In: Forma urbis, les plans généraux de Lyon XVIe-XXe siècles. Archives Municipales, Lyon. Les dossiers des Archives Municipales, Archives municipales, 2e édition revue, corrigée et augmentée. http://www.archives-lyon.fr/static/archives/contenu/sommaires/forma_urbis.html

Vanario M (1999) Lyon, du XVe siècle à nos jours: l'évolution de la ville à travers les principaux plans. In: Forma urbis, les plans généraux de Lyon, XVIe–XXe siècles. Archives Municipales, Lyon (Les Dossiers des Archives Municipales n. 10). http://www.archives-lyon.fr/static/archives/contenu/sommaires/forma_urbis.html

**Enali De Biaggi** is a professor in The Department of Geography and Spatial Planning since 2002 at the Université Jean Moulin Lyon 3. After a Masters degree at the University of São Paulo in 1996, she joined the Institut des Hautes Etudes de l'Amérique latine at the Université Sorbonne Nouvelle Paris 3 for her PhD named "Cartography and spatial representations in Brazil" (2000).

**Fernanda Padovesi Fonseca** is a professor in the Department of Geography since 2008 at The Universidade de São Paulo. She has obtained both Masters and Doctorate degree at the Universidade de São Paulo, the latter on the subject "The inflexibility of the cartographic space, a question for Geography: analysis of the discussions about the role of cartography", presented in 2004.

# Colonial Sydney: A Cartographic Record of Change

**Robert Clancy and Alice Tonkinson**

**Abstract** This review of the cartographic record of colonial Sydney illustrates change from an isolated English penal colony of 1000 inhabitants, to an independent international city of half a million. Patterns of map production and distribution reflect changes in society and the structural evolution of a 'world city.' Histiocartographic analysis illustrates the:

1. transition from a make-shift village to a world centre,
2. evolution from a penal colony (1788–1840) via a 'nodal city' controlling exports, to an international city, important in its own right,
3. and the switch from a small, 'walking city' with colonial outposts to an urban/suburban complex absorbing its rural hinterland (the Cumberland Plain).

Printed maps and their publication and distribution characteristics both illustrate change, and add to the dynamic with their focus on colonial printing history. Maps in the First Fleet Journals (1789–1792) show early hope, then disappointment while those of French visitors (Baudin, 1802) depict a composite and busy village. The impact of Macquarie is seen in a map by Thomas Mitchell (Surveyor General), published with the "NSW Calendar & Directory" (1831), a list of residents connected with 'the General Post Office.' It was the first local, commercially lithographed map, and represented a watershed moment when Australian printing could replace cheaper printing in Europe. Traditional European sources (e.g. printed atlases, Hansard, published journals) contained few large-scale maps of Sydney—from this time nearly all important maps were published by local printers.

R. Clancy (✉)
University of Newcastle, Callaghan, NSW, Australia
e-mail: clancy_robert@hotmail.com

A. Tonkinson
State Library of New South Wales, Sydney, NSW, Australia
e-mail: alice.tonkinson@gmail.com

 
M. Altić et al. (eds.), *Dissemination of Cartographic Knowledge*, Lecture Notes in Geoinformation and Cartography, DOI 10.1007/978-3-319-61515-8_3

## 1 The Evolution of Colonial Sydney

The cartographic record of colonial Sydney documents an unique evolution of a city—from an English colonial convict camp of 1000 people (1788) to a bustling independent and international city (Federation and independence in 1901) of half a million. Contemporary maps illustrate the chaotic structural and functional impact of this explosion in population, with Sydney lacking the 'buffer' of earlier planned development found with most European and American cities. Many European cities with their classical post-Renaissance traditions combined the aesthetic with function and contemporary ideas of defence. Whilst in the New World, cities were built either according to Spanish colonial concepts of large central squares according to rules codified in the 'Laws of the Indies' (1573), or based on ideas that combined the influence of the enlightenment with the pragmatics of commerce, as modular 'gridded squares,' with both residential and commercial blocks around a central square. An example of the latter is Savannah in Georgia, planned by James Oglethorpe. Thus, to an extent the post-industrial -revolution population explosion in both the old and new worlds—often in the order of 100's fold—occurred within a built environment that did not exist in Sydney.

After 1870, Sydney began to evolve from being a 'nodal city' handling the wool and mining exports transported from its hinterland, to a major commercial and political centre within the Asia-Pacific region. Steam travel made possible a subdivision of the early land grants given to encourage agriculture for food, with the development of 'railway suburbs' the result. Despite this decompression of a congested Sydney, the commercial imperative now driving development meant a balance between the home and workplace fell well short of the ideal. As can be seen in Fig. 1, the population of Sydney continued to increase. The number of sheep in NSW also continued to increase. Whilst this reflects the dominant wool economy, it is more reflective of economic status, with sheep numbers halving in the 1890s, a time of major economic depression, with many from the rural population, re-locating to Sydney.

The end of the colonial period came in 1901, with Federation in Australia. This was a watershed between the previous dominant commercial interests, and ideals of liberalism with concern for individual living conditions and those who were disadvantaged. Enlightened industrial philanthropists saw the advantages of decentralisation and healthy living spaces with congenial work conditions, as captured in Ebenezer Howard's 'Garden City Movement' (1898). This movement advocated satellites surrounding a central hub, all linked by rail and road, with wide boulevards and open spaces giving a balance of residential, industrial and agricultural zones. However, in Sydney by 1900, 35% of inner city homes were dilapidated, unsanitary and unfit for habitation. It took the arrival of Bubonic Plague around Darling Harbour in 1900 and a subsequent Royal Commission (1909) to demand change. Daceyville was gazetted as a 'Garden Suburb' (1912) and the Cumberland Plan based squarely on Howard's ideas would follow, but too late, as post WWII immigration and individual commercial priorities denied any serious opportunity for planned development. From the time of Governor Arthur Phillip, planned development would be compromised by individual greed, driving Sydney upwards and outwards, with its population in 2016 approaching five million (Australian Bureau of Statistics 2016).

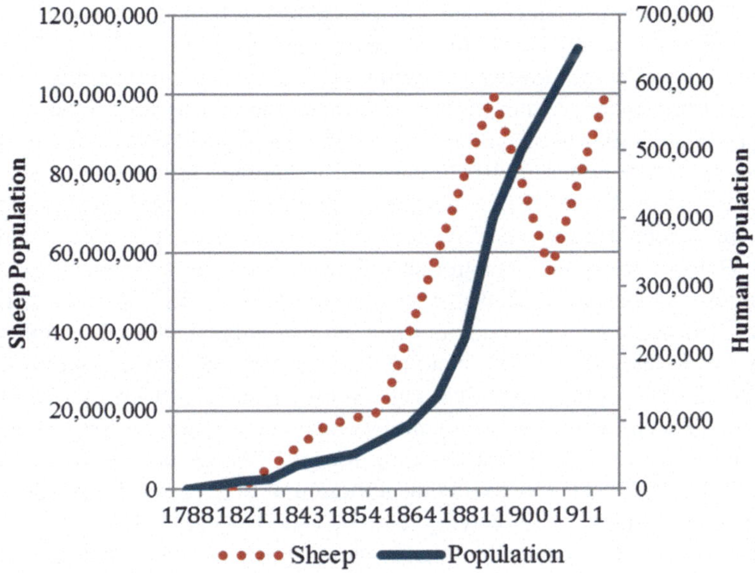


**Fig. 1** Population of NSW and number of sheep (1788–1926), compiled by the authors from official sources

## 2 The Cartography of Urban Development

Several key milestones of change are relevant to the cartography of urban development. First, was an 'escape' from the physical confines of the Cumberland Plain, the basin area between Sydney and the mountainous Great Divide to the west, which was crossed in 1813, unlocking a vast fertile hinterland that would generate wealth from wool and mineral exports. Sydney evolved into a nodal city, acting as a conduit between European markets and its rural hinterland, always remaining the main location of population growth.

Second, Governor Lachlan Macquarie's tenure (1810–1821) with his interest in urban planning as well as a related attention to social and economic issues, impacted on the built environment and street structure. Third was the need for a City Council (established 1842) to control urban development. Fourth, the discovery of gold to the west of Sydney in 1851, and the subsequent doubling of its population in the 1860s rising to 400,000 by 1890, put enormous physical and social pressures on urban development. Fifth, steam travel (1855), through the creation of 'railway suburbs,' enabled a level of decentralisation. Sydney was no longer a 'walking city' with adjoining suburbs coping with an overflowing population now expanding to cover Sydney's original rural hinterland over the Cumberland Plane. Although by 1890, 60% of Sydney's population lived in suburbs, the inner city area never adjusted with continued congestion and disconnect between the population and services. Sixth, waves of bubonic plague (1900–1920) with their associated terror and disruption, lead to a Royal Commission charged with providing guidelines to re-model an urban environment that had outpaced both services and efficiency.

The following maps illustrate changes related to these milestones. The first Governor of the colony, Arthur Phillip, had a clear plan for Sydney. Johnson argues this plan was a composite of symbols representing a magical code, tying the city to a prosperous future (Johnson 1992: 39).[1] Phillip's plan "Sketch of Sydney Cove, Port Jackson in the County of Cumberland," published in 1789 in his journal cites Albion (his original name for 'Sydney') on a fixed point (the Observatory) at 45 degrees from north, in a design that reflected views in the Spanish text "Laws of the Indies." (Johnson 1992: 39)[2] A second sketch by Phillip 4 years later is more realistic.[3] The settlement is a convict town, now orientated along the Tank Stream flowing into Sydney Cove, with quarters for marines and convicts, gallows and a prison, and a road south to the brickfields. Maps by senior officers William Dawes and Watkin Tench identify early urban development adjoining a sparse hinterland limited in the west, by the Great Dividing Range.[4] Change occurred with government grants (1791–1831) encouraging agricultural and pastoral development to feed the colony. Maps accompanying texts by David Mann (1811) and J. Burr and G. Ballisat (1814) document this phase, blurring any boundary between Sydney Town and its surrounds.[5] Details of early urban development are captured in Charles Leseuer's Plan de la Ville de Sydney (1802) published with the printed journal of the French expedition led by Nicolas Baudin (1800–1803) (Fig. 2). Trappings of a busy urban development include Mr. Campbell's store, a hospital

[1]Johnson describes Governor Phillip's 'Albion' (Sydney) as "conceived more as a diagram of hope than as a practical affair... by gematria, the substitution of numbers for letters" (1992: 38–44).

[2]Arthur Phillip's journal, *The Voyage of Governor Phillip to Botany Bay* (1789) was published just 1 year after the colony was established at Sydney Cove. It includes Phillip's idea of how he wished Sydney to be developed: Sketch of Sydney Cove. Port Jackson in the County of Cumberland.

[3]Immediately prior to Phillip's departure, he commissioned a second map to record how plans were put into place [*A survey of the Settlement in New South Wales. New Holland* (1792)]. The manuscript is held by State Records, Sydney and includes written comment by Phillip that no land to the west of the boundary can be granted or leased—a forlorn hope!

[4]Six volumes written by officials arriving in 1788 and known as the First Fleet Journals, were published between 1789 and 1798. Three contained important maps of early Sydney: Phillip (see footnote 2 above); *A Map of all those Parts of the Territory of New South Wales. . .belonging to the Settlement established at Port Jackson.*by William Dawes (1792) published in *An Historical Journal of the Transactions at Port Jackson and Norfolk Island. . .*(1793) by John Hunter; and *A Map of the hitherto explored country contiguous to Port Jackson. . .* by Captain Watkin Tench, published in his *A Complete Account of the Settlement of New South* (1793). The maps by Dawes and Tench are the first to record the immediate hinterland to the colony at Sydney Cove—areas later to be incorporated into modern Sydney.

[5]At the time of crossing, the Great Divide had confined Sydney to the Cumberland Plain for 25 years. Two general maps identifying the then limits of settlement were published. The first by David Dickenson Mann (1811) in *The Present Picture of New South Wales* [*A new Plan of the Settlement of New South Wales* (1810)] identifies the districts of settlement, while *Plan of the Allotments of Ground, granted from the Crown in NSW* (1814), J. Burr and G. Ballisat sold with a guide for immigrants, details specific grants in this region.

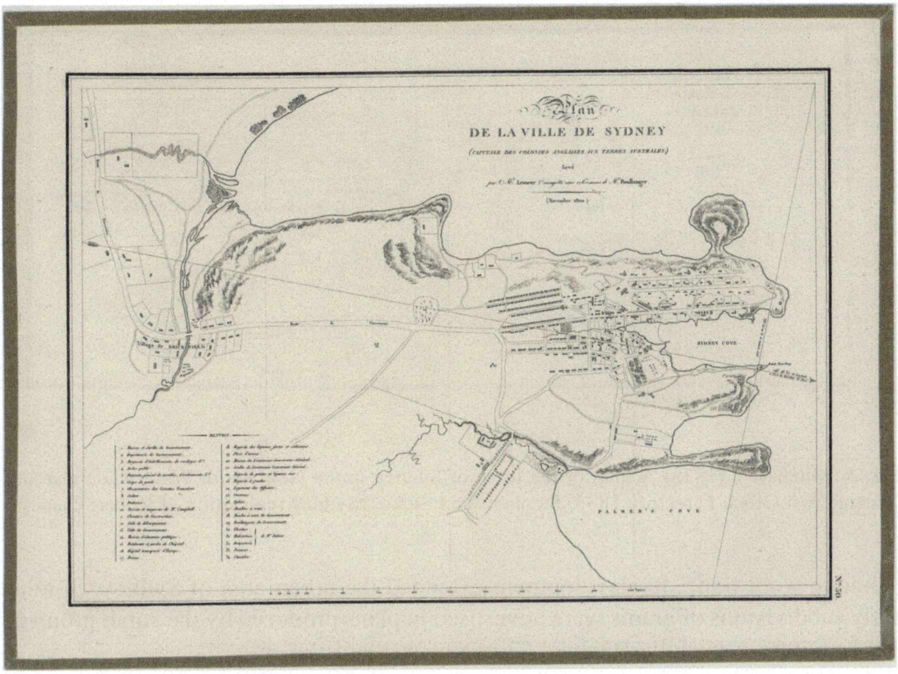


**Fig. 2** Lesueur CA (1802) Plan de la Ville de Sydney. Capitale des Colonies Anglaises in (1824) Voyage de Découvertes Aux Terres Australes, Arthus Bertrand, Paris (collection of Robert Clancy)

wharf, a school, salt works and hints of early extension to the south, though gallows and a jail are darker reminders that this remains a convict settlement.

*A Map of the Town of Sydney* (1832) by Thomas Mitchell (Surveyor General) published by Stephens and Stokes in *NSW Calendar and Directory*, was the first locally printed, commercial map (see Fig. 3 for the 1833 edition). It shows the dramatic change in urban development that followed the opening of the western plains and Macquarie's tenure, with a population now 22,000. Sydney is a busy port, building ships, handling rural product and servicing whalers and sealers. Military and convict sections are replaced by commercial and domestic expansion, which, however, remain blurred. A 'green-zone' extends from the harbour to Hyde Park, and the beginnings of suburbs can be seen. It is a blueprint for the future city of Sydney. The fourth edition of the *Calendar* with an updated map (1836) was published by Ann Howe—daughter-in-law of George Howe, who was the first commercial printer in Sydney (1805).

The subdivision of land adjoining central Sydney (while the town was still confined as a 'walking city') to accommodate both commercial and domestic needs, is anticipated in a map by J. Basire (*Plan of Sydney with Pyrmont*..., 1837) published in the British Parliamentary Papers. The property on land adjoining Darling Harbour was owned by the Macarthur family, known for initiating the now

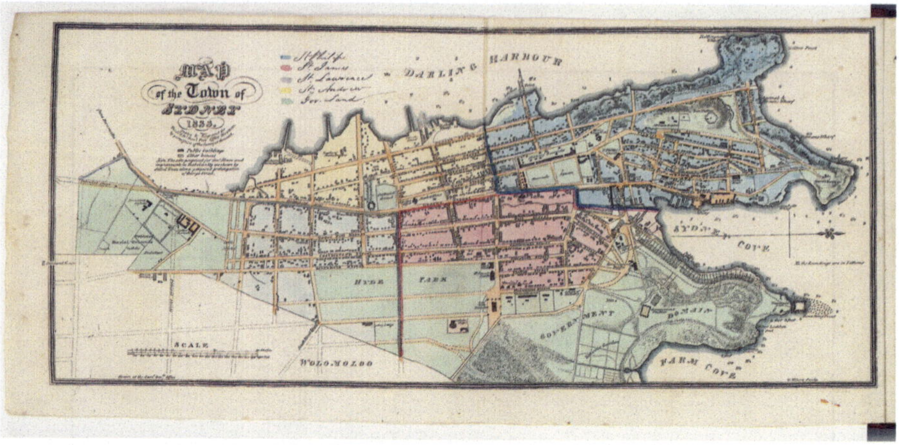

**Fig. 3** Mitchell T (1833) A map of the town of Sydney in the New South Wales Calendar and General Post Office Directory 1833. Stephens and Stokes, Sydney (collection of Robert Clancy)

lucrative wool trade. Its development extended the urban area of Sydney. Similar early subdivisions of grants were advertised in plans produced by the small group of local printers, especially Raphael Clint and William Baker.

Locally produced maps of Sydney increased in the 1850s after the discovery of gold. A new group of small companies such as Woolcott and Clarke (*Map of the City of Sydney*, 1854) and Smith and Hinton (*Map of Sydney and Suburbs*, 1854) are prominent, together with the continued appearance of almanacs, directories and guides. These maps show an increasingly congested city with busy ports but limited transport. Water reticulation from Busby's Bore and gas works at Darling Harbour, mark the changes.

After 1870, maps by small-sized printers are few. Major changes brought about by steam transport and increasing population can be seen with the appearance of a new breed of large-scale printers, many also with outlets in Melbourne. The major producers of these maps in Sydney were John Sands Ltd. and Gibbs, Shallard and Co. Maps from publications such as the *Sands Directories* (1858–1933) included comprehensive documentation of changes in Sydney as it cannibalised land previously part of large land grants. For example the map from the 1894 *Directory* (*Sydney and Suburban Municipalities*) includes tabulated demographic data of the new municipalities, details of trades etc., documenting the late colonial evolution of urban Sydney, while the 'new railway suburbs' represent a beginning of separation of urban and commercial Sydney (Fig. 4). Details of urban Sydney can be seen in the fire-insurance maps of Percy Dove (*A New and Complete Wharf, Street & Building Directory…of Sydney*, 1880) with 119 blocks at a scale of 1:480, and printed by John Sands Ltd. Risks and building materials are colour coded. By 1900 Sydney was a congested city, with services disconnected from those living within it, and health at risk. A map accompanying the *Report of the Outbreak of Plague at Sydney*, (1900) published by the Government Printer shows the distribution of

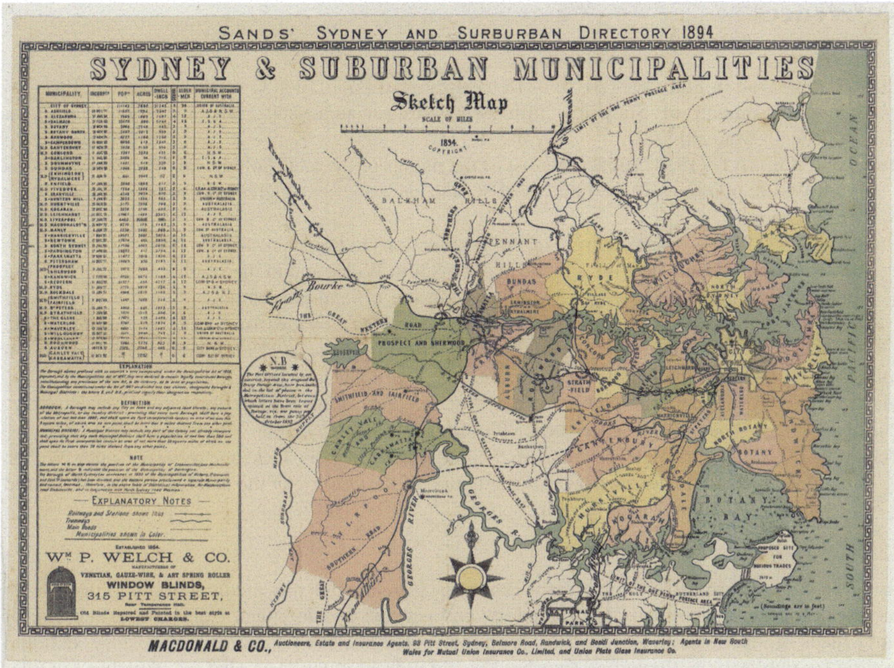


**Fig. 4** John Sands (1894) Sydney and Suburban Municipalities, in John Sands, Sydney and Suburban Directory, Sydney (collection of Robert Clancy)

houses clustered around Darling Harbour, where cases were identified, reflecting transmission via the rat-flea from ships moored in the area.

Four major printers (defined as employing more than 60 staff) dominated map production between 1860 and 1890 in urban Sydney. They were W.C. Penfold; John Sands Ltd.; S.T. Leigh and Co.; and Gibbs and Shallard (which sold out to McCarron and Stewart in 1892). Specialist map publishing in Sydney was pioneered by H.E.C. Robinson who began in Sydney in partnership with Higinbotham and Harrison (1882–1887), then as a sole trader from 1895. The development of a strong commercial publication of maps reflected changing needs, with a general population now requiring urban maps as street directories, bicycle routes as well as for tourism and land sales.

Throughout colonial NSW, the government was a source of cadastral, planning and subdivision maps. The Surveyor General's Department produced manuscript maps largely for survey purposes until 1829, when a lithographic press was acquired. The first printed map was *The Hundred of Parramatta* by Peter Berni (held at the State Library of NSW). Once commercial printing of maps began in 1832 (see above), few detailed maps of urban Sydney containing primary material appeared from European sources.

## References

Australian Bureau of Statistics (2016) 3218.0 – Regional Population Growth, Australia, 2014–15. Australian Bureau of Statistics. http://www.abs.gov.au/ausstats/abs@.nsf/mf/3218.0. Last accessed 15 Mar 2017

Burr J and Ballisat G (1814) An acurate list of the names of the land-holders, in the colony of New South Wales; pointing out the number of aces in each district, as granted from the Crown. Burr & Co, London

Hunter J (1793) An Historical Journal of the Transactions at Port Jackson and Norfolk Island with the discoveries which have been made in New South Wales and in the southern ocean since the publication of Phillip's voyage, compiled from the official papers, including the journals of Governors Phillip and King and of Lieut. Ball, and the voyages from the first sailing of the Sirius in 1787, to the return of that ship's company to England in 1792. John Stockdale, London

Johnson PA (1992) In the beginning: from the first fleet to Macquarie. Globe 37:38–44

Mann DD (1811) The present picture of new South Wales. John Booth, London

Phillip A (1789) The voyage of Governor Phillip to Botany Bay. John Stockdale, London

Tench W (1793) A complete account of the settlement at port Jackson, in new South Wales: including an accurate description of the situation of the colony; of the natives; and of its natural productions. G. Nicol and J, Sewell, London

**Robert Clancy** AM FRS(N), Professor Emeritus. Career academic medical immunologist, with research programme in mucosal immunology and vaccine development. Collector and curator of early maps of Australia and Antarctica, with maps recognised as documents of history, defining criteria for publications (including four books) and presentations. Awarded membership to Order of Australia for contributions to both immunology and historic cartography.

**Alice Tonkinson** BA is a postgraduate student, studying a Master of Museum and Heritage Studies at the University of Sydney. She works part-time at the State Library of NSW and also administers a private map collection.

# Invariable Image of Cyprus: Reproduction of Siege Maps of Nicosia and Famagusta

**Merve Senem Arkan**

**Abstract** The island of Cyprus had witnessed a clash in 1570–1571 between the two renowned nations; the Venetians and the Ottomans. This prominent conflict between the East and the West brought about the creation of the siege maps. While depicting the war between the Venetians and Ottomans, the maps were also given information about the cities themselves. The cartographers depicted this conquest with the maps of the two most significant cities of the island; Nicosia; the capital city and Famagusta; the main port city. The maps of the conquest of the island played a role of telling and creating the image of what is going on in the Eastern Mediterranean Sea. These urban maps allow us to witness the turning point in history of Cyprus.

These sixteenth century siege maps were used to represent these two cities even after the conquest of the island by the Ottoman Empire. Consequently after the siege, the contents of these maps were changed and reproduced by the map makers in the later centuries. They became only material to represent the two main cities of Cyprus for next 300 years.

The paper focuses on the siege maps as informative materials about the two cities and the situation on the island for the European audience. It studies the reproduction of these maps and the changes that were made by the map makers accordingly to the time period and reasons behind using the sixteenth century maps to represent these two cities in the later centuries.

## 1 Introduction

The island of Cyprus had witnessed a momentous event in the sixteenth century; a clash between the Venetians and Ottomans to command over a small yet important island in the Eastern Mediterranean Sea. The Ottoman troops started the conquest of the island in 1570 under the orders of Sultan Selim II. The Venetians confronted the Ottoman threat after 80 years of dominance over Cyprus. The sieges of two cities; the capital city, Nicosia and the main port city, Famagusta were crucial in

M.S. Arkan (✉)
Eötvös Loránd University, Budapest, Hungary
e-mail: m.senemarkan@gmail.com



M. Altić et al. (eds.), *Dissemination of Cartographic Knowledge*, Lecture Notes in Geoinformation and Cartography, DOI 10.1007/978-3-319-61515-8_4

this takeover. The Ottoman troops marched towards Nicosia first. After the capture of the capital city, the troops moved on to Famagusta. The siege of Famagusta lasted for almost a year which ended with the victory of the Ottoman Empire (Gunnis 1947: 19–20). This turning point in Cypriot history was illustrated on several maps.

In the sixteenth century, because of the power of the Ottomans and their threat to Europeans, a war between the West and East was noteworthy. An important event such as the war in Cyprus turned many eyes to the island which initiated a new era of urban maps and boosted the map production. The priority laid with the town plans of Nicosia and Famagusta as to illustrate the war between the Venetians and Ottomans (Navari 2003: 26; Marangou 2002: 124).

Within these maps, Giovanni Francesco Camocio's and Georg Braun and Frans Hogenberg's maps were perhaps the most significant ones and they were copied by many other cartographers during as well as after the conquest. The sixteenth century maps of Famagusta and Nicosia became the main source for representation of these cities until the start of the British Era in 1878.

The paper focuses on the siege maps as informative materials about the two cities and the situation on the island for the European audience. The paper studies the reproduction of these maps and the changes that were made by the map makers accordingly to the time period and reasons behind using the sixteenth century maps to represent these two cities in the later centuries.

## 2 Famagusta

Giovanni Francesco Camocio produced many isolarii, besides being a map maker he was also a publisher. His maps normally illustrate wars and so, during this period, the conflict between the Venetians and the Ottomans was dominant (Stylianou and Stylianou 1980: 36). As a result of this, Camocio depicts several maps of Cyprus including both Famagusta and Nicosia to show the conquest of the island (Fig. 1).

The map of Famagusta shows the city and siege together. It illustrates the city's fortification and bastions with the citadel. In the centre, the main public buildings (the Palace and the St. Nicholas Cathedral and another church near the Palace) and private buildings were depicted without giving an accurate layout of the inner city. Only two main streets can be seen; street from sea gate to the main square and another one from the main square to the south side of the city, with determined main piazza near the Palace. In addition, the harbor is given in detail as well. In the inner harbor, he depicted four ships along with ships surrounding the city outside of the harbor. On the land, outside the walls, there are the Ottoman troops, in the background Ottoman tents, and in the foreground troops some on horses, some walking, and some are depicted with canons, advancing on the doomed city.

Besides Camocio's map, a similar representation of the city was used on Braun and Hogenberg's map *in Civitates Orbis Terranum*. The *Civitates Orbis Terrarum* included mostly European city maps but also those in Africa, Asia, the Middle East,

**Fig. 1** Giovanni Francesco Camocio, "Famagosta", in *Isole famose porti, fortezze, e terre maritime sottoposte alla Ser.ma Sig.riadi Venetia, ad altri Principi Christiani, et al Sig.or Turco, nouamēte poste in luce. In Venetia alla libraria del segno di S. Marco.*, Venice, ca.1575 (Courtesy of Sylvia Ioannou Foundation, Vaduz)

Peru and Mexico, first volume was published in 1572 (Stefoff 1995: 61). As part of the project, there is a plan of Famagusta as well (Stylianou and Stylianou 1980: 67). The engraving of the map was admitted to Camocio in ca. 1568. The inner part of the city is dominated by private houses while the same three public buildings were depicted differently (the Palace and two churches) to distinguish them from other buildings and to show their importance. Especially the Palace's features are given in detail; larger than other buildings with two storey so the building stood out in the map. On the map, there is no written name of the buildings. Again, the city walls, the citadel and the bastions are well depicted, as well as the harbor. In the 18-line text beside the map, information is given about the political situation of the city, the conquest and the siege of 1571 (Stylianou and Stylianou 1980: 68; Navari 2003: 82), and at the top of the map is a written account of Famagusta's strong defense and the Turks domination (Fig. 2).

Several other cartographers reproduced these maps during and after the siege. There are a few additions and changes according to preferences of the map maker; reproducing with colour or replacing the Ottoman troops with the landscape. Balthasar Jenichen adds movement to the Camocio's map, by including smoke

**Fig. 2** Georg Braun and Frans Hogenberg, "Calaris/Malta/Rhodus/Famagusta", in *Civitates Orbis Terrarum*, Cologne, 1572, (Courtesy of The National Library of Israel, Eran Laor Cartographic Collection, Shapell Family Digitization Project and The Hebrew University of Jerusalem, Department of Geography—Historic Cities Research Project, Jerusalem)

coming from the canons, both on the city walls and from the Ottoman batteries[1] (Stylianou and Stylianou 1980: 57), or Simon Pinargenti's map from 1572–1573 which is a colored version of Camocio's map (Stylianou and Stylianou 1980: 47) (Fig. 3). Other than these map makers; Donato Bertelli has published the same map in 1575 as well[2] (Stylianou and Stylianou 1980: 46) where he made few changes on the troops and ships at the surroundings of the city but kept the city's representation same.

In addition, the image of Braun and Hogenberg's map without the siege was used by many cartographers as well; Sebastian Münster published a map of Famagusta in *Cosmographey oder beschreibung aller Länder* in 1574[3] (Stylianou and Stylianou 1980: 26; Navari 2003: 94); Francesco Valegio portrayed the same image in *Raccolta di le piu illustri et famose citta di tutto il mondo* which was published in

[1]Balthasar Jenichen, "Famagosta. . .BI", Germany, 1571, M. Michaelidas Collection, Nicosia.

[2]Donato Bertelli, "Famagosta", in *Civitatum Aliquot. . . Disegni di alcune piu illustri citta, et fortezza del mondo. . . Venetiis MDLXXIIII, Donati Bertelli formis*, 1575, Bibliothéque Nationale, Paris.

[3]Sebastian Münster, "Famagusta", in *Cosmographey oder beschreibung aller Länder*, Basel, 1574.

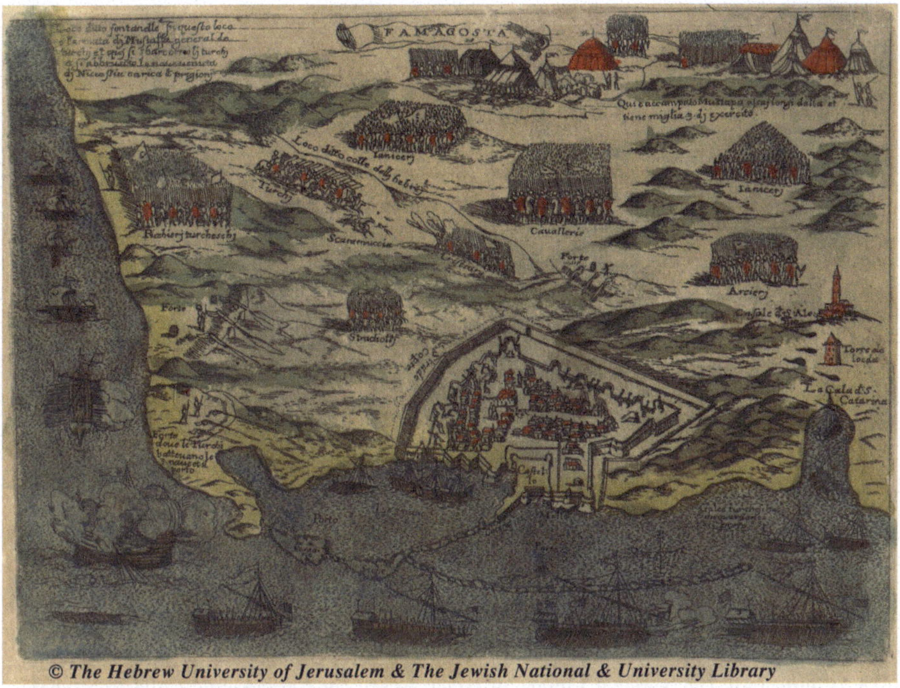


**Fig. 3** Simon Pinargenti, "Famagosta", in *Isole che son da Venetianella Dalmatia*. . ., Venice, 1572–1573, (Courtesy of The National Library of Israel, Eran Laor Cartographic Collection, Shapell Family Digitization Project and The Hebrew University of Jerusalem, Department of Geography—Historic Cities Research Project, Jerusalem)

1579[4] (Stylianou and Stylianou 1980: 74). Later on, Henry de Beauvau used the same reproduction in his book *Relation Journaliére du Voyage de Levant* in 1615 (Stylianou and Stylianou 1980: 89) with a small change of depicting less number of private buildings (Fig. 4).

In *Atlantis Maioris Appendix* same representation of the city was published by Joannes Janssonius (publisher), and Iodocus Hondius (author of the map) in 1630[5] at the frame of the map of Asia (Stylianou and Stylianou 1980: 93) and also it was used as an inset in John Speed's map of the Turkish Empire in 1626. The inset is located on the top left corner[6] (Stylianou and Stylianou 1980: 90). In these two images, the representation of the fortification and the inner city is vague since they are a small part of a bigger map. On these maps the appearance of the city mostly stayed same; the same main public buildings are identifiable which surrounded by

[4]Francesco Valegio, "Famagusta... ciuitas Cypri", in *Raccolta di le piu illustri et famose citta di tutto il mondo*, Venice, ca.1579.

[5]Joannes Janssonius, Iodocus Hondius, "Famagusta", in *Atlantis Maioris Appendix*, Amsterdam, 1630, Bibliothéque Publique et Universitaire, Genéve.

[6]John Speed, "Famagusta" (inset in The Turkish Empire), in *Prospect of the Most Famous Parts of the World...*, London, 1626, Bibliothéque Publique et Universitaire, Genéve.

**Fig. 4** Henry de Beauvau, "Famagosta", in *Relation Journaliére du Voyage de Levant,* Nancy, 1615, (Courtesy of The National Library of Israel, Eran Laor Cartographic Collection, Shapell Family Digitization Project and The Hebrew University of Jerusalem, Department of Geography—Historic Cities Research Project, Jerusalem)

private buildings. The important identifications of the city; fortification, bastions and the citadel were portrayed, and for the outside of the city, landscape took place of the Ottoman troops.

The fortifications and the inner city of the maps show the sixteenth century version of the city. From the written accounts, we know effects of the war and the details of the city transformation from a Christian city to a Muslim one with rising of the new Ottoman style buildings over the city. The siege was one reason for the slow demolishing of the city as well as the decreasing population, which in return damaged the city. Houses were left empty and churches were not in use. The population was excessively gathered around the Palace area (Uluca and Akın 2008: 179). It seems clear that the city was a ruin and almost empty, brought about by war, natural disasters, ill-health and economic decline. Frederic Hasselquist was a traveler who visited the island around the 1750s, he writes,

> Famagusta ought to be seen, less for itself than because it was formerly the strongest place in the island, and much talked of in history . . .. The town is in far worse condition than the fort; all the houses built by the Venetians are utterly demolished or deserted. There are but three hundred inhabitants, chiefly Turks, who occupy the miserable remains of the famous city of Famagusta. (Cobham 1908: 307)

During the 100 years of Ottoman domination, the city went through both dire and prosperous situations. The Ottoman administration's first work was to restore damaged structures caused by the siege. The fortifications were the most damaged which was the priority for the Ottomans. During the process of converting the city, the first alteration started with the conversion of St. Nicholas Cathedral into a mosque (Gunnis 1947: 88–89). Besides the main Cathedral, new constructions started to rise inside the walled city; baths, fountains, schools, and khans were constructed during the 300 years of the Ottoman rule (Uluca and Akın 2008: 184). However, the altering conditions of the city during the Ottoman Era or changes on the appearance of the city were not reflected on the maps.

In the later centuries, Famagusta had appeared on maps other than these copies of sixteenth century maps such as nautical charts and plans of the city. Even though the city had plans from the Ottoman Era, these plans depicted the city walls and the harbour but omitted information about the interior of the city such as buildings, streets and squares. Moreover, some of the plans were erroneous in their depiction of even the main features of the city. Descriptive and informative accounts from foreign countries were not detailed in their description of the inner town. The plans depicted the inner city empty due to restrictions against Christian population entrance into Famagusta during the Ottoman Era. When the Ottoman Empire took over the island, they closed down the city to Christians and foreign visitors which caused the Christians to move out of the city and form the town called "Varosha". Varosha appeared on some of these plans as small village or gardens outside the city.

During the Ottoman era of the island, the port cities of Cyprus appeared only as an appendage on the charts. Particularly French ships and other foreign ships were visiting Cyprus because of the privileges acquired from the Ottoman Empire, for commercial reasons or for refitting (Bulut 2002: 198–199). As the port city, Famagusta appeared on nautical charts that show the depths of the harbour and give an image of the walled city. Usually, these images were symbolic; showing the fortification and few buildings inside. An imaginary view of Famagusta on the charts was depicted because the city itself was no longer significant and the harbor was in a poor condition and not being used to its full capacity. The only information about the cities' socio-political change was the addition of the crescent symbols to identify the Muslim occupation of the island.

In the later centuries of the Ottoman rule, the prohibition for entering the city was removed, whereupon travel texts offering information about the inside of the city reappeared. But the city, now dead, still did not gain the map makers' interest. Charts were created that only focused on the harbor because even if the port had left its glorious days behind, it was still in use. For their audience of mainly sailors, nautical charts that focused on the harbor were suitable.

These maps of Famagusta have represented the city until the start of the British Era in the nineteenth century. Besides the main port city, the capital city Nicosia was represented by the sixteenth century siege maps in the later centuries.

## 3 Nicosia

One siege map stands out for Nicosia; Camocio's map of the city (Stylianou and Stylianou 1989: 26) was reproduced several times by various map makers. On the map, outside the walls, he illustrates the troops of the Ottoman army that surrounded the city. Within the walls, he chose to illustrate and label the major and the most significant buildings of the city in the middle, St. Sophia Cathedral. Besides that, there are illustrations of other churches that stand out, however, none of them are marked. The symbolic images of these churches are that of classical designs with single towers as opposed to the St. Sophia Cathedral which has a different architectural shape with an additional tower. The rest of the inner city was drawn with private buildings spread without an accurate plan of the city. We cannot really determine the streets, or squares on the map. He depicted the diagrammatic shape of the city and focused on the 11 bastions that are one of the most significant features of the city. During the Venetian era, the outline of the city changed to create better and stronger fortifications against a possible Ottoman attack which is the circular shape of the fortification today; the perfect military architecture as mentioned by Cosmescu "The perfect shape of Venetian Nicosia is unique: a circle, the most perfect of the geometrical forms, with its center quite close to the cathedral of S. Sophia" (Cosmescu 2016: 46). On this map, the river is also depicted, as flowing from one side, filling the moat and continuing to the other side as it was after it was diverted. Two bridges are depicted; one towards the north-west and the second one towards the south-east direction. He only showed two gates instead of three; Famagusta (Giuliana) and Paphos (S. Domenico), and even then, their names are not specified. There are other inaccuracies about the fortification of the city; for example; the Paphos gate should be between adjacent bastions at the south. Another example is that the bastions were drawn as symbolic rather than as a realistic representation because their edges are too sharp-cornered (Fig. 5).

Simon Pinargenti published a colored version map of Nicosia in his isolario in 1573[7] (Stylianou and Stylianou 1989: 36). Another map of the city which is based on Camocio's map without the siege is from the town book by engraver Francesco Valegio, *Raccolta di le piu illustri et famose citta di tutto il mondo*[8] (Stylianou and Stylianou 1989: 42). There are other maps that present the same image of the city with only a few changes. Marco Sadeler's work *Viaggio da Venetia…*, after 1610 which was first published by Giacomo Franco in his work *Descrittione… Viaggio da Venetia à Costantinopoli*, in 1597 (Stylianou and Stylianou 1980: 77; Navari 2003: 42) was based on Camocio's map (Fig. 6).

Another copy of the same map of Nicosia was from a French booklet under the title of *Relation iournaliere du voyage du levant faict and descrit par haut et*

---

[7]Simon Pinargenti, "Nicosia", in *Isole che son da Venetia nella Dalmatia*, Venice, ca. 1572–1573, the Hebrew University of Jerusalem, the Jewish National and University Library, Jerusalem.

[8]Francesco Valegio, "Nicosia", in *Raccolta di le piu illustri et famose citta di tutto il mondo*, Venice, 1579.

**Fig. 5** Giovanni Francesco Camocio, "Nicossia", in *Isole famose porti, fortezze, e terre maritime sottoposte alla Ser.ma Sig.ria di Venetia, ad altri Principi Christiani, et al. Sig.or Turco, novamente poste in luce, Venice, alla libraria del segno di S.Marco*, Venice, ca. 1575, (Courtesy of Sylvia Ioannou Foundation, Vaduz)

*puissant Seigneur Henry de Beauvau, Baron du dict lieu et de Manonville, Segneur de Fleuille, Sermaise, Domepure etc* which covers Henry de Beauvau's journeys in the Levant in 1615[9]. The booklet also has a small map of Cyprus and a plan of Famagusta (Stylianou and Stylianou 1980: 89). Angelicus Maria Myller's book describing his voyages to Jerusalem, under the name *Peregrinus in Jerusalem, Fremdling in Jerusalem, Oder Ausfuhrliche Reisz = Beschreibungen* also used this map which was published in 1729[10] (Navari 2003: 251).

The representations of the inner city stayed the same on these maps. On few maps, the gates were removed completely and on others, only one of the gates, usually the Paphos gate was depicted similar to Camocio's map, in the wrong place. The main difference on all these maps is the absence of the siege which was transformed to the landscape. The siege maps of the city were replaced with urban maps that only focused on the city without showing the battle. On the other

[9]Henry de Beauvau, "Nicosie" in *Relation Journaliére du Voyage du Levant*, Nancy, 1615, the Hebrew University of Jerusalem, the Jewish National and University Library, Jerusalem.

[10]Angelicus Maria Myller, "Nicosia", in *Peregrinus in Jerusalem, Fremdling in Jerusalem, Oder Ausfuhrliche Reisz = Beschreibungen*, 1735, Bank of Cyprus Cultural Foundation, Nicosia.

**Fig. 6** Giacomo Franco, "Nicosia", in *Viaggio da Venetia à Costantinopoli Per Mare, e per Terra...*, 1597/1598 (Courtesy of Sylvia Ioannou Foundation, Vaduz)

hand, the maps did not illustrate changes that the city was going through under the new government.

The new government was reconstructing old buildings as well as creating new and oriental-style ones around the city. With these innovations, the city had a new look which would last for centuries; a mixture of Christian and Muslim cultures and designs where the lower part is in Gothic style and the upper part is in the Ottoman style with cumbas (Gürkan 2006: 33–34). New buildings constructed according to the Muslim tradition included mosques (mescit), mausoleums (türbe), khans, baths, fountains and others (Gürkan 2006: 35). The city preserved its oriental appearance. After the siege and under the Ottoman domain, the city appeared more like the beautiful oriental city that the travelers wrote about rather than a city in ruins. Dixon describes the city:

> This little sister of Damascus is a labyrinth of lanes and alleys; winding under minarets, towers and fruit-trees; round about khans, kiosques, and fountains; in and out among brown walls, running waters and broken grounds. Only a native of the town can find his way from gate to mosque, from coffee-house to khan. (Dixon 1879: 141)

The map maker only used information available from the previous century, and it ended up as a replica of the sixteenth century maps of Nicosia without the siege. It does not give any additional information about the city or any new information about the Ottoman rule and the changes the city had gone through in this century. The inaccuracies and the lack of information that started with Camocio's map were continued with other maps, such as; the number of gates, their locations or layout of the inner city, etc.

As a result of the siege, the city's fame continued even after its conquest, although, in the next two centuries the number of maps of the city decreased drastically. The city did not have a problem with restrictions, however, as the inland city on Ottoman territory, Nicosia did not appeal to the foreign cartographers anymore. The city's fame came from its 'ideal city' architecture that the cartographers replicated from previous maps, or they only focused on its fortifications. Even though travelers wrote about the city and its new ruler, these did not appear on the maps. The image of the city was limited to the loose sixteenth century representation without giving any changes or developments that city went through during the Ottoman Era.

## 4 Conclusion

After the sixteenth century, the maps on the later years were not detailed or informative about the Cypriot cities or its events. The nautical charts were created for the port cities and the only inland city that appeared on the maps was Nicosia. After Famagusta and its harbor became an enemy territory, Nicosia as an inland city disappeared from the maps. Therefore, both cities and the once-famous harbor lost their value in the eyes of the Europeans and the European map makers. When a city loses its importance, it effectively disappears.

The conquest had a huge impact on the Cypriot history and cartography. The sieges turned the eyes of the European map makers towards the island however during the Ottoman era this interest were lost. With the decline of number of maps and the lack of Ottoman maps, these maps became the main source of cities' representation. The maps were used over and over again, repeating same mistakes and same images of the cities without showing any improvement or changes.

The siege maps were used for 300 years that represent these main cities of the island marking them as important assets in cartography. Under the Ottoman Empire, some changes were made within the walls, such as the transformation of buildings, conversion of churches into mosques, or additions or changes to buildings according to the new lifestyle and aesthetics of the Ottoman Empire. But these changes were not reflected in the maps.

Although these maps preserve the knowledge of the sixteenth century Famagusta and Nicosia, they also create invariable images of two major cities of the island for centuries which show the power of the siege maps over the Cypriot cartography. The siege maps of Famagusta and Nicosia from this period determined Cypriot urban cartography for the next three centuries.

## References

Bulut M (2002) The role of the Ottomans and Dutch in the commercial integration between the Levant and Atlantic in the seventeenth century. JESHO 45(2):197–230

Cobham CD (1908) Excerpta Cypria: materials for a history of Cyprus: with an appendix on the bibliography of Cyprus. Kraus Reprint, New York

Cosmescu D (2016) Venetian renaissance fortifications in the Mediterranean. McFarland, North Carolina

Dixon WH (1879) British Cyprus. Chapman and Hall, London

Gunnis R (1947) Historic Cyprus: A guide to its town and villages, monasteries and castles. Methuen, London

Gürkan HM (2006) Dünkü ve bugünkü Lefkoşa. Galeri kültür yayınları, Lefkoşa

Marangou AG (2002) The harbours and ports of Cyprus. Laiki Group Cultural Centre, Nicosia

Navari L (ed) (2003) Maps of Cyprus: from the collections of the Bank of Cyprus Cultural Foundation. Bank of Cyprus Cultural Foundation, Nicosia

Stefoff R (1995) The British Library companion to maps and mapmaking. British Library, London

Stylianou A, Stylianou JA (1980) The history of the cartography of Cyprus. Cyprus Research Centre, Nicosia

Stylianou A, Stylianou JA (1989) The history of the cartography of Nicosia. Leventis Municipal Museum of Nicosia, Nicosia

Uluca E, Akın N (2008) Gazimağusa Limanı: 13. Yüzyıldan 20. yüzyıla bir Akdeniz limanının gelişimi. İTÜ Dergisi/A Mimarlık 7(1):62–73

**Merve Senem Arkan** has graduated from Archaeology and Art History Major at Eastern Mediterranean University in Cyprus. In 2007, she completed MA in Art and Design History at the Kingston University, UK. She holds a Master's degree in Eastern Mediterranean Studies from Eastern Mediterranean University in Cyprus. Her MA thesis was "The Cartography of post-Medieval Famagusta: from the 16th to the 19th centuries" which was awarded in 2011. She started her Ph.D. in the History of Cartography Department at Eötvös Loránd University, Budapest where she works under the supervision of Dr. Zsolt Török. Her thesis "The Urban Cartography of Cyprus: Between the 16th and 20th Centuries" was awarded in December 2016. Since July 2015 she became the honorary secretary of The International Society for the History of the Map (ISHMap).

# Part II
# Territory, Sovereignty and Borderlands

# Between Secrecy and Silent Cooperation: The Dissemination of Knowledge About the Republic of Dubrovnik in the Context of the Ottoman–Venetian and Napoleonic Wars

**Mirela Altić**

**Abstract** Despite its exceptional political, commercial and naval strength, the Republic of Dubrovnik was remarkably restrained regarding the public availability of its maps. Only two original maps of the Republic of Dubrovnik are preserved to this date—one created between 1718 and 1746, and the other from the earliest nineteenth century. What links them is that both remained in manuscript form, and both were expressly banned by the Republic's authorities from being either copied, published or shown to foreign nationals. The turning point in this regard was "Stato di Ragusi", the first commercial map of the Republic compiled by Vincenzo Maria Coronelli in 1688, according to the information provided by Dubrovnik authorities. Created under the most unusual circumstances, the map remained the only printed detailed map of the Republic of Dubrovnik up until its fall in 1808. The paper analyzes how the Republic of Dubrovnik was represented on maps made by Dubrovnik authorities and how the same space was seen by foreign cartographers. Furthermore, the paper evaluates the impact of the Republic's diplomats on the dissemination of knowledge about the Republic in the context of the Ottoman-Venetian and Napoleonic wars.

## 1 The Republic of Dubrovnik and Its Cartography

The Republic of Dubrovnik was a maritime republic centered on the city of Dubrovnik (Ragusa in Italian and Latin).[1] The Republic existed from 1358 to 1808. Like Venice, it is a typical example of a maritime city-state which, with the help of certain tributary relationships with the Ottoman Empire, managed to

[1]Medieval Dubrovnik originated from a late antiquity settlement. From the early Middle Ages until 1205 Dubrovnik recognized the Byzantine rule and it developed as the centre of the bishopric and administrative centre of a Byzantine theme. From 1205 until 1358 the city recognized the Venetian rule. From 1358 until 1526 the city developed under the rule of Croatian-Hungarian

M. Altić (✉)
Institute of Social Sciences, Zagreb, Croatia
e-mail: mirela.altic@gmail.com



M. Altić et al. (eds.), *Dissemination of Cartographic Knowledge*, Lecture Notes in Geoinformation and Cartography, DOI 10.1007/978-3-319-61515-8_5

retain sovereignty over four and a half centuries.[2] It reached its commercial peak in the fifteenth and sixteenth centuries, before being conquered by Napoleon's French Empire and formally annexed by the Napoleonic Kingdom of Italy in 1808.

The Republic of Dubrovnik had a small surface area but, politically and as a naval force, it was an extremely powerful state in the Mediterranean with colonies and diplomatic missions from Portugal and Spain in the west to Africa in the south and Asia in the east.[3] The Republic, which highly valued its territorial sovereignty and had a significant naval fleet in the Mediterranean, realized early on the need for cartography but was extremely reserved towards the public use of maps. The Republic of Dubrovnik did not use maps in a way most countries did, which, in order to promote their power, disseminated information about their realms by means of commercial cartography, thus also heralding their political power. The authorities of the Republic of Dubrovnik decided on exactly the opposite strategy: geographical data on its territory was considered highly sensitive information which had to be carefully protected from possible abuse, particularly by numerous informers and spies who came to Dubrovnik by sea and land.[4] The confidentiality of information was to contribute to the security of Dubrovnik but also the neighboring Ottoman Empire. No less important, the lack of real information had to further support the myth of wealth and power of the Republic. Situated between the possessions of the Venetian Republic and the Ottoman Empire, the Republic of Dubrovnik skillfully had to balance between these two empires, carefully controlling the dissemination of information about Dubrovnik and the maintaining a powerful image in the minds of the surrounding world (Kunčević 2012: 16, 25).

The few maps produced by the Republic of Dubrovnik served the needs of the administration of the Republic, and only a small circle of the ruling aristocracy had access to those maps. This resulted in the production of a very small number of maps that were kept only in original manuscripts, and were never copied or shown to foreigners. This certainly does not mean that the citizens of Dubrovnik were not well acquainted with the achievements of cartography of that time. On the contrary, a look into the libraries of wealthy Dubrovnik families confirms that many highly valued cartographic editions were represented in their collections. One of the most prominent Dubrovnik writers and historians, Mavro Orbini (mid of sixteenth century—1614), mentions that, while writing his famous work "Il Regno degli Slavi" (Pesaro, 1601), he used the works of Abraham Ortelius, Giacomo Gastaldi,

---

kings and from an autonomous town community it turned into an aristocratic republic. This period was the most prosperous in Dubrovnik's development (Foretić 1980).

[2]In 1433, Dubrovnik won the exclusive right to trade with countries ruled by Islamic rulers, and after that they began paying an annual tribute to the Sultan.

[3]Consulates of Dubrovnik among others were located in London, Antwerp, Brussels, Cadiz, Mallorca, Rome, Naples, Messina, Corfu and Goa (Mitić 1973).

[4]For the same reason, the Republic of Dubrovnik avoided building modern roads so as to, in the event of war of some European power against the Ottoman Empire or Dubrovnik itself, aggravate the enemy raids by land, that is, limit potential attacks exclusively from the sea side which was well defended.

Sebastian Münster, and Wolfgang Lazius.[5] Maps were of course well-known to Dubrovnik seafarers as well. In one of Dubrovnik sales documents from 1318 it is mentioned that a ship had a compass in its equipment. The Dubrovnik family Gučetić preserved the manuscript (probably a transcript) about an astrolabe (between 1428 and 1460). The poet Mavro Vetranović (sixteenth century) made an astronomical atlas, and the painter Vlaho Držić (sixteenth century) a relief globe with the specially marked equator and the prime meridian. According to a record from 1759 the Algerian pirates seized one map on a Dubrovnik ship on its way to Genoa, while three compasses, a map of the Adriatic Sea, and two navigation maps were found on another seizure in 1769 (Bašić 2006: 150). So, maps were definitely a part of Dubrovnik's everyday life, whether they were used to navigate merchant ships or as a source of information in the works of Dubrovnik writers and scientists. No less important, two internationally recognized cartographers came from Dubrovnik, Vicko Dimitrije Volčić (1563–1607), a prominent portolan maker, and Ruđer Bošković (1711–1787), one of the most eminent scientists of the eighteenth century, the author of the first science-based map of the Papal States (Altić 2014: 71–89). Significantly, none of the mentioned two was cartographically active in Dubrovnik.

## 2 Dubrovnik Land Records and the Emergence of Cadastral Maps

It is not exactly known when the people of Dubrovnik decided to compile the first map of the entire Republic, but we know when they started by preparing cadastral records. Their emergence was prompted by the expansion of the territory of the Republic. After Dubrovnik expanded by including the nearby Pelješac peninsula in 1326–1333, in 1336 Dubrovnik authorities, with the purpose of a planned division of the land, organized the first register of land for their new possession on Pelješac (Lučić 1980: 57–89). A document from 1336 represents one of the oldest land records created by a state government in the South Slavic regions, but it did not contain cartographic material.[6]

The first known cadastral plan was created in 1358 when, with the aim of the planned construction of Mali Ston (also located on the Pelješac peninsula), the Republic prepared a plan of the above mentioned settlement. The cadastral plan of Mali Ston from 1358 is the oldest preserved cadastral plan in the area of today's Croatia. Also preserved is another plan for the land west of Ston dating from 1359.

[5]Orbini himself probably also tried to draw maps. Namely, in one manuscript copy of his work, four hand-drawn maps of the Adriatic coast were attached. These were rough sketches that were created by copying from some Venetian template.

[6]State Archives of Dubrovnik, Zemljišnik Dubrovačke Republike, XII seria, Cathastricum No. 2a.

The people of Dubrovnik later also continued to compile land registry books (in the form of a lists of real estate). Each time they expanded the territory of the Republic (in 1357 they acquired parts of Župa, Šumet and Rijeka, in 1399 the coast around Slano, in 1419–1427 Konavle and Cavtat), but even then the cadastral maps appeared only exceptionally (Foretić 1947: 94–96). Dating from the fourteenth century, the emergence of only one other cadastral plan is known, namely that of Cavtat (not preserved in the original but in a transcript from the seventeenth century). At the same time, for the area of the city of Dubrovnik itself, from the thirteenth century onwards, only real estate listings were compiled, without cartographic representations (Benyovsky Latin and Zelić 2007).

## 3 Dubrovnik in Venetian and French Cartography Until the Appearance of the First Commercial Map of the Republic

Venetian maps during the fifteenth and the first half of the sixteenth centuries show the area of the Republic of Dubrovnik exclusively on the portolan charts of the Adriatic, or geographic maps based on Ptolemy's sources. The first significant efforts of Venetian cartographers to improve the representation of Dalmatia occurred during the War of Cyprus (1570–1573), when the first series of panoramas of individual cities and forts were created, but with no significant mapping of a wider area.[7] The panoramas made on this occasion, mostly by the pen of Croatian cartographers in Venetian service (Natale Bonifacio, Martin Rota Kolunić), were published during the 1570s in the Venetian isolarii (Camocio, Pinargenti, Rosaccio, Lafreri). In that period the first printed panoramas of Dubrovnik were created.

The interest of foreign powers for mapping Dubrovnik increased particularly during the so-called anti-Ottoman coalition at the beginning of the seventeenth century that was also supported by some citizens of Dubrovnik. Namely in 1607 Archduke Charles Emmanuel of Savoy encouraged, with the Spanish and Austrian support, a general uprising against the Ottomans, which counted on Dubrovnik as a place from which military forces would penetrate into the interior of the Balkans. During 1607 and 1608 the spies of Charles of Savoy were in Dubrovnik, trying to gather supporters of the uprising. As part of intelligence work in this period, two of the agitators, Imberto Saluzzio and Filiberto Provona, secretly created a precise map of the city and a plan of the city walls and the port (Ćosić 2011: 43–44, Principe 1991: 191–202). To date, these plans remain the oldest known exact cartographic presentation of the city of Dubrovnik.

The first Venetian attempts at compiling modern regional maps of Dalmatia, including the territory of the Republic, based on field measurements, originate from

[7]The exception in this regard are the only map of Istria made based on the template by Petar Kopić and the map of Zadar and Šibenik Counties made based on the map by Martin Rota Kolunić.

**Fig. 1** Detail of Republic of Dubrovnik on a 1645 Venetian map of Dalmatia appeared during Ottoman-Venetian War for Candia (1645–1669) (courtesy of Biblioteca Nazionale Marciana, Venice)

the time of the Candian[8] War (1645–1669). Then, for the purpose of warfare in the Dalmatian hinterland and later because of the need of demarcation of respective territories between the Venetian Republic and the Ottoman Empire, the first map series was compiled that included a regional map of Dalmatia and thus the territory of the Republic of Dubrovnik. The map of Dalmatia prepared in 1645 shows that the quality of the display of the Republic lags far behind the display of Dalmatia (Fig. 1).[9] The most obvious error is certainly the representation of Pelješac peninsula as an island, while in the entire area of the Republic only two settlements are marked, Dubrovnik and Ston (not accidentally, these were the two strongest fortresses of the Republic). A surveying the Dubrovnik area had not been conducted because the city authorities did not grant approval. The Ottoman Empire with which the Republic, due to its commercial interests on the Balkans and in the East, nurtured very close relations, would not look favorably at a the cooperation of Dubrovnik and Venice, especially not in the midst of an Ottoman-Venetian War, which was fought in the Dalmatian hinterland, and which in this area mostly went to the detriment of the Turks. The Turks themselves were also unwilling to map their possessions in Europe because they feared that maps could get into enemy hands. They expected from its allies to act alike. As the Republic of Dubrovnik was on the mainland and on all sides surrounded by the territory of the Ottoman Empire, it was

---

[8]Candia was the name of Crete when it was under Venetian from 1205 to 1669.

[9]The Biblioteca Nazionale Marciana, Carte manoscritti, IT.VII.200(10050), Tav. 19bis [Carta della Dalmazia], c. 1645.

regarded that the publication of data on the topography of the Republic would represent a potential threat to the security of the Ottoman territory.

Thanks to good trade cooperation between France and the Ottoman Empire, the Turks looked with less reprimand at displaying the Republic on the French maps. Although they also were not allowed to map nor were they provided with data on the topography of the Republic. It is noticeable that during the seventeenth century the French regularly included the representation of the Republic of Dubrovnik in their maps that were based on older Venetian templates, which they then complemented with the data from a number of travelogues.

Among these stood out especially Pierre Du Val, who produced a map of the Christian-Ottoman borders in 1663 that also included the Republic of Dubrovnik.[10] Although relatively small in scale, the display includes the drawn borders of the Republic and a number of settlements that are not found in the Venetian maps. In accordance with the absence of field mapping, the display of the territory is extremely deformed, and the land border of the Republic is drawn wrongly, including the part of Herzegovina which was under Ottoman rule at the time of creation of the map.[11] Based on the aforementioned map, in the same year Du Val also published a separate map of the Republic in his Geography with which he published also a short description of the Republic.[12] This map would serve as a template for most printed maps of the Republic published in the European atlases and geographies in that time.[13]

It is quite certain that, at that time, the Republic of Dubrovnik possessed its own map of its territory. Although to date the original of it has not been found, its existence in the mid seventeenth century is expressly mentioned in written documents. The map was kept safe as the original manuscript on the premises of the Dubrovnik Senate, and only the selected individuals of Dubrovnik nobility could see it. The prohibition of copying and distribution of the map of the Republic of Dubrovnik, as well as the general lack of significant Dubrovnik cartographic production negatively affected the circulation of knowledge about the geography of the Republic of Dubrovnik. As a result of such an attitude by the Dubrovnik authorities, in all European cartographies of that time there was a slower progress of representing the Republic than in the adjacent Venetian Dalmatia.

---

[10]Les Confins des Chrestiens et des Turcs en Terre Ferme. Pierre du Val, Paris: 1663. Copperplate in color; 37 × 50.5 cm.

[11]The representation of the land border of the Republic might suggest that Du Val was familiar with the fact that, in 1431, Dubrovnik managed to annex Trebinje, Vrm and Klobuk, but as already next year, the Ottoman rule overturned that decision, the controversial settlements remained part of the Ottoman Empire (Stulli 1989: 53).

[12]Estat de la Republique de Raguse, Pierre du Val, [ca 1:1,400,000], Paris: Pierre Du Val, 1663. Copperplate; 12 × 10 cm. Published in: Le monde Ou La Geographie Universelle contenant les descriptions, les cartes, & le blason des principaux pays du monde (Paris, 1663), vol. II, pp. 364–367.

[13]Nicolas Sanson used Du Val's map as source for his map of Dalmatia and the Republic of Dubrovnik (Sindik 1931: 281–288).

## 4 Vincenzo Maria Coronelli and the Creation of the First Detailed Commercial Map of the Republic of Dubrovnik (1688)

During the following Ottoman-Venetian War (1684–1699) the Venetians once again renewed their interest in the mapping of Dalmatia and the Republic of Dubrovnik. Compiled on the basis of relatively limited field measurements, new maps were created during the war for all parts of Dalmatia. However, the Venetian mapping to the south could spread only to the mouth of the Neretva River (west of the border with the Republic). Because of its obligations to the Turkish side, the Republic of Dubrovnik even then could not be officially mapped by Venetian cartographers. But then, Dubrovnik diplomacy saw a new opportunity: to secretly support the development of a first detailed commercial map of the Republic that was to be made for them by the most famous Venetian cartographer, Vincenzo Maria Coronelli. Documents are preserved in the Dubrovnik archives that bear witness to the 1685 discussion on forwarding information about the Republic to Coronelli by unofficial channels, as well as on offering a pecuniary award to the Venetian cartographer. Accordingly, in March 1685, Coronelli was sent a text and a geographical map of the Republic through an intermediary, with the insistence of the Dubrovnik authorities that the initiative should be presented exclusively as a private matter, and by no means an official interest of the Republic.[14] Coronelli thus came into possession of an unidentified text and map, which he would use for the purposes of making a map of the Republic and the pertaining geographical description.

Why did the people of Dubrovnik, at the height the Ottoman-Venetian War (1684–1699) that raged directly around their territory, support the compilation of the first detailed printed map of the Republic by the pen of a cartographer who came from the country with which Dubrovnik had extremely tense relations? This decision was made when it became obvious that the Turks were losing this war as well (in 1683, the Turks were severely defeated at the end of their second siege of Vienna) and that, due to Ottoman territorial losses in the hinterland of Dubrovnik, the Republic of Dubrovnik might have a direct border with the Venetian Republic.[15] In addition, in 1667 Dubrovnik survived a catastrophic earthquake that substantially had weakened the Republic, and from which it only began to recover.

---

[14]State Archives of Dubrovnik, Consilium Rogatorum, 127, 1684–1685, 138–139.

[15]After the Venetians joined the Christian Holy League in 1684, the Dubrovnik Republic reacted with a decisive change in its foreign policy orientation by leaning toward the Habsburgs. An agreement was signed under which the obligation to protect the Republic was transferred to the Habsburgs, for which the Republic would pay a tribute. Because of the threatening Ottoman danger, the agreement was supposed to remain secret, and the tribute payment to begin only after the Ottoman defeat and the liberation of Bosnia, Herzegovina, Slavonia and Hungary. However, after the French declaration of war to Austria in 1690, it became clear that the Ottoman Empire was to survive, and Dubrovnik's rapprochement to the Ottomans was resumed. They paid the total residual amounts of tribute to the Ottomans in 1695, and continued to pursue a policy of neutrality.

All this prompted the Dubrovnik authorities to a silent cooperation with the Venetians who, by publishing the map, contributed to the trade re-empowerment of the Republic.

At that time, based on Venetian field measurements carried out during the war, Vincenzo Maria Coronelli was compiling new maps of Dalmatia for his atlas, so the publication of the first detailed map of the Republic of Dubrovnik was less prominent among other maps of Dalmatia. Although the original text and the map sent to Coronelli is not preserved, judging by the flattering description of the Republic published in his atlas "Mari, Golfi, Isole", Coronelli published the integral text which he had received from Dubrovnik. The same can also be concluded for the map that contains hitherto completely unknown details of the Dubrovnik territory, and which certainly was based on the Dubrovnik cartographic template. Coronelli supplemented that template, compiling both the Venetian and Dubrovnik information into his map which, entitled "Stato di Ragusi", would be printed in Venice and Paris (Fig. 2).[16] The map was completed for printing in 1688.[17]

Coronelli's map contained data on the Republic that had never been seen before (Škrivanić 1959: 1999–204). Earlier numerous errors in the contour of the coast and particularly the Pelješac peninsula were now corrected. For the first time, the map also contains a significant number of place names that were first mentioned on a publicly available map. The settlements are mainly located accurately, for some he also inserted statistical data on the number of houses or the number of inhabitants.[18] The place names are generally not Italianized, thus further confirming that the data from Coronelli's map originated from Dubrovnik. In addition to numerous settlements, the map records numerous churches and monasteries but also maritime information such as names of bays and capes, as well as data about fresh water sources and fish catching areas. The boundaries of the Republic are not drawn, either on land or at sea, but an indication "di Ragusi" is written along the islands that belonged to the Republic (Vis, Mljet, Elaphite Islands), while the symbol of the Venetian lion drawn on the islands under Venetian rule, so the sea border is indicated indirectly. Quite another case is the land border which, given that the map was created at the height of the Ottoman-Venetian war in which the Venetians

[16]Stato di Ragusi, Bocca di Fiume Narenta, Isole di Lesina e Curzola nella Damatia possedute dalla Serenissima Republica di Venetia. Vincenzo Maria Coronelli, [1:305,000], Venice: [1688]. Copperplate in color; 58 × 43 cm.

[17]Coronelli's map of the Dubrovnik Republic does not have the original dating, but we learn from a note on the adjacent map sheet titled "Corso delli Fiumi Drino e Boiana ..." that the map of the Republic was compiled in March 1688. In fact, Coronelli wrote on the aforementioned map "Dello Stato di Ragusi il P. Coronelli ne publicara una carta particolare nel Mese di Marzo 1688".

[18]On the island of Šipan, for example, he stated that there were 100 houses in Luka Šipanska, whereas 40 houses in the neighboring Suđurađ. On the Pelješac peninsula (Sabioncello o Penisola di Stagno), he stated that there were 30 houses in Kučište, 30 houses in Orebić, 12 houses in Borje, 10 houses in Obuće, 10 houses in Dingač, 8 houses in Trstenica, and 60 houses in Žuljana. For Dubrovnik, there was a note that it was destroyed in an earthquake in 1667, but that it was later restored. For the port of Dubrovnik, he noted that it closed by means of big chains, and that it could accommodate large ships, whereas for the port of Gruž, that it could accommodate 100 galleons.

**Fig. 2** Stato di Ragusi by Vincenzo Maria Coronelli produced in 1688 as first detailed commercial map of the Republic of Dubrovnik made with silent support of Dubrovnik authorities (courtesy of National Library of Zagreb)

conquered part of the Turkish territory in the Dubrovnik hinterland, could not have been marked before the new demarcation. However, Coronelli did not fail to draw the Venetian lion in the area of the mouth of the Neretva River, which the Venetians conquered in 1685, and in the area of the Bay of Kotor, respectively, where the Venetians won Herceg Novi in 1687.

The map thus designed was published in Coronelli's work *Corso geografico universale* (Venice, 1692), and then also in his isolarii. The map is a careful artistical design, too. It is embellished with two cartouches, one title cartouche with St. Blaise, the patron saint of the Dubrovnik Republic,[19] and the other with the coat of arms of Dalmatia and a scale. It was the first detailed map of the Republic intended for public use. Its publication with silent cooperation by Dubrovnik authorities was a precedent that Dubrovnik would not repeat. Coronelli's map of the Republic would therefore remained the only published (detailed) map of the Republic until its downfall in 1808.

[19]The area below the title cartouche, apparently intended for the coat of arms of the Republic, remained empty on all preserved copies of the map.

**Fig. 3** Official map of the Republic of Dubrovnik probably drawn by Miho Pešić, between 1718 and 1746 and kept in Dubrovnik Senate house for administrating the Republic (courtesy of National Archives in Dubrovnik)

# 5 Manuscript Map of the Republic (1718/1746)

The only original map of the Republic made in Dubrovnik and preserved until the present day was created during the first half of the eighteenth century (Fig. 3). According to the inscription on the map, we know that in 1746 Miho Pešić donated that map of Dubrovnik to the priestly association of St. Peter in Dubrovnik, of which he himself was a member.[20] It is not known whether Miho Pešić was also the author of the map, but it is quite possible in view of his work and biography. The same handwriting in both the dedication and the toponymy on the map testifies to the fact that Miho Pešić, even if he was not the author of the original, created that map as a copy of another, original Dubrovnik map.

The priest Miho Pešić was a prominent Dubrovnik painter and diplomat who certainly possessed the skills and knowledge to make a copy, perhaps even to produce an original map (Lupis 2002: 123–134). Pešić also had a direct reason for creating this manuscript map. As secretary to the Dubrovnik Consul, in 1718, he took part in the peace conference in Passarowitz. The Treaty thus signed established a new border between the Ottoman Empire and the Venetian Republic, which also was directly related to the border of the Republic of Dubrovnik. Accordingly, this map accurately reflects the land borders between the Republic and the Ottoman Empire. After the Ottoman-Venetian War of 1684–1699, the Treaty of Karlowitz (1700) created a new line of demarcation in the Dubrovnik hinterland between the Ottoman Empire and the Venetian Republic, but also the Republic of Dubrovnik.[21]

[20]Hanc Legum Ragusinarum Collectionem/Venerabili Congregationi Presbyterorum/Sancti Petri in Cathedra/Michael Pessich eorundem Socius, & Consors/In perenne Sui amoris, & obsequij testimonium/Libens donavit Anno Sal; [utis] MDCCXLVI, Miho Pešić. Manuscript in color; 37 × 114.5 cm.

[21]Since the Venetians conquered the immediate Dubrovnik hinterland, which hitherto had been under Ottoman rule, that would have meant that, for the first time, Dubrovnik would have a direct land border with Venice, which would seal Dubrovnik's fate. There was a swift diplomatic action in which Dubrovnik persuaded the Ottomans and the Austrians to force the Venetians to return part

The Passarowitz peace conference of 1718, held after another Ottoman-Venetian War (1716–1718), confirmed the state of borders in the Dubrovnik hinterland that was established in 1700. On that occasion, Pešić quite probably used some map in the negotiations, possibly that particular copy. The map had no original dating (only the year of its donation is known, 1746), so we can only assume that it was probably created for the purpose of peace negotiations of 1718, or immediately thereafter, but certainly prior to 1746 when it was donated to the Dubrovnik monastery (Škrivanić 1958: 260).

Besides the land borders to the Ottoman Empire and sea borders to the Venetian Republic, the map shows the territorial structure of the Republic consisting of captaincies and principalities (Conte, Governo). The representation of the internal territorial and judicial system of the Republic certainly confirms that the map was created in Dubrovnik for their domestic needs and that it was not publicly available. A map of such content had to be one of the maps for which it is mentioned that they were kept in the Dubrovnik Senate for the purposes of administering the Republic. In support of this are also the prominent state attributes of the Republic: its patron saint drawn in color, sitting on the title cartouche and blessing the city, and a colored coat of arms engraved in the center of the map.

It is interesting that, in its base, Pešić's map does not rely on Coronelli's map at all. The contours of the coast and the islands on Pešić's map are entirely different, often more accurate, especially in the representation of Pelješac and Konavle. Coronelli has the largest number of place names, in particular those related to navigation, bringing us primarily the view of the Republic from the sea. In contrast, although technically considerably poorer drawn than Coronelli's, Pešić's map contains detailed information about the borders and the structure of the Dubrovnik territory, allowing us a different view of the Republic, a view of its structure from the inside through the eyes of Dubrovnik's inhabitants themselves.

## 6 Venetian, French and Austrian Efforts in the Eighteenth Century

The problem of the circulation of topographical knowledge about the Republic of Dubrovnik continued to be clearly visible in the Venetian maps of the eighteenth century. Although the knowledge about Venetian Dalmatia progressed significantly in the meantime, the representation of the Republic progressed much slower. The problems with the representation of the Dubrovnik territory are best illustrated by a

---

of the conquered area along Dubrovnik's border to the Ottomans. A narrow corridor of Ottoman territory was thus created between the Dubrovnik Republic and the Republic of Venice, on Sutorina and Klek, that served as a buffer zone. That ad hoc created corridor survived all political changes on the Balkans in the next three centuries to provide the todays Republic of Bosnia and Herzegovina its narrow access to the Mediterranean in the areas of Klek and Sutorina.

Venetian map of Dalmatia from 1732, created for the purpose of demarcation.[22] The quality of the representation of the part of the Republic included on the map is significantly different from the representation of the neighboring Venetian territory. The representation of the Pelješac peninsula only slightly improved compared to Coronelli's time, while the depictions of Mljet and Vis remained almost unchanged. Significant progress can be seen on Venetian manuscript maps from the late eighteenth century. But those rare maps of Venetian Dalmatia were not available to commercial cartographers, so their impact on the content of printed maps intended for commercial use was quite limited. Thus the representation of the Republic included on Thomas Salmon's map of Dalmatia from 1753 completely relied on Cantelli's map from 1689.[23] The maps of the Dubrovnik area that French hydrographer Nicolas Bellin compiled for the purposes of his 1771 pilot book of the Adriatic were also based on Coronelli's templates.[24] A certain advancement can only be seen on the maps by Pietro Santini from 1780,[25] and Melchiori and Frane Zavoreo from 1787, respectively,[26] which primarily resulted from improvements in nautical charts of Venetian and French provenance (Fig. 4). However, the ban on the mapping of territory of the Republic continued to hamper the production of larger scale commercial maps. Also, it is significant that none of the cartographers of the eighteenth century managed to publish a printed plan of the port of Dubrovnik.

However, at the very end of the eighteenth century, there would be certain changes in relation to the cartographic policy of the Republic of Dubrovnik. On the eve of the downfall of the Republic of Venice and all the more certain expansion of the French Empire, Dubrovnik was turning to cartographic cooperation with the Austrians. The Ottoman-Austrian wars (1737–1739, 1788–1791), in which Dubrovnik provided intelligence to the Austrians, had strengthened the cooperation of the Republic with the Habsburg Monarchy. Interestingly, even then Dubrovnik refused to send a copy of the map of the Republic to the Austrians. However, the lack of an adequate plan of the city and its immediate surroundings was a big problem in the waiting, especially for the Austrians who, under the treaty of 1684, had an obligation to protect the Republic. Therefore, in 1783, for the first time, the city

---

[22] Descrizione topografica della Dalmazia, [ca 1:200,000], 1732. Manuscript in color; 140 × 53 cm.

[23] Carta geografica della Dalmazia/Giovanni Thomas Salmon. Venecija: Albrizzi, 1753. Copperplate; 33.5 × 43.5 cm. Published in Salmon's work "Lo stato presente di tutti i paesi e popoli del Mondo" (Venetia, Giambattista Albrizzi, 1753).

[24] Decription Géographique du Golf de Venise et de la Morée. Avec des remarques pour la Navigation et des Cartes et Plans des Côtes, Villes, Ports et Mouillages, Paris, Didot, M.DCC. LXXI.

[25] Nuovelle carte de la Partie Occidentale et Orientale de Dalmatie dressee sur les lieux par P. Santini, 1:350,000, Venezia, 1780. Copperplate in color, two sheets.

[26] Nuova carta topografica della provincia di Dalmazia, divisa ne suoi territorii delineata dalli Sig [no]ri Ingegneri Melchiori e Zavoreo e da Lodovico Furlanetto dedicati a sua eccellenza Angelo Emo. . .[1:350,000]. Venezia: Lodovico Furlanetto, 1787. Copperplate in color; 118 × 47 cm.

**Fig. 4** Republic of Dubrovnik on a section of a 1787 Venetian map of Dalmatia by Melchiori and Frane Zavoreo (courtesy of War Archives in Vienna)

authorities allowed the survey of the city by Austrian military topographers. The work of Austrian surveyors resulted in a city map in a scale of 1:14,400, the first modern map of the city of Dubrovnik[27]while, by 1790, a map was compiled of the immediate surroundings of the city.[28] Of course, both maps remained in manuscript and were not copied. Interestingly, even then, the people of Dubrovnik did not show greater interest for these maps, so both in their only copies remained in Vienna.

## 7 Austrian Survey of the Republic (1804)

With the downfall of the Venetian Republic (1797) and the establishment of Austrian rule in Dalmatia and the Bay of Kotor, the Austrian General Staff began to prepare a topographic survey of the newly acquired territories. As the Republic of Dubrovnik was located between two parts of the Austrian possession, in 1803, the Austrian Foreign Minister suggested to the Dubrovnik authorities to permit the survey of the Republic as well. In support of this request a need was emphasized to connect the triangulation network, which Austrian surveyors were to develop in Albania, with the triangulation network they were to set in Dalmatia, and that it had

[27]Plan von Ragusa 1:14,000, 1783. Manuscript in color; 33 × 23 cm.

[28]Plan von Ragusa mit umliegender Gegend, letstere landschaftlich dargestelt, 1:5400. Manuscript in colour; 57 × 40 cm.

to be done over the territory of Dubrovnik. The Austrian side offered to cover all the costs of the surveys carried out on the territory of the Republic and not to create a separate map of the Republic for public use. Finally, in 1804, Austrian military cartographers Catinelli, Sponville and Geppert arrived in Dubrovnik. However, when they asked for an interpreter from the Dubrovnik authorities to visit the terrain with them, the Senate did not approve it. Dubrovnik requested a postponement of works, to bring the issue to the Turkish side, and to get their approval for the proposed field surveys that the Austrians would carry out in the Republic.[29] Apparently dissatisfied with Dubrovnik's cooperation with the Austrian authorities, the Turkish side did not reply to Dubrovnik's inquiry concerning the survey of the terrain. Only after Catinelli sent a sharp note of protest to the Dubrovnik Senate, the works began in May 1804 (Muljačić 1958: 260). The field works lasted until August of the same year during which the Austrian officers measured the base at Konavle and then developed a trigonometric network on the whole territory of the Republic. At the end of August, on the basis of the conducted survey of the terrain, they compiled a topographic map of the Republic, a copy of which was sent to the Dubrovnik Senate on 25 August 1804 (Fig. 5). In order to diplomatically thank the three Austrian officers, the Dubrovnik Senate sent to each a golden sword as a gift.

The map of the Republic of Dubrovnik compiled in 1804 has the characteristics of a topographic map.[30] Made in a scale of 1:115,200, it contains a representation of the entire territory of the Republic without the island of Lastovo (Lagosta). When comparing this map with all previous representations of the Republic, it is clear that it is an original work. For the first time we see all the settlements of the Republic,

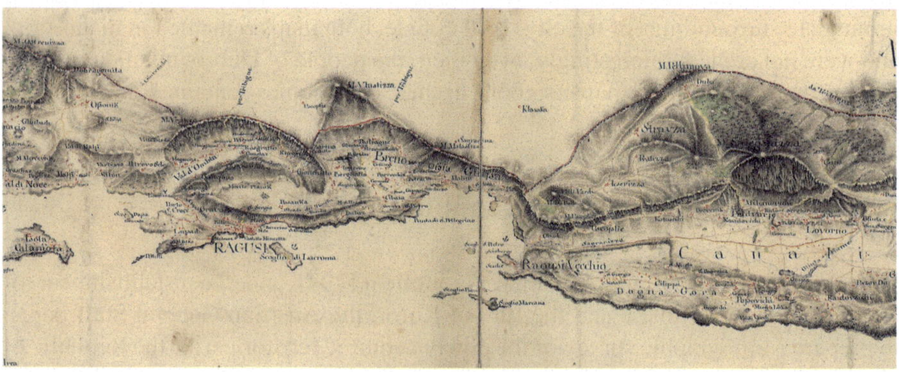

**Fig. 5** Part of a topographic map of the Republic of Dubrovnik created by the Austrian General Staff in 1804 with approval of Dubrovnik authorities (courtesy of War Archives in Vienna)

[29]State Archives of Dubrovnik, Consilium Rogatorum 210, VI, 42.

[30]Karte des Gebietes der Republik Ragusa. Nach dem Viertel-Mil. Maßstab aufgenommen durch die Oblte. Geppert und Catinelli des GQM Stabes und dem zugetheilten Lt. Sponville v. Hohenlohe Bartenstein Inf. Reg. Gez. durch Oblt. Geppert. 1:115,200.- Manuscript in color; 137 × 40 cm.

the hydrographic and transport networks, and the orography of the surrounding terrain. However, the quality of the representation lags significantly behind Austrian topographical maps of the time. Because of the pace of the survey and a number of other restrictions imposed by Dubrovnik, the relief could not be shown exactly by hatching but only generalized by shading. From the omission of the island of Lastovo it is obvious that the Austrian cartographers did not have time to visit all the islands that belonged to the Republic. Also, the map has no coordinate grid. In fact, in a letter the surveyors sent to the Dubrovnik Senate, they quite strikingly apologized to the people of Dubrovnik because the map had not a drawn grid since "they were unfamiliar with latitudes and longitudes for either of the major settlements of the Republic".[31] Given the tone of the excessively friendly letter, it is clear that the lack of measurements of the positions of the settlements themselves was a result of restrictions imposed on them by the people of Dubrovnik. In fact, even during the survey, the Dubrovnik authorities insisted that all trigonometric marks in the field be urgently removed.[32]

After the survey was completed the Austrians kept their promise. The compiled map of the Republic of Dubrovnik was not published nor were the data collected through the survey included in other printed maps of the Adriatic coast. Thus, for example, the printed topographic map, published in Vienna in 1804, completely omits the representation of the territory of the Republic (the map reaches eastward merely to the mouth of the Neretva River, while the Bay of Kotor is represented as an inserted additional map).[33] Interestingly, a stormy debate took place in the Senate regarding the Austrian manuscript map of the Republic of Dubrovnik. Namely, after the French occupied Dalmatia and the Bay of Kotor in 1806, French troops quickly found themselves in front of Dubrovnik. Immediately before entering the city in April 1806, the French authorities requested the people of Dubrovnik to send them a copy of the topographic map of the Republic, which was held in the office of notary public but, as far as it is known, the Dubrovnik authorities did not send the map to the French (Lučić 1988: 131). Just a month later, Dubrovnik surrendered to French troops without resistance, but the Republic was formally abolished only in January 1808.[34]

[31]State Archives of Dubrovnik, Acta et diplomata, nineteenth century, 278.

[32]State Archives of Dubrovnik, Consilum Rogatorum. 210, VII, 70.

[33]Charte des nordwestl. u. südöstl. Theiles v. Dalmatien, mit dem österreichischen Anthaile von Albanien, nach den besten bisher bekannten Hülfsmitteln u. astronomischen Bestimungen entworfen. Scale 1: 535,000. Wien: 1804.

[34]Under the terms of the Peace of Pressburg (1805) the French took over from the Austrians Dalmatia and the Bay of Kotor. During 1806 the French military occupied Dalmatia and the Russian the Bay. Despite their policy of neutrality, the Dubrovnik Senate found itself in a hopeless situation. Threatened by Russian occupation, Dubrovnik surrendered to the French troops without resistance on May 27, 1806.

## 8 French Survey of the Dubrovnik Littoral After the Fall of the Republic

Immediately after the establishment of French rule over the eastern Adriatic coast, Napoleon commissioned extensive mapping to encompass the territory of the Republic of Dubrovnik as well. Because of numerous naval operations of the French against the British in the Adriatic, as early as 1806, Napoleon ordered a hydrographic reconnaissance of the entire Croatian Adriatic coast (Altić 2012: 18). The survey, conducted under the leadership of the French cartographer Charles-François Beautemps-Beaupré, took three years. After the survey of the northern Adriatic (1806), during 1808 and 1809, aided by Lieutenant Tician, Beautemps-Beaupré conducted a hydrographic survey of the area between Dubrovnik and the Bay of Kotor.[35] Eleven manuscript charts were created based on that survey, and were subsequently published under the common title *Reconnoissance hydrographique du Golfe de Cattaro et des environs de Raguse…*.[36] The largest part of the Dubrovnik territory was represented on a sheet titled "Reconnoissance hydrographique des environs de Raguse", depicting the area from Zaton to Cavtat, while detailed port plans were compiled for Gruž (Gravosa), Slano, Zaton (Malfi), and Molunat (Molonto). The French hydrographic survey thus resulted in the first modern port plan of Dubrovnik with marked depths and a modern nautical chart of the entire local waters. In addition, a topographic plan of the suburban surroundings was also compiled in a scale of 1:4710 with a detailed representation of all Dubrovnik's fortifications, settlements outside the fortifications, and of traffic communications connecting them, in other words, everything that the French military required to control the city on land and sea.[37]

Immediately after establishing French sovereignty in order to assess the possibilities for moving his troops in the coastal hinterland, Napoleon ordered a topographic survey as well. Already in 1806 General Mathieu Dumas issued an order to conduct a topographic survey of Dalmatia. As the matter was urgent, the first topographic map was created by joint efforts of Italian and French engineers. In the next months, four engineer-geographers were singled out from the Italian topographic office and placed under the command of General Poitevin, a military engineer commander based in Split. The urgency of the work and the proximity of

---

[35] Beautemps-Beaupré's charts of the Northern and Central Adriatic Sea were bound into a single atlas, also known as the Napoleonic Atlas, in 1806 (only few copies exist).

[36] As a result of the survey of 1808 and 1809 the following charts and maps were created: chart of southern Adriatic from Ston to Budva, chart of the Port of Gruž and the mouth of Ombla river, general map of Dubrovnik and the surrounding area, three charts of the Dubrovnik sea area, chart of the Bay of Kotor, map of Kotor and the surrounding area, chart of the harbor of Molunat, chart of the harbor of Slano, and chart of the harbor of Zaton. On 1 May 1810, in addition to the charts of southern Adriatic, Beautemps-Beaupré compiled a comprehensive military report.

[37] Manuscript charts for southern Adriatic stayed unpublished till 1825, when the selected charts, eleven in total, were published in *Neptune de la Méditerranée*, the maritime atlas of the Mediterranean published in Paris.

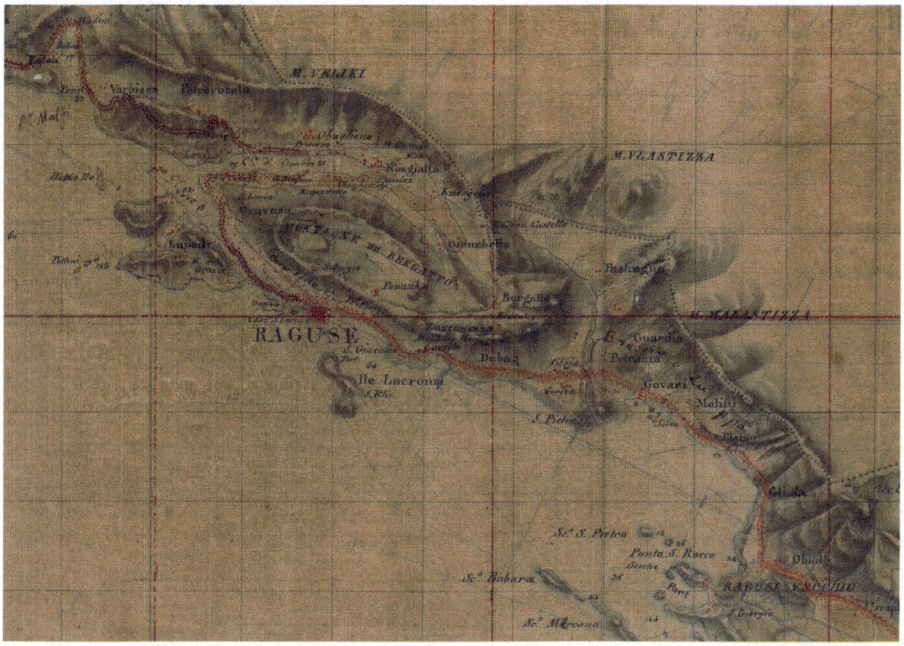


**Fig. 6** Republic of Dubrovnik on a section of 1807 French topographic map of Dalmatia (courtesy of War archives in Vienna)

the enemy in certain areas did not allow for engaging in detailed mapping. On-site inspections were carried out in a hurry, especially those in the vicinity of Zadar and Dubrovnik. As early as 1807, the topographic map of Dalmatia and Dubrovnik was completed in a scale of 1:80,000 and included 48 sheets.[38] The territory of the (former) Republic is shown on sheets 35 and 40 (Fig. 6). We can see that it is a modern military topographic map on which special attention is paid to the representation of the relief. In contrast to the Austrian topographical map, where the depiction of the terrain configuration is one of its weaknesses, the relief here is very concisely represented by hachures. Due to needs of movements of the military, roads were mapped in particular detail and categorized into four brackets. In addition, arable land was specially marked, so to some extent the map also shows the land use. Also, the engineers managed to measure the depths of harbors and passages, the positions of islands, and their exact distance from the mainland. The position of single measuring points was determined using a plain table and compass, because there was unfortunately not enough time to wait for the completion of coastal height measurements that were carried out by the hydrographer Beautemps-Beaupré (Altić 2012: 21–22). The maps were copied by hand in several clean copies; however, they did not appear in print.

[38]Carte militaire et marine de la Dalmatie comprenant la Dalmatie ex-Venitienne la Republique de Raguse les bouches de Cattaro et les iles adjacentes...

# 9 The First Printed Topographic Map of the Dubrovnik Area

The first commercial maps that also included the area of the (former) Republic were published by several Viennese publishers as early as in the times of French rule over Dubrovnik. That is the topographic survey of Dalmatia, completed by the Austrian General Staff in 1804, finally gave a mathematically based and more or less unified representation of the entire coastal area. Commercial editions of these sheets soon appeared, using the material of the topographic map of Dalmatia. Not accidentally the authors of the first printed topographic maps of Dalmatia came from the ranks of cartographers who worked for the General Staff as well. In this regard, a decisive step was made by Maximilian de Traux, a military cartographer in the service of the Habsburg Empire. During the first period of Austrian administration in the area (1797–1806) he had carried out mapping in Dalmatia and Austrian Albania.[39] When the material from the 1804 topographic survey became available to him, he compiled the first printed topographic map at a scale of 1:180,000, which was published in Vienna and Budapest in 1810 (Fig. 7).[40]

**Fig. 7** Dubrovnik area on the first printed topographic map of Dalmatia by Maximilian de Traux (Vienna, 1810) based on the Austrian military survey of 1804 (courtesy of War archives in Vienna)

[39]Already in 1804 Maximilian de Traux compiled a manuscript atlas of Dalmatia and Austrian Albania in which, along with extensive textual description of the said provinces, he published 27 manuscript maps (4 regional maps and 23 detailed plans, mainly ports and forts). Significantly, the atlas did not contain the representation of the Dubrovnik Republic.

[40]Carte von Dalmatien und dem Gebieth von Ragusa aus ächten Quellen. Maximilian de Traux, 1:180,000, Wien und Pest, Kunst und Industrie Comptoir, 1810. Copperplate in 8 sheets. 58 × 43.6 cm each.

It was the first publicly available topographic map of Dalmatia and the late Republic of Dubrovnik in a relatively large scale. The map included all former knowledge of the topography of that area. It was also recognized by other publishers. For example, the map was also published in the redaction of A. Arrowsmith in London as early as 1812, while in 1829, the same map was published by the Viennese company Artaria. The appearance of the printed map of Dalmatia and the Dubrovnik area in 1810 marked a turning point which ushered in a new era in the development of cartography in the area of the passed away Republic of Dubrovnik. When in 1815 the Austrian Empire once again assumed authority over the Croatian Adriatic coast, the production of maps depicting that area would grow rapidly. After 1815, for the first time and in a short period a significant number of plans of the city of Dubrovnik, its fortifications, topographic maps of the surrounding area and detailed nautical charts of the local waters appeared. While the key role in the mapping of that area would continue to be played by the Austrian General Staff, the data obtained by mapping would no longer be treated secretively. The results of the cadastral survey of Dubrovnik (1837), as well as the next topographic survey of the same area, in a scale of 1:28,800, thus quickly echoed on maps of commercial publishers. The secrecy that for centuries characterized the cartographic production of the Dubrovnik Republic would thus eventually give way to new trends of commercial needs.

## References

Altić M (2012) Die hydrographische und topographische Vermessung Dalmatiens durch die Franzosen zur Zeit Napoleons [French topographic and hydrographic surveys of Dalmatia in the Napoleonic period]. Cartographica Helvetica 46:17–24

Altić M (2014) Exploring along the Rome Meridian—Roger Boscovich and the first modern map of the Papal States. In: Liebenberg E, Collier P, Török Z (eds) History of cartography: lecture notes in geoinformation and cartography: IX. Springer, Heidelberg

Bašić Đ (2006) Pomorstvo Dubrovnika od XII. do početka XX. stoljeća [Shipping in Dubrovnik from the 12th century to the beginning of the 20th century]. Pomorski zbornik 44:139–177

Benyovsky Latin I, Zelić D (2007) Knjige nekretnina dubrovačke općine (13-18. st.) [The books of communal real property of Dubrovnik (13th to the 18th centuries)]. Hrvatska akademija znanosti i umjetnosti and Zavod za povijesne znanosti u Dubrovniku, Dubrovnik

Ćosić S (2011) Mavro Orbini i raskol dubrovačkog patricijata [Mavro Orbini and the factions within the Ragusan Patriciate]. Radovi—Zavod za hrvatsku povijest 43:37–56

Foretić V (1947) O izmjeri zemlje u Dubrovačkoj Republici [On the land survey in the Dubrovnik Republic]. Geodetski list 2–3:94–96

Foretić V (1980) Povijest Dubrovnika do 1809 [History of Dubrovnik until 1809]. Nakladni zavod Matice hrvatske, Zagreb

Kunčević L (2012) Dubrovačka slika Venecije i venecijanska slika Dubrovnika u ranom novom vijeku [Ragusan image of Venice and Venetian image of Ragusa in the early modern period]. Anali Dubrovnik 50:9–37

Lučić J (1980) Najstarija zemljišna knjiga u Hrvatskoj—dubrovački zemljišnik diobe zemlje u Stonu i Pelješcu iz god. 1336 [The oldest cadastral book in Croatia—Dubrovnik cadastral book of the land partition on Ston and Pelješac from the year 1336]. Anali Dubrovnik 18:57–89

Lučić J (1988) O kartografiji u Dubrovačkoj Republici [On cartography in the Republic of Dubrovnik]. Anali Zavoda za povijesne znanosti JAZU u Dubrovniku 26:125–134
Lupis V (2002) Don Miho Pešić: dubrovački slikar 18. stoljeća i njegovo doba [Don Miho Pešić: a Dubrovnik painter of the 18th century and his time]. Peristil 45:123–134
Mitić I (1973) Konzulati i konzularna služba starog Dubrovnika [Consulates and Consular Service of Old Dubrovnik]. Historijski institut JAZU u Dubrovniku, Dubrovnik
Muljačić Ž (1958) Prva moderna geografska karta dubrovačkog područja [The first modern map of Dubrovnik region]. Naše more 4/5:260–264
Principe I (1991) Tri neobjavljene karte Dubrovnika iz XVI-XVII. stoljeća [Three unpublished maps of Dubrovnik from the 16th to 17th centuries]. Dubrovnik 1:191–202
Sindik I (1931) Dubrovačka Republika na kartama XVII vijeka [The Republic of Dubrovnik on the maps of the 17th century]. Rešetarov zbornik:281–288
Škrivanić G (1958) Karta Dubrovačke Republike od Mihajla Pešića [The map of the Republic of Dubrovnik by Miho Pešić]. Istorijski časopis VIII:255–261
Škrivanić G (1959) Koronelijevi podaci o Dubrovačkoj republici i njegovoj izradi karte republike [Coronelli's information on the Republic of Dubrovnik and his work on the map]. Istorijski časopis IX–X:199–204
Stulli B (1989) Povijest Dubrovačke Republike [The history of the Dubrovnik Republic]. Arhiv Hrvatske, Zagreb

**Mirela Altić** is a chief research fellow at the Institute of Social Sciences in Zagreb, Croatia. In the Department of History, University of Zagreb, Dr. Altic holds the rank of full professor and lectures on the history of cartography and historical geography. Besides her specialization in South Eastern and Central European map history, last few years she publishes extensively on the Jesuit cartography of Americas and conducts research in European and American archives and libraries. She is the author of twelve books, numerous scholarly papers and a contributor to The History of Cartography Project. For the academic year 2013–14 she was awarded the David Woodward Memorial Fellowship at the University of Wisconsin, Madison. In 2016 she was invited by American Geographical Society Library to give an annual talk on Maps and America: The Arthur Holzheimer Lecture Series with the title "Encounters in the New World: Jesuit Cartography of the Americas". She is currently Vice-chair of ICA Commission on the History of Cartography.

# Cartographic Exchange and Territorial Creation: Rewriting Northern Japan in the Eighteenth and Nineteenth Centuries

**Edward Boyle**

**Abstract** The Tokugawa era (1603–1868) witnessed a dramatic expansion in the creation and circulation of maps, which moved from being comparatively rare items at the beginning of the period to banal objects of mass-circulation at the end. Yet the shape of Japan being represented on these maps was greatly altered over the course of this period, particularly with regards to the amorphous area north of Japan, known as Ezo. This transformation in geographical representation similarly affected visions of Japan held beyond its shores, which were partially the product of an increasingly, if punctuated and inconsistent, global circulation of geographic materials in comprehensible forms.

The geography of these northern areas of Japan was gradually clarified by the early nineteenth century, as European efforts at mapping the region were combined with the results of a succession of Tokugawa state-sponsored exhibitions that explicitly aimed to increase the state's knowledge of its diffuse northern reaches. It will be argued here that the relational aspect of cartographic exchange is crucial to the demarcation of this territory as being Japanese and under the authority of the Tokugawa state. Greater appreciation for the exchange involved in cartographic territorial creation not only allows for the transnational process of state demarcation to be recovered, but also hints at the inherently relational nature of the imperial sovereignty that came to literally remap vast areas of the globe during the nineteenth century.

## 1 Introduction

> In the morning we were visited by a new party of Japanese, superior to the others in dress, and equally so in behaviour. We derived not only pleasure, but information also, from their society...They permitted me to copy a large chart of the islands to the North of Japan, and promised to bring me one of their own doing the next day. (Broughton 1804: 100–1)

E. Boyle (✉)
Center for Asia-Pacific Future Studies, Kyushu University, Fukuoka, Japan
e-mail: tedkboyle@gmail.com

 
M. Altić et al. (eds.), *Dissemination of Cartographic Knowledge*, Lecture Notes in Geoinformation and Cartography, DOI 10.1007/978-3-319-61515-8_6

The above encounter, narrated by William Broughton in the published account of his voyage in the *Providence*, occurred on 25 September 1796, at a place Broughton termed 'Endermo Harbour', known today as Muroran in Hokkaido. The *Providence*, a 420-ton sloop of 16 guns with a full complement of 115 men, had just sailed across the Pacific and up the coast of the main island of Japan with the specific aim of charting those 'islands to the North of Japan' represented on the map that Broughton was permitted to copy. What Broughton embarked upon in the Providence was a voyage of discovery, one that promised to 'open new sources of knowledge and trade' (Broughton 1804: iii). The area which Broughton was intent on 'discovering' was one widely considered to still be unsatisfactorily represented upon European charts. Broughton's voyage was self-consciously conducted in the image of Cook, and the fact that the area around the Okhotsk Sea had remained uncharted by the latter in the course of his explorations added to the allure of the region for Broughton himself (on the wider background to British interest in Japan, see King 2010). Broughton's voyage of discovery should be understood as both following in well-worn tracks while entering uncharted waters.

In seeking to definitively chart the lie of the land for the curiosity of the nascent geographers and astronomers of European capitals, Broughton ran into the presence of Japanese upon these islands, who were similarly exercised by questions of trade and discovery as they sought to clarify the geographical outline of this space north of Japan proper. This study will examine the cartographic exchange that surrounded the meeting of two individuals, William Broughton and Kato Kengo, at this harbour of Endermo in order to highlight how the 'horizon of possibilities' represented on their maps had come to be developed. In doing so, it will demonstrate how central such examples of mutual exchange were to the emergence of the geography of the northwestern Pacific, which was successively rediscovered and rewritten multiple times over the course of the eighteenth and nineteenth centuries.

## 2 Broughton's Voyage of Discovery

### *2.1 William Broughton's Mapping of 'Insu'*

> [The Providence's officers concurred] to survey the coast of Asia, commencing at the island of Sakhalin. . .my intention was also to complete the survey of the adjacent islands, viz. the Kuriles, and those of Jeso and Japan, left unfinished in Captain Cook's last voyage. (Broughton 1804: 65)

In October 1793, William Robert Broughton was granted command of HMS *Providence*, and instructed to proceed to the west coast of North America to re-join the expedition of George Vancouver. Born in 1762, Broughton had served in the Royal Navy since 1774, and would therefore have been well used to the vagaries of fitting-out, delays in which meant that he was ultimately unable to depart the Channel until February 1795. Broughton was in fact retracing his own steps, having served as the commander of a brig, the *Chatham*, which had accompanied

Vancouver's *Discovery* on the latter's voyage to the American northwest. Vancouver's mission had been twofold; to accurately chart the northwest coast of the continent while also overseeing the First Nootka Sound Convention that had been signed between Britain and Spain in 1790 (Clayton 2000: 330–337). Having arrived in early 1792, Vancouver's negotiations with his Spanish counterpart had stalled, and Broughton had been dispatched back to Britain for further instructions at the end of that year.

Putting in at Rio de Janeiro, Tasmania, Tahiti, and Hawai'i on the voyage, Broughton had returned to Nootka Sound by March 1796, and found that Vancouver had long since departed. Correctly anticipating that Vancouver would already have surveyed the south-eastern part of South America, Broughton instead resolved to turn his attention to the coast of Asia, the survey of which he believed remained as yet 'unfinished'. Broughton sailed into 'Endermo Bay' on 15 September of that year and remained there for 2 weeks, provisioning and repairing his ship, and seeking to open lines of communication with both the 'natives' and Japanese (Broughton 1804: 96). On 1 October he rounded Erimo-Misaki and headed towards the Kuril Islands, reaching them 4 days later and cruising among them until the 18 of that month, before heading south and then west in order to strike the east coast of Japan just north of Edo (modern Tokyo). Sailing through the Ryukyu (Okinawa) Islands and wintering in Macao, he returned to the Ryukyus, where the Providence was wrecked on a reef near the Miyako Islands on 17 May 1797. Having fortunately acquired a 35-man tender while in Macao, he subsequently proceeded up the east coast of Japan in the smaller vessel and returned to the harbour at Endermo on 12 August 1797. Once again provisioning and refitting here, the tender weighed anchor on the 21 and left the bay on the 23. After plying back and forth for several days, waiting for the wind, he passed through the 'Straights of Sangaar', separating the main island of Japan from that of 'Insu', on 29 August 1797, passing the Japanese town of Matsumae on the 30.

Under the, as we shall see, mistaken impression that he was the first European in these waters, Broughton charted the west coast of the main island before continuing onwards to the 'extensive island north of Insoo' (Broughton 1804: 291) on 8 September. By 11 September, Broughton was struggling to 'reconcile the extent of this land with the island, North of Insoo in the Japanese chart' (Broughton 1804: 295). Three days later he was convinced that 'the extent of this land was much greater than we could suppose the Japanese island to be, as laid down in their chart' (Broughton 1804: 299). On 16 September, having been halted in a bay, Broughton confirmed to his satisfaction that there was no passage to the sea to the north, and began to run south down the opposite coast. He exited what his chart would subsequently term the 'Gulf of Tartary' on 23 September of that month, and reaching the island of Tsushima, located between Japan and Korea, on 12 October. After provisioning at the Korean port of Pusan and exploring the islands off the southwestern coast of Korea, he returned to Macau via the Ryukyus and Formosa (Taiwan) on 27 November.

As this account of his voyage shows, Broughton's aim was to 'finish' that which Cook had not, the charting of lands at the eastern end of Asia. According to the account of Broughton's voyage, what he had accomplished was a survey of not just the 'eastern coast of Tartary', but also that 'of Chica, Jesso or Insu', an island that was certainly 'an object of curiosity to Europeans' (Broughton 1804: vii). This object was a semantically slippery one, for having resolved to survey 'Jeso', Broughton subsequently found himself referring to the territory in question by an alternative designation. On his approach to 'Volcano Bay', within which the harbor of Endermo was located, he had questioned three boats of fishermen, who indicated that 'Matsmai' was located to the west and that they called 'their own land *Insu*' (Broughton 1804: 89). While Broughton correctly noted that this was the 'famous land of Jeso', in general he continued to utilize this new moniker of Insu/Insoo in the remainder of his account, as the quotations above indicate, adding another possible designation for the island to a list that already included 'Jesso, Matsumay, or Matmay, and Chica' (Kruzenshtern 1813 II: 49). The following year Broughton felt he had received some clarification when he returned to Endermo and came to understand that this was the name of the island utilized by its inhabitants.

The linguistic mapping being practiced here by Broughton, through seeking to understand the geographical designations adopted by the various inhabitants of the region being discovered, accords with the drawing of 'lines of commensurability' conducted by the earlier expedition of Lapérouse to this region, as highlighted by Michel Bravo. This 'ethnonavigation' was utilized in order to discern the geographical identity of a given territory. Nevertheless, such a process depended upon an image of territory which could be subsequently rewritten in accordance with the 'geographic gifts' granted in the course of 'orchestrated encounters' with native peoples (Bravo 1999: 217–226). The development of such an image, and role of the material exchange of maps within it, were essential to Broughton's ability to seek the discovery of this 'ill-defined' stretch of Asian coastline.

## 2.2 *Tracing 'Insu' on Broughton's Map*

> These people informed us that the proper name of this extensive island was Insu or Insoo, and universally called so by the natives: Matzmai applying only to the town and district inhabited by the Japanese, situated opposite to the coast of Nipon, in the straits. (Broughton 1804: 272)

There is little doubt that the sounds Broughton transcribed as 'Insu' were identical to those of 'Jeso', a land that had been mentioned in European reports since before Japan had been visited (Boscaro and Walter 1994: 84). Maps of Japan had featured an island with similar designations (such as Yezo, Jesso, etc.) since at least the 1617 map of Christophorus Blancus in Rome, almost certainly based upon an original by Ignacio Moreira (Schütte 1962; Hubbard 1994; plates 28–35 & 57–60 in Walter 1994). While the map by the Jesuit Girolamo de Angelis has subsequently become famous as an early representation of this island immediately to Japan's north, and influenced the work of Robert Dudley (Schütte 1969), it was unusual in attempting to show the entire area of the island. Given the paucity of

European information about Japan, it was far more common for European maps to represent Yezo as 'enframing' the rest of Japan, in the style of Japanese maps of the period (Yonemoto 2003). This is most obviously seen in maps that were produced on the basis of Ishikawa Ryusen's, including those of Adrien Reland (Yonemoto 2016). It was on the basis of another of Ishikawa's maps transported to Europe that a young Swiss scholar, Johann Caspar Scheuchzer, would produce a map to accompany his translation of Engelbert Kaempfer's famous work on Japan in 1727 (Fig. 1). The map selected by Scheuchzer showed a small island of Matsumae off the coast of the larger island. While almost certainly not the map Kaempfer originally envisaged as accompanying the text (Massarella 1995), it accords with the author's description in the book. There, the island of *Jezo* 'was invaded and conquer'd...by Joritomo, the first Cubo, or secular monarch, who left it to the Prince of Matsumai, (a neighboring Island belonging to the great Province Osiu) to be by him govern'd and taken care of' (Kaempfer 1906: 106).

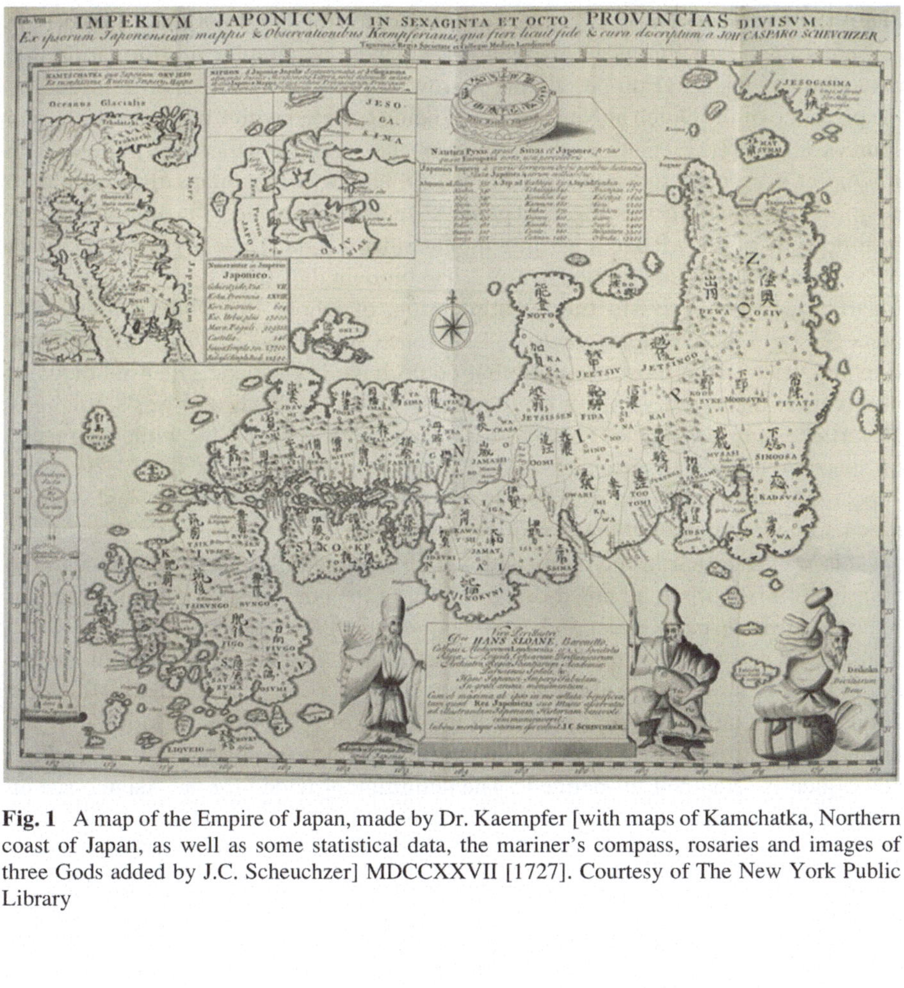


**Fig. 1** A map of the Empire of Japan, made by Dr. Kaempfer [with maps of Kamchatka, Northern coast of Japan, as well as some statistical data, the mariner's compass, rosaries and images of three Gods added by J.C. Scheuchzer] MDCCXXVII [1727]. Courtesy of The New York Public Library

Scheuchzer's decision to adopt this mode of representation speaks to the naturally insular nature of the political imagination, where the perceived separation between an area of direct Japanese rule and that of the native *Ezo* inhabitants came to be literally represented on this map of Ryusen's (On insular representations, see Steinberg 2005).[1] As Kaempfer admitted on the page following the description above, 'As to [Jeso's] figure, I could not gather anything positive, neither from the accounts I had from the Japanese, nor from the Maps, I met with in the country, they differing much from each other' (Kaempfer 1906: 107). This absence of definitive geographic information was indicated by the two alternative possibilities for the geography of northern Japan shown. The work Kaempfer had probably intended to base his map of Japan on was shown on Scheuzer's map in a cartouche that provided an alternative representation of the relation between Japan and 'Jesogashima' (Jeso island(s)) to its north (see Walter 1994, plates 74–76a). A second cartouche represented Yezo as being an extension of Kamchatka, and was taken 'From a recent map of the Russian Empire'.[2]

As Scheuzer's offering of alternative geographies suggests, attempts to position these 'islands to the North of Japan' in relation to 'the coast of Asia' would long cause confusion for European mapmakers. Japanese geographical information that made its way back to Europe, via the narrow channel of transmission afforded by the presence of the Dutch at Nagasaki, was being collated with Chinese and Russian sources in the representation of this part of the world. As the map published in Kaempfer's account indicates, a central problem was that maps that sought to position Japan within this more expansive spatial vision had ran up against the insufficiency of knowledge regarding the far side of Asia. The encyclopaedic work of Nicholas Witsen, first printed in 1692, had brought the 1643 Dutch voyage led by Maarten Gerritsz. Vries to Europe's attention, of whom Broughton's immediate predecessor in these waters, comte de Lapérouse, had observed, 'The navigation of Captain Vries is the most accurate that could have been made given methods of observation that were very crude' (quoted in Coen's journal of Vries's Voyage). Nevertheless, by the time of Lapérouse's own voyage in the 1780s, while the results of successive Russian expeditions and the published charts of Cook's voyages appeared to have displaced a geography of these regions based upon Vries' voyage, they had not yet put anything definitive in its place.

This was reflected in the maps with which Broughton himself travelled. In the margins of his journal, he noted on approaching the coast that 'we had no account relating to these parts excepting Cooke's 3d voyage & general chart, in which Matsmai is laid down as an island North of Japan, which we suppose this to be' (Gould 2011: xli). As already noted, however, one of the primary motivations for Broughton's fixation with this stretch of the 'coast of Asia, from the latitude of 35° north to the latitude of 52° north' was that it had not been surveyed by Cook, and consequently remained 'ill-defined'. The definition provided for the Asian coast on

[1]In what follows, Ezo is a geographical designation, while *Ezo* refers to the people who inhabited this land of Ezo.

[2]'Ex recentissima Russici Impery Mappa'.

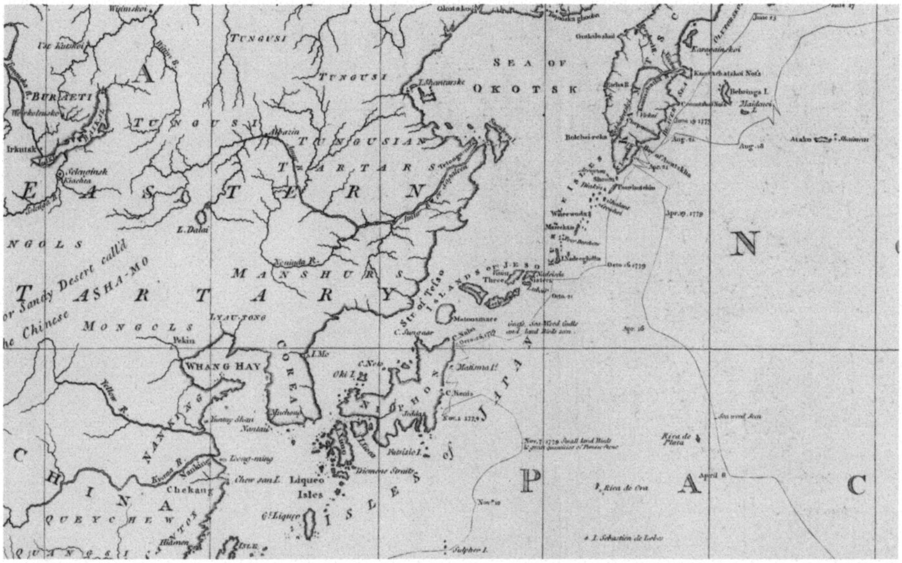


**Fig. 2** Representation of the islands north of Japan on Roberts' map of Cook's voyage, from A general chart: exhibiting the discoveries made by Captn. James Cook in this and his two preceding voyages, with the tracks of the ships under his command. By Lieut. Henry Roberts of his Majesty's Royal Navy. W. Palmer sculp. (London, G. Nicol and T. Cadell, 1785). Courtesy of the David Rumsey Map Collection

this chart, the 1785 'General Chart of Cooks Voyages' by Henry Roberts, was the result of an earlier exchange of cartographic material that had occurred during the course of Cook's third voyage (Fig. 2). According to the introduction to the account of that voyage, the representation of the coast of Asia on the map was on the basis of 'a Russian MS chart' (Cook and King 1785: lxxxii), which had been acquired by the expedition from an Ismayloff during a 3-week stay on 'Oonalashka' (Unalaska in the Aleutians) in 1778 (Kippis 1788: 226).[3] The vision of these islands north of Japan which Broughton carried with him to Asia was already one based upon a process of cartographic exchange that had occurred in the extremes of the northern Pacific. This had created an image of Japan's north where the 'Islands of Jeso' immediately above the main Japanese island gave way to the Kuril Islands stretching away towards Kamchatka.

The origin of such a representation of these islands ultimately leads back to the general map of the Russian Empire that had been published by the Academy of Sciences in St. Petersburg in 1745. This 'Complete Map of the Russian Empire' relied for its representation of this part of the world on the charts of Martin

[3]This was a Gerasim Gregorievich Izmailov, who many years earlier had been abandoned on one of the Kuril Islands by the Hungarian adventurer Benyowsky during the course of the latter's escape from Kamchatka (Akitsuki 1999: 172).

**Fig. 3** Representation of the islands north of Japan on the *Mappa Generalis Totius Imperii Russici*, Academy of Sciences, St. Petersburg 1745. Courtesy of the David Rumsey Map Collection

Spangberg (Fig. 3). As a member of the Second Kamchatka Expedition (c. 1733–1743), Spangberg had visited the southern Kurils and eastern coast of Hokkaido in 1739. On his chart, Spangberg represented the Kurils and Jezo as a number of small islands located to the north of the main island of Japan (Postnikov 2007, 81). These were incorporated on the general map together with a representation of Sakhalin based upon that adopted by the Qing in the so-called Kangxi atlases, a copy of which had been sent to Peter the Great in 1721 (Hostetler 2002, 71). The Complete Map amended the representations of both these islands and Sakhalin, with the Kuril Islands leading up to Kamchatka being located some distance from an island of 'Matuszma' immediately to the north of the main island of Japan, while Sakhalin was represented in the 'arrowhead' form characteristic of its appearance in the Kangxi atlas.

The publication of a history of Kamchatka and the Kuril Islands by another member of this Expedition, Stepan Petrovich Krasheninnikov, in 1755, and the rapid dissemination of its translation and accompanying maps throughout Europe (there was an English translation by 1764 and one into French by 1767) meant that the results of these Russian explorations came to be widely circulated.

Krasheninnikov's map differed from the one drawn up in 1745, positioning the island of 'Matma' or 'Matsumay Island' much closer to both that of 'Kunasir' and the Asian mainland, while noting that Kunasir and the islands immediately adjacent were referred to as 'Jeso' by the inhabitants. Krasheninnikov's text noted that the natives of 'Eturpu' and 'Urupe' are also included by the Japanese as being part of 'Jeso', together with those of Kunasir and Matsumay (the map is located between pages 53 and 54 of Krasheninnikov 1764).

The representation adopted by Krasheninnikov was also that of Gerhard Friedrich Müller, who like Spangberg and Krasheninnikov was another veteran of the Second Kamchatka Expedition, and who had apparently provided much of the information on the Kurils recounted by Krasheninnikov. Müller's map, initially published in the same year of 1755, offered the same description and representation as on Krasheninnikov's maps. Müller's map was also disseminated in the West, being published by the Academy in French in 1758 and affixed to his 'Voyages et découvertes faites par les Russes' of 1765. This latter text was also widely-known and repeatedly referenced by Cook during his voyage. The representation common to the maps of Müller and Krasheninnikov came to form the basis for Thomas Bowen's chart of Cook's discoveries that accompanied the 1780 publication of Joseph Banks' *New System of Geography* (Fig. 4).

**Fig. 4** Thomas Bowen (1780): *A new & accurate chart of the discoveries made by the late Capt. Js. Cook and other distinguished, modern navigators made between the latitudes of 80° north and 50° south and extending to 260° east long. from the meridian of Greenwich exhibiting Botany Bay, with the whole coast of New South Wales in New Holland, also New Zealand, Norfolk and the various other islands situated in the Great Pacific Ocean, & the Northern & Southern Hemispheres.* Courtesy of the National Library of Australia, MAP NK 1584

Tracing this history of representations of the eastern end of Asia in this fashion can be interpreted as showing a steady increase of geographical knowledge leading inexorably to more accurate maps. This increasing geographical knowledge allowed for the territory of little-explored parts of the earth to be successively rewritten on the basis of each subsequent voyage. Nevertheless, as we have already seen with the example of Kaempfer's account, rewriting was not necessarily a guarantee of greater accuracy, and was sometimes undertaken for reasons far from opaque. A comparison of the two charts representing Cook's voyages is an example of this, where the 1785 chart with which we began, and which Broughton took on his voyage, is significantly different from not only the chart of 1780, but also the material on which it was based, stretching back to the incorporation of Spangberg's discoveries on the 1745 chart. Compared to these maps, it alters the representation of the Asian coast, shrinks and adjusts the orientation of Sakhalin, and repositions the island of 'Matoosmae' so that the latter is closer to the main island of 'Niphon' and further from the Asian coast. Such 'rewritings' take on an arbitrary air, whether based on Russian manuscript maps or not.

## 2.3 *Mapping the World in Enlightened Fashion*

> the condition of the ships, of the sails and cordage, made it unsafe to attempt, at so advanced a season of the year, to navigate the sea between Japan and Asia; which would otherwise have afforded the largest field for discovery; that it was therefore advised to keep Eastward of that island, and in our way thither to run along the Kuriles, and examine more particularly the islands that lie nearest the Northern coast of Japan, which are represented as of a considerable size, and independent of the Russian and Japanese governments (Cook and King 1784 III: 385)

> there are, however, parts of the coasts, both of Asia and America, which would yet afford employment for the labourers in the science of DISCOVERY. The Asiatic coast, from about the latitude of 35° to 52° North, is at present very ill defined… (Vancouver 1798: 397)

The latter half of the eighteenth century is frequently considered to have witnessed the triumph of a positive geography, and an end to such geographic arbitrariness. The voyages of discovery undertaken by the likes of Cook, Lapérouse and Broughton have served as both examples and metaphors of 'science in action'. In Bruno Latour's famous formulation, 10 years prior to Broughton's arrival among these islands, a French expedition led by the comte de Lapérouse had landed on Sakhalin and been told by the natives they encountered there that Sakhalin was, indeed, an island. This information was communicated to Lapérouse through drawings made both on the sand and on paper (Latour 1987: 216). Despite the loss of the expedition, Lapérouse had fortunately conveyed this information back to the expedition's 'center of calculation' in Paris through the 'immutable mobiles' carried by his interpreter across Russia. The incorporation of the information contained on these mobiles within contemporary scientific knowledge enabled

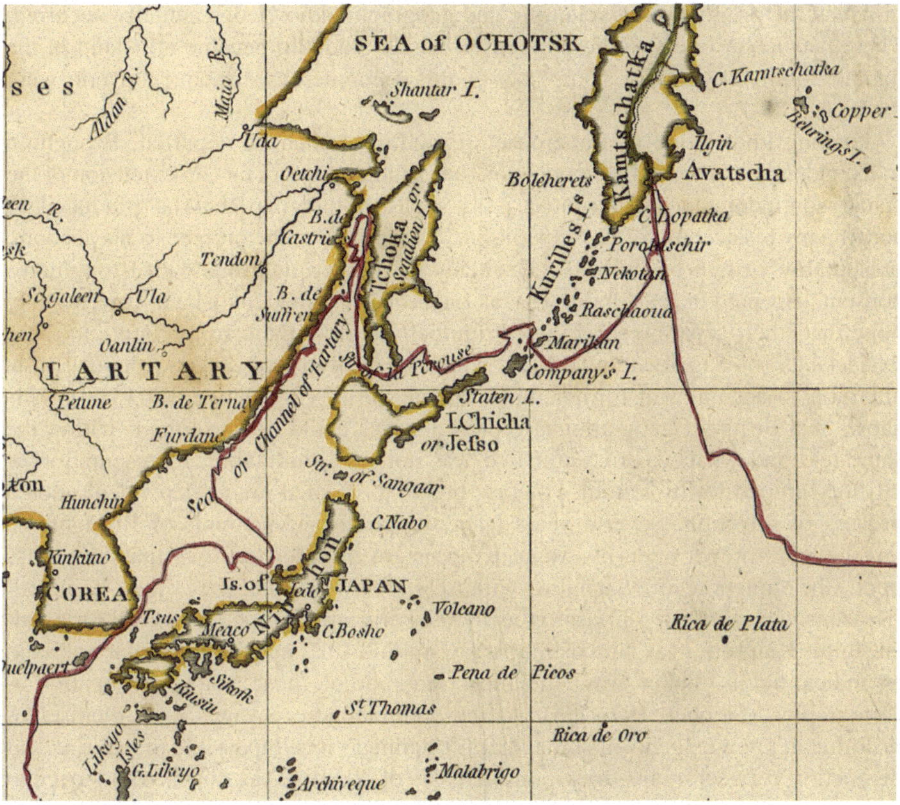


**Fig. 5** Lapérouse's voyage in the islands north of Japan, from the *Map of the World, or General chart of the known parts of the globe, illustrating the voyage of La Pérouse in 1785, 1786, 1787 & 1788*, G. G. & J. Robinson, Paternoster Row 1799. Courtesy of the John Carter Brown Library at Brown University

subsequent expeditions to the region to travel with a more accurate picture of the territory to be encountered than that possessed by earlier explorers (Fig. 5). The moment at which Sakhalin is shown to be insular is able to be communicated to Europe and thus able to be represented on all future maps made of the area.

Latour's account offers a brilliant exposition of the process by which a specific piece of information was communicated, represented in immutable yet mobile form, transported back to Europe and subsequently disseminated through a variety of channels as scientific knowledge. However, it overstates the centrality of this moment on the shores of de Langle Bay, in part because of the drama inherent in the manner in which it appears that two distinct modes of geographical consciousness come into contact on the beaches of Sakhalin itself. As the example of Cook's two maps above indicates, enlightened geography was not necessarily able to proceed through the simple accumulation of additional knowledge on the same map, and its acceptance and re(-)presentation continued to occur in arbitrary fashion. Latour's 'center of calculation' is too totalizing a metaphor, standing outside of the manner in which the production of scientific and geographic knowledge actually occurred. There was never one specific moment at which Sakhalin became an island in the manner implied by Latour's emphasis on this encounter, as subsequent events were to show.

Lacking knowledge of Lapérouse's expedition when he departed, Broughton independently arrived at the conclusion that Sakhalin was in fact an extension of the Asian continent, and represented it as such on the map that he produced to accompany his account of the expedition.[4] Nevertheless, the preface to his account, presumably written by its editor, Broughton's brother, and thus with Broughton's consent, accepted that Sakhalin was, as Lapérouse claimed, an island. It went on to hope that 'their separate surveys will mutually correct the errors relative to these two islands [of Sagaleen and Jesso], which have been laid down with such little attention to accuracy in former maps of the world' (Broughton 1804, vii). This shows that different geographical representations could be 'inscribed' within the same text and what was disseminated was not the 'immutable' representation of territory argued for by Latour's theory, but an abstracted image dependent upon a process of selection. Several years later, the published accounts of Broughton's voyage were drawn upon by Adam Krusenstern in order to prove that 'my ideas upon the junction of Sachalin with Tartary were perfectly well founded' (Kruzenshtern 1809 II: 182). Into the 1820s, Paris and Europe's foremost Orientalist, Julius Klaproth, was lambasting Arrowsmith's 1822 map of an insular Sakhalin as indicating its maker was 'the most ignorant of those whose occupation is cartography' (Stephen 1971: 35). As these emblematic voyages were themselves resulting in knowledge itself somewhat 'ill-defined', it is impossible to make a clear distinction between maps drawn on the basis of 'positive' as opposed to 'speculative' geographical knowledge. There is no dividing line whereby modes of representation suddenly became 'scientific' and 'accurate'.

One reason for this, as the above highlights, was because of the varied individuals and institutions involved in attempting to bring these territories onto the map, the collective products of which could confuse, rather than stabilize, representations of parts of the globe. A second, however, is the assumption, obvious in Latour's account, ascribing the incorporation of geographic knowledge as only occurring at one 'center', identified with here with Paris, and more generally Europe. Yet the presentation of the cartography of these voyages as 'an empirical, objective and unproblematic science concerned only with the presentation of geographic information' (Edney 1994: 101)[5] assumes its maps as resting on the claims to science

[4]In his application of Latour's notions to the Japanese expedition of Mamiya Rinzo, Brett Walker asserts that Broughton was seeing Sakhalin a "second time" (Walker 2007: 294), but it seems unlikely that this was the case. Barry Gough concludes that "[d]etails of [la Perouse's] geographic discoveries, which had been sent back to Paris in September 1787 from Petropavlovsk, were apparently not known to Broughton, though if they were (and it seems unlikely) Broughton gives no hints" (Gough 2010, "Introduction": xxii).

[5]Edney notes this is what Harley calls the "illusion of cartographic mimesis" and Wood and Fels the "cartographic myth".

made by its producers, rather than in the shifting sources of knowledge out of which that geographic information was formed. Yet despite European claims, it did not retain a monopoly on the production of the knowledge utilized within this cartography, as the remainder of this article will highlight.

# 3 Situated Knowledge

## *3.1 Geographical Genealogies*

> I acquired from him a very compleat map of the Japanese islands, with strong injunctions not to acknowledge from whom I procured it; as they explained the parting with it would bring them into disgrace and punishment, were it known. (Broughton 1804: 272)

Among the officials that Broughton encountered while moored in 'Endermo Bay' during September 1796 was the Matsumae official Kato Kengo. Born in 1762 as the eldest son of a clan retainer, Kato had entered the Shogunal Academy in 1785 and subsequently served as an official and then physician for the Matsumae, the feudal family granted responsibility for managing Ezo. He was already considered something of a diplomat, having been in the party that had met with Adam Laxman when the Russian envoy had sailed down the Kuril Island chain in order to return three Japanese castaways and request the opening of Japan to trade with Russia. The merchants residing in Nemuro had acceded to the Russian request that they be allowed to winter there, giving time for the wheels of the Shogunate's response to this unwanted incursion to slowly grind into gear. As a result, the likes of Kato had spent some time getting to know these Russian visitors, before the latter were granted permission the following spring to proceed along to the coast to the town of Matsumae itself. This time was partially spent in an exchange of geographical knowledge between the Japanese hosts and their Russian guests (see 'Adam Laxman: Journey of Laxman's Embassy to Japan' in Wells 2004: 37–38).

The 'large chart of the islands to the north of Japan' which Broughton refers to is almost certainly a version of a map known as the 'Matsumae Chizu' (Fig. 6). The map in question was one that Kato had first drawn in either 1791 or 1792, and of which numerous versions were made over subsequent years, with over 70 manuscript maps based on this design known to be in existence in Japan today (Takagi 2011: 114). Kato had already handed a version of this map over to Laxman's Russian expedition in exchange for the opportunity of copying the latter's maps, and now took advantage of his being sent to deal with the troublesome arrival of another group of foreigners to once again display his knowledge of the islands. He apparently made quite the impression on the visitors, for when Broughton was returning to Volcano Bay in his tender in the summer of 1797, after having wrecked the *Providence* on the reef in the Ryukyus that May, word reached the Matsumae that the English had been asking whether 'Kato was well?'. The clan once again

**Fig. 6** Copy of a map of Ezo, showing Karafuto, the main island of Ezo, and the Kuril Islands out to Etrorofu, Kato Hisashi (Kengo) *Matsumae Chizu* (original c. 1791). Courtesy of the Northern Studies Collection, Hokkaido University, MAP 645

dispatched him to deal with these unwanted yet knowledgeable guests (Tsuko Ichiran 6 1913: 365).

The representation Kato offered of these islands to Japan's north provided ample evidence of the recent surge of interest in this region. The name by which the map is known, 'Matsumae map', reflects the fact that Kato was an official of the clan, but is also indicative of the geographic lineage from which the map emerged. The construction of the map shows the traditional priority of the Matsumae as having been control over a number of distinct points situated near the coast of this collection of amorphous islands, with the map listing the members of the clan to whom each village or *Ezo* settlement was assigned. This means of comprehending space resulted from the unique way in which Matsumae authority had come to legitimated, through trade with the barbarian inhabitants, the *Ezo*, who had given their name to the geographic designation of the territory over which the Matsumae's rule extended. This means of defining the territorial extent of their authority was distinct from all of the other domains that made up the Tokugawa state, which were allocated both control over and responsibility for the collection of revenue, and thus had a defined territorial basis, although one which remained notably unrepresented on national maps of the period (Berry 2006). By contrast, the Matsumae were effectively granted licence to define the territorial extent of their rule through their monopoly over trade relations with the *Ezo*. As is obvious from the maps submitted by the Matsumae during the Shogunate's official mapping projects, the clan failed to scale or produce maps of their lands to the standards mandated by the Tokugawa regime (Kawamura 1984). Consequently, the extent of land claimed by the Matsumae was represented in a severely truncated fashion on official maps of Japan during this period (Kuniezu Kenkyukai 2005).

The lands of the Matsumae on these maps incorporated the main island of Ezo, today's Hokkaido, and extended out to both Sakhalin and the Kuril Islands, although this assertion of control was largely on the basis of trade being conducted with the *Ezo* living in those regions (Fig. 7). The actual extent of direct Matsumae authority, an area at the southern tip of Hokkaido that was officially mandated for Japanese residence, and through which Shogunal inspectors passed on their official tours, was rendered comparatively accurately. The rest of the map, though, was largely drawn on the basis of information derived from *Ezo* traders coming to Matsumae, rather than Japanese surveyors heading out into this ill-defined space known as the barbarian islands. Over the course of seventeenth and eighteenth centuries, the system whereby specific trading sites within the lands of the *Ezo* were allocated to the clan's retainers was extended over the whole of the main island of Ezo, while simultaneously being hollowed out through the sale of such trade privileges to merchants. The map of Kato noted above shows the vestiges of this system through its list of clan retainers officially responsible for villages within the area of official Japanese habitation and for the trading posts across the area officially inhabited by the *Ezo* people.

While it is often asserted that the maps submitted by the Matsumae sought to deliberately conceal the extent of the area they claimed to rule over (Takagi 2006: 144), this seems to be merely an assumption of historians. Significantly, in the history of his clan that the scholar and official Matsumae Hironaga wrote a mere decade before Kato produced his map, he chose to illustrate the text with a version of the map that the Matsumae had submitted to the Shogunate in its last two surveys (Takakura 1987: pl. 9). In the same text, Hironaga gave comparatively accurate latitude readings for the southernmost and northernmost points on the island (at Matsumae and Soya, see Matsumae 1781: 113), and also estimated the circumference of the main island of Ezo as being about 600ri "rather than the 800 it was thought in the past". This figure was only slightly less than the 650ri estimate offered almost a century later, in 1884's *Hokkaido-Shi [Hokkaido Almanac]*. In other words, while the Matsumae had a reasonable idea of the extent of the territory claimed under their authority, they were unable to accurately represent it on the map. This is why Matsumae accounts of the land under their control were characterized by lists of Japanese villages and native settlements extending out in either direction from the Matsumae base of Fukuyama Castle, in the manner reproduced by Kato.

Some seventeenth and eighteenth century images of Ezo produced by other administrations or for popular consumption did position its main island, whether

**Fig. 7** Extent of 'Matsumae' on the Complete Map of Japan assembled from the Genroku-era provincial surveys (1697–1700), located to the northeast of the main Japanese island of Honshu. Karafuto is represented as a very small island to the north of Ezo's main island, with a desultory scattering of islets showing the Kuril Islands to the east. Courtesy of the Ashida Collection, Meji University Library

together with Karafuto and the Kuril Islands or not, in relation to the continental coast of Tartary. However, references to the continent remained geographical expressions that joined with other more fantastic ones, like the Isle of Women, in locating this space in a context defined by Japan. As a Chinese-derived notion of Japan's place in the world being shaped by a civilized-barbarian order with Japan at its centre came to be popularly expressed over the period of Tokugawa rule, the *Ezo* came to serve as one of Japan's 'familiar foreigners' (Yonemoto 2003: 33). The land these *Ezo* inhabited therefore 'enframed' the core of Japan, associated with representation of the traditional notion of 66 provinces and 2 islands as being Japan's natural boundaries (Unno 1999). This is shown most literally in the *Wakan Sanzai Zue*, an encyclopaedia written in stages around the turn of the eighteenth century (Fig. 8). In it, the *Ezo* are listed as 'near' rather than 'distant' foreigners, and a map of their lands offered, yet this same land is literally located outside of the frame of the map when positioned in relation to Japan itself (on the significance of the description of the *Ezo* given in this Chinese-derived encyclopedia, see Morris-Suzuki 1998: 13–23).

While Broughton's desire to map this 'ill-defined' area of the Asian coastline reflected the hitherto marginal position of this region in a global pattern of European exploration, the area of the *Ezo* that Kato was seeking to mark was hardly less marginal within a history of Japanese mapping, despite the enormous difference in relative distance from their respective 'centers'. However, this absolute difference in geographical distance was in fact somewhat misleading, in that both of these surveyors were seeking to write this territory of Ezo onto the maps of their worlds.

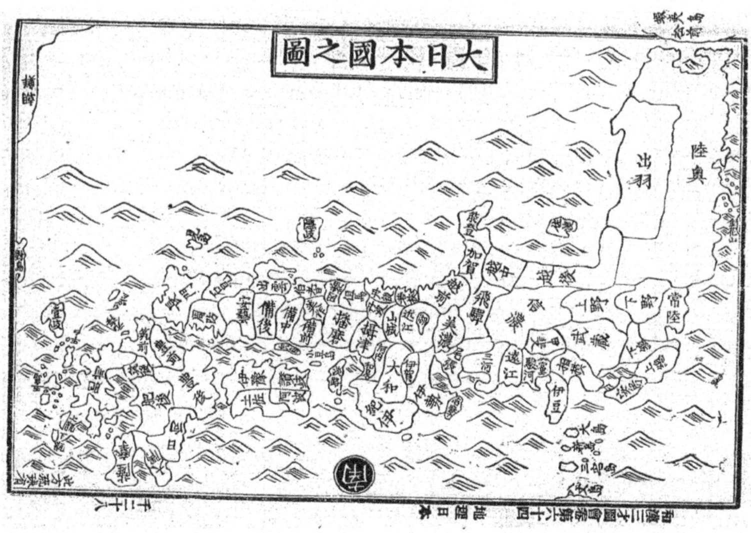


**Fig. 8** Great Japan being literally 'enframed' by 'Matsumae' and the 'Ezo Islands', placed beyond the border in the top right corner of this 'Map of Great Japan'. Terajima Ryoan, *Wakan Sanzai Zue*, Book 64, original 1713, 1888 reprint. Osaka: Chukindo. Courtesy of the National Diet Library

## 3.2 *Writing 'Insu' into the World*

> Our Japanese friends joined us for dinner, and presented me with a chart of their own doing; in return I gave them Captain Cook's general chart of the world, which gratified them extremely. (Broughton 1804: 101)

The traditional map of the Matsumae's own representation of their area of authority, which Hironaga adapted for his *Matsumae-shi*, reflected this general understanding of the space of the *Ezo* as existing on the edge of, and distinct from, the islands of Japan itself. Nevertheless, his account prefigured the dramatic changes that subsequent decades would bring through the incorporation of his own map of the Matsumae domain. This map largely adopted a traditional understanding of the space of Ezo, with its representation of the area held to be under the authority of the Matsumae being remarkably stable, but there was one important distinction (Fig. 9). This latter map illustrated a new awareness of the wider geography within which this space of Ezo was situated, with the Kuril Islands now being drawn as stretching out as a series of stepping stones towards a Kamchatka-like peninsula that was marked with the label *oroshia-koku* (the country of Russia, see also Takakura 1987: pl. 3).

The appearance of Russian lands coming to 'enframe' the islands of Ezo in the same manner in which the latter had traditionally been situated at the edge of Japan resulted from increasing concern over reports of Russian advances in Ezo and rumours of a secret foreign trade being conducted to Japan's north. The call to occupy Ezo in the face of this Russian threat is generally understood to have swelled from the mid-1780s (Harrison 1953; Keene 1969), advocating the 'opening' of the Ezo, the introduction of agriculture on the islands, and the incorporation of this new province into Japan. These warnings showed that both Japan as a whole and those places that served to 'enframe' the nation were coming to be incorporated into a more global understanding of space. The territory of the nation came to be defined through not only the accumulation of geographic knowledge but also situating it within the context provided by the acquisition of information from overseas.

The insertion of this amorphous space claimed by the Matsumae into more global networks is perfectly demonstrated within the leaves of the *Matsumae-shi* itself, where these two distinct representations of the Matsumae's own area of rule are followed, a few pages later, with a map of Bering's voyage.[6] The incorporation of this map of the North Pacific, together with the amendments that Hironaga had made to the traditional Matsumae map of their territory, underscores how this space of Matsumae rule was now beginning to be positioned not only in relation to Japan, but to other states as well (see Funakoshi 1986: pl. 86 for a reproduction). This shift

[6]It is unclear whether the map was originally part of the text or was added later, but it seems more likely that Kato drew this map of Bering voyages on the basis of discussions with, or maps received from, Laxman at Nemuro. This would mean the map was drawn and added to the *Matsumae-shi* at least a decade after the latter was originally written. This is the interpretation accepted by Japanese historians. Note, though, that the representation of Ezo adopted on this map is far closer to the maps Hironaga drew earlier for the *Matsumae-shi* than to Kato's understanding shown on the 'Matsumae-Chizu'.

**Fig. 9** Copy of the New Outline Map affixed to Matsumae Hironaga's *Matsumae-shi*, (1781), 'based on an old map revised with new information'. The main island of Ezo is divided in two, with 'Karato-shima' (Karafuto) to its north. The Kuril Islands stretch out northeast from Ezo to a Kamchatka taking up the right-hand edge of the map, labelled as the 'country of Russia'. Courtesy of the Northern Studies Collection, Hokkaido University, SCROLL 99

indicates a distinct change in how the spatial extant of Matsumae rule came to find representation upon the map, especially when we consider the identity of the mapmaker, as the chart of Bering's Expedition inserted into the *Matsumae-shi* notes that it was copied from a map also made by one Kato Kengo.

The manner in which Ezo appears upon this chart of Bering's discoveries itself speaks to the developing cartographic bricolage, for the mode of representation adopted for the area of Ezo on this map is one that combines distinctive elements from both Hironaga's maps of his clan's area of control and Kato's own 'Matsumae map'. The North Pacific chart appears to show Ezo and Karafuto as two large islands, as adopted by Hironaga and in contrast to the majority of early maps of the region in which the main Ezo island was represented as far larger than a subservient Karafuto. The similarly outsize impressions created by the Kuril Islands closest to Ezo, those of Kunashiri, Etorofu and the cluster of islets standing for Shikotan and the Habomais, is one that is also visible on Kato's map. While the map of the Pacific includes the island beyond Etorofu, known in Japanese as Uruppu, among those represented in a distinctively oversized way, Kato's 'Matsumae Chizu' map does not, with its representation of Ezo essentially halting at the edge of Etorofu and petering out in a desultory representation of a number of islands beyond.

While similarly seeking to account for new geographic knowledge, therefore, these two maps of Kato, likely drawn a few months apart, seek to abstract the space known as Ezo in two distinct ways. That produced for the Matsumae sought to provide an abstract image of the extent of land under the governance of the Matsumae, with very little reference to its surrounding geography, merely noting a small section of 'Santan' to the west of the island of Karafuto, and simplified representations for the remainder of the Kuril Islands out beyond Etorofu. By contrast, the other map sought to position a rapidly changing picture of the lands of the *Ezo* within a new context provided by the image of the Northern Pacific, one presumably made possible by Kato's encounter with Adam Laxman and his officers in the winter of 1792. The fluid cross-currents of geographic knowledge found themselves represented on these maps in a manner amenable to there being subsequently exchanged on the shores of Endermo Bay.

## 4 Rewriting the World Again and Again

> In the year 1802 there appeared in St. Petersburg a chart of the Russian discoveries in the north-eastern parts of the Great Ocean...what particularly distinguishes this chart is an island between Jesso and Sachalin, called Karafuto, or Schisa. The west coast of Jesso, as well as this island, were inserted on the authority of a Japanese chart ... (Kruzenshtern 1808 II: 28–29)

Kato's presentation of a 'chart of the islands to the north of Japan' to Broughton was a 'geographical gift', but in this case, one that earned its own 'geographical reward' in return, a copy of Cook's chart of the world. This is in contrast with the 'Chinese' that Lapérouse encountered on Sakhalin. There, the natives were 'not all that interested in maps and inscriptions', for these are only 'intermediaries for their exchange' and 'not considered important in themselves' (Latour 1987: 218). By contrast, for Kato, the inscription on the map was of crucial importance, and recognized as such. Kato made clear to Broughton that he ran the risk of punishment in granting a copy of this map of 'the islands to Japan's north' to the interlopers, a risk seemingly confirmed by the famous Siebold incident which occurred some three decades after the encounters between Kato and Broughton at Endermo Bay. Nevertheless, the geographical knowledge represented by such inscriptions was adjudged worth the risk by both sides, in their efforts to accurately map this area to Japan's north. It is within this process of cartographic exchange that the 'ill-defined' territory of this part of the world came to be successively rewritten until its appearance on the map became fixed.

Eight years after Broughton's second appearance at Endermo, a Russian expedition led by Adam Johann von Krusenstern was unknowingly following in his footsteps up the west coast of the island of Yezo, stopping at Soya near to northern tip of that island. Asking a local Japanese officer about the island of Karafuto, Krusenstern was informed it was just to the north, being 'distinctly perceived as soon as the weather began to clear up, separated from Jesso by a channel' (Kruzenshtern 1813 II: 47). According to the chart issued to Krusentern for his voyage, this island of Karafuto was a small indented island orientated from east to west and located between the islands of 'Matmai' and Sakhalin. While the shape of the island of Sakhalin followed that used on Lapérouse's map, that of 'Matmai' was orientated from north-east to south-west, in the same manner as on the chart that Kato had gifted Broughton. In fact, the representation of both 'Matmai' and this island of Karafuto that nestled between it and Sakhalin were taken directly from Kato's map. Kato had also shared this map with members of Laxman's expedition in 1792, and copies of it had subsequently been sent back to Moscow. By the time of Krusenstern's arrival, the spatial representation of these islands north of Japan had long since been rewritten by the Japanese themselves, but Kato's work still defined the territory created for Krusenstern at the outset of his voyage.

## 5 Conclusion

> I could not reconcile the extent of this land with the island North of Insoo, in the Japanese chart; neither did it continue the same easterly direction: however, to clear up the matter, we had only to continue our course, keeping well in with the land. (Broughton 1804: 295)

By the tail end of the eighteenth century, the representation of the area under Matsumae authority, which had long encompassed the islands today known as Hokkaido, Sakhalin and the Kuril Islands, was being rapidly and repeatedly rewritten on the map. What had altered was the manner in which this 'ill-defined' place began to be incorporated into a more globalized conception of space, one in which the boundaries of Ezo came to have meaning not only in the context of its relation to Japan, but with regards to other polities as well. The growing awareness towards this change in Japan is shown not only in the well-documented spate of works advocating a more direct form of rule over the *Ezo* and their lands, but in the person of Kato Kengo himself. Kato not only provided this representation of Japan's Ezo to myriad groups of foreign explorers arriving in the region, but also sought to reposition and remap the region in the context of this newly-globalized understanding of space.

This study has traced how this geographical region to the north of Japan, which by the nineteenth century was at the intersection of the territorial extents of the three Asian powers of Russia, China and Japan while serving as an area of increasing fascination to European and American expeditions, was being represented on the map. Its attention has therefore been less on the immutable mobiles through which geographic knowledge was communicated back to centres of calculation, and instead on the endlessly mutable representation of this geography that was a created in the course of these cycles of accumulation. Such representations were liable to be repeatedly rewritten through processes of cartographic exchange like that which this article has begun to recover.

**Acknowledgements** Attendance at the Dubrovnik Conference was possible thanks to a JSPS Grant-in-Aid for Young Scientists (B) 'Scaling borders: Tension and deterritorialization, Japan and beyond', 2016–2019 (Project No. 16K17071).

## References

Akitsuki T (1999) Nihon Hokuhen No Tanken to Chizu No Rekishi [A history of the exploration and cartography of the Northwest Pacific]. Hokkaido Daigaku Toshokankokai, Sapporo

Berry ME (2006) Japan in print: information and nation in the early modern period. University of California Press, Berkeley

Boscaro A, Walter L (1994) Ezo and its surroundings through the eyes of European cartographers. In: Walter L (ed) Japan: a cartographic vision: European printed maps from the early 16th to the 19th century. Prestel, Munich

Bravo MT (1999) Ethnographic navigation and the geographical gift. In: Livingstone D, Withers C (eds) Geography and enlightenment. University of Chicago Press, Chicago

Broughton W (1804) A voyage of discovery to the North Pacific Ocean: in which the coast of Asia, from the latitude of 35° north to the latitude of 52° north, the island of Insu (commonly known under the name of Jesso), the north, south and east coasts of Japan, the Lieuchieux and adjacent islands, as well as the coast of Corea, have been examined and surveyed, Performed in His Majesty's Sloop Providence and her tender, in the years 1795, 1796, 1797, 1798. T. Cadell & W. Davies, London

Cook J, King J (1784) A voyage to the Pacific ocean. Undertaken, by the command of His Majesty, for making discoveries in the Northern hemisphere, to determine the position and extent of the west side of North America; its distance from Asia; and the practicability of a northern passage to Europe. Performed under the direction of Captains Cook, Clerke, and Gore, in His Majesty's ships the resolution and discovery, in the years 1776, 1777, 1778, 1779, and 1780. Vols I and II written by James Cook; vol III by James King. W & A Strahan, London

Edney MH (1994) Mathematical cosmography and the social ideology of British cartography, 1780-1820. Imago Mundi 46:101–116

Funakoshi A (1986) Sakoku Nihon ni kita 'Koki zu' no chirigaku shiteki kenkyu [The acceptance of K'ang-hsi maps in Japan in the age of her isolation; a study of geographical history]. Hosei University Press, Tokyo
Gough B (2010) Introduction. In: David A (ed) William Robert Broughton's voyage of discovery to the North Pacific 1795–1798. Hakluyt Society, London
Harrison JA (1953) Japan's northern frontier: a preliminary study in colonization and expansion, with special reference to the relations of Japan and Russia. University of Florida Press, Gainesville
Hubbard JC (1994) The Map of Japan engraved by Christopher Blancus, Rome, 1617. Imago Mundi 46:84–99
Kaempfer E (1906) The history of Japan, together with a description of the Kingdom of Siam, 1690-92. J MacLehose, Glasgow
Kawamura H (1984) Edo Bakufusen Kuniezu no Kenkyu [Research into provincial maps produced by the Edo Bakufu]. Kokon Shoin, Tokyo
Keene D (1969) The Japanese discovery of Europe, 1720-1830. Stanford University Press, Stanford
King R (2010) 'The long wish'd for object'—opening the trade to Japan, 1785-1795. Northern Mar/le marin du nord 20(1):1–34
Kippis A (1788) The life of Captain James Cook, vol 2. JJ Tourneisen, Paris
Krasheninnikov SP (1764) History of Kamchatka and the Kurilsky Islands, with the countries adjacent. R Raikes, Glocester
Kruzenshtern IF (1813) Voyage round the world in the years 1803, 1804, 1805, & 1806: by order of His Imperial Majesty Alexander the First, on board the ships Nadeshda and Neva, under the command of Captain A.J. von Krusenstern of the Imperial Navy. In two volumes (trans. RB Hoppner). John Murray, London
Kuniezu Kenkyukai (2005) Kuniezu No Sekai [Kuniezu: province maps of Japan made by the Tokugawa government]. Kashiwa Shobo, Tokyo
Latour B (1987) Science in action: how to follow scientists and engineers through society. Harvard University Press, Cambridge, MA
Massarella D (1995) A history of *The History*: the purchase and publication of Kaempfer's *History of Japan*. In: Bodart-Bailey B, Massarella D (eds) The furthest goal: Engelbert Kaempfer's encounter with Tokugawa Japan. Japan Library, Folkestone
Matsumae H (1781) Matsumae-shi. In: Terasawa H, Wada T, Kuroda H (eds) (1979) Hoppo Mikokai Kobunsho Shusei [Collection of previously unpublished documents on Japan's North]. Sobunsha, Tokyo
Morris-Suzuki T (1998) Re-inventing Japan: time, space, nation. ME Sharpe, New York
Postnikov AV (2007) Cartographic solution to the problem: are America and Asia joined? (from the early 18th century to the 1740s). In: Postnikov AV (ed) The history of mapping of Moscow Tsardom and the stories of Russian cartography. ICA/Russian Academy of Sciences, Moscow
Schütte JF (1962) Ignacio Moreira of Lisbon, Cartographer in Japan 1590–1592. Imago Mundi 16:116–128
Schütte JF (1969) Japanese cartography at the court of Florence; Robert Dudley's maps of Japan, 1606–1636. Imago Mundi 23:29–58
Steinberg PE (2005) Insularity, sovereignty and statehood: the representation of islands on portolan charts and the construction of the territorial state. Geografiska Annaler Ser B Hum Geogr 87(4):253–265
Stephen JJ (1971) Sakhalin: a history. Clarendon, Oxford
Takagi T (2006) Edo Bakufu no Kuniezu Sakusei to Matsumae-han no Taiou [The role of the Matsumae in creating provincial maps under the Bakufu]. In: Hokkaidoshi Kenkyu Kyougikai (ed) Hokkaido Rekishi to Bunka—Sono shiten to tenkai [Hokkaido's history and culture—perspectives and prospects]. Hokkaido Shuppan Kikaku Sentaa, Sapporo
Takagi T (2011) Kinsei Nihon no Hoppō-zu Kenkyū [Research on Northern mapping in early modern Japan]. Hokkaido Shuppan Kikaku Sentaa, Sapporo

Takakura S (1987) Hokkaido Kochizu Shusei [Collection of historical maps of Hokkaido and adjoining regions]. Hokkaido Shuppan Kikaku Sentaa, Sapporo

Tsuko Ichiran 6 (1913) Kokusho Kankokai, Tokyo

Unno K (1999) Chizu Ni Miru Nihon [Japan in maps: Wakoku, Zipangu, Dainippon]. Taishukan, Tokyo

Vancouver G (1798) A voyage of discovery to the North Pacific ocean, and round the world; in which the coast of north-west America has been carefully examined and accurately surveyed. Undertaken by His Majesty's command, principally with a view to ascertain the existence of any navigable communication between the North Pacific and North Atlantic oceans; and performed in the years 1790, 1791, 1792, 1793, 1794, and 1795, in the Discovery sloop of war, and armed tender Chatham, under the command of Captain George Vancouver. Vol 3. G. G. & J. Robinson, and J. Edwards, London

Walker BL (2007) Mamiya Rinzō and the Japanese exploration of Sakhalin Island: cartography and empire. J Hist Geogr 33(2):283–313

Walter L (1994) Japan: a cartographic vision: European printed maps from the early 16th to the 19th century. Prestel, Munich

Wells D (2004) Russian views of Japan, 1792-1913: an anthology of travel writing. RoutledgeCurzon, New York

Yonemoto M (2003) Mapping early modern Japan: space, place, and culture in the Tokugawa period, 1603-1868. University of California Press, Berkeley

Yonemoto M (2016) The European career of Ishikawa Ryusen's map of Japan. In: Wigen K, Sugimoto F, Karacas C (eds) Cartographic Japan: a history in maps. University of Chicago Press, Chicago

**Edward Boyle** is an Assistant Professor at the Faculty of Law and the Center for Asia-Pacific Future Studies, Kyushu University, Japan. He holds a BA from the School of Oriental and African Studies in London and an MA from the Faculty of Law at Hokkaido University. He is a PhD candidate at Hokkaido University. Currently, he is charged with the task of establishing Japan's first interdisciplinary Center for Border Studies at Kyushu University.

In English, he has published or pending articles and reviews in *Imago Mundi, Europa Regional, Eurasia Border Review*, *Journal of Borderlands Studies*, *Geographical Review of Japan (Series B)*, and others. He has translated a book by Akihiro Iwashita (*Japan's Border Issues: Pitfalls and Prospects*, Abingdon: Routledge, 2016) and co-authored a number of chapters. He runs the Kyushu University Border Studies website (http://cafs.kyushu-u.ac.jp/borders/), is editor-in-chief of *Border Bites*, and tweets at @BorderstudiesRM.

His doctoral research concentrates upon the incorporation of Japan's north into the space of the state during the seventeenth to nineteenth centuries, looking at the history of the cartography of the region as well as the concepts of territory that underpinned them. More recently, he has been looking at the comparative history of early modern imperial mapping, contemporary practices of bordering and the multiscale nature of borders under globalization. As well as Japan, he is engaged in research on Georgia and Northeast India. His work intersects with political science, geography, history, and scholarship on international relations.

# Change of Sovereignty and Cartographic Advance: Cartographic Implications of the Spanish-American War of 1898

**Eric Losang and Imre Josef Demhardt**

**Abstract** Rarely there is a greater booster to advancing cartographic coverage of an area than becoming a theatre of war. While the topography of the European nation states experienced a—by and large—steady improvement since the French Revolution, the colonial peripheries often experienced such significant updates only during times of contention or change of sovereignty. This paper discusses the cartographic build-up and fall-out of one such colonial clash, the Spanish American War of 1898. That conflict involved territories scattered around half the globe from Cuba and Puerto Rico in the Caribbean to the Philippines in the north-eastern Pacific Ocean. This case study evaluates various aspects of how becoming part of the "imperial consciousness" is transforming the cartographic coverage of hitherto peripheral colonial territories like the Philippines and Puerto Rico from—by and large—private and small scale maps to official and detailed large scale map (series). This paper inspects the development of cartography of the said territories from the 1850s to the early 1900s, emphasizing administrative requirements, public demand, forms of publication and changes in content.

## 1 Cartographic Production in the United States Between Civil War and Imperialism

On the eve of the Civil War (1861–65), few detailed maps existed of areas in which fighting was likely to occur. Thus, when hostilities started there was a lack of maps for a public in sudden need of more detailed geographical information. This turned into an opportunity for publishers in New York, Boston, and Washington to establish a profitable market. Another vital yet inexpensive source for maps were the major newspapers. These were mainly produced by journalists and artists, rather

E. Losang (✉)
Leibniz-Institut für Länderkunde, Leipzig, Germany
e-mail: e_losang@ifl-leipzig.de

I.J. Demhardt (✉)
University of Texas at Arlington, Box 19529, Arlington, TX 76019, USA
e-mail: demhardt@uta.edu

 
M. Altić et al. (eds.), *Dissemination of Cartographic Knowledge*, Lecture Notes in Geoinformation and Cartography, DOI 10.1007/978-3-319-61515-8_7

than skilled cartographers. They tried "to convey a sense of battlefields and their general geographic situation" (Schulten 2001: 22) rather than to strive for topographic accuracy.

As Susan Schulten observes, the origins of the modern world atlas lie in the strong Post-Civil War desire for information (Schulten 2001: 26). To meet this demand the publishing of almanacs, encyclopedias and atlases became a profitable industry at the beginning of the 1870s, with urbanized and industrialized hotspots like Chicago, New York and Philadelphia. Here an unprecedented competition between publishing houses of different background for the information market began and initiated the evolvement of commercial cartography. These publishing companies were able to meet the increased demand by taking on innovations that allowed for faster, cheaper, and easier printing and production, mainly due to new engraving and printing techniques. Wax engraving, invented separately in America (1830s) and England (1840s), revolutionized the American commercial map and atlas publishing.[1] Thus, the American atlas emerged as a profitable and informative way to visualize America's transformation. The development of the atlas narrative (arrangement and content of maps) in the second half of the nineteenth century can be connected to the ongoing transformation and development of the United States driven by industrialization and globalization.

Throughout the 1870s atlases remained parochial with a strong focus on the transformation of the contiguous United States, emphasizing the area's transition from the unorganized 'Wild West' via territorial status to statehood. Eighty percent of the U.S. atlases completely ignored individual overseas countries by using continent maps. Only since the mid-1880s the narrative changed as the predominant hemispheric partition of the world[2] was replaced by a U.S. centered view as a new link between Europe and Asia, emphasizing the connections rather than the isolation of the Monroe doctrine Americas.

It is not exaggerated to state that the public demand for geographic information, especially in times of war, strongly influenced the modes and the style of cartographic production. The change from a highly sophisticated craft to a demand-driven mass production mirrors the evolving era of industrialization and can be taken as a first but unconscious effort to democratize geographic knowledge, which gained importance in the Spanish-American War. In that way, imperial mapping in the United States was the prelude to public internalization of imperialism by reflecting America's newly defined responsibility in geo-politics at the turn of the century.

[1]For an in-depth discussion on the changing aesthetics, workflows and influence of wax engraving on cartography in the U.S. cf. Woodward (1977): 45; Schulten (2001): 26.

[2]For the ideological implications of hemispheric world regionalization see Lewis and Kären (1997).

## 2 The Spanish-American War of 1898: Turning Chance into Destiny

In hindsight the military clash between Spain, the old but declining power in the Western hemisphere, and the United States of America, making its rapid ascent from a regional to a global economic and military power in the course of the nineteenth century, seems as inevitable as the outcome predictable. After the major setback by losing the mainland colonies in Latin America to independence movements in the 1810s and the three devastating civil or Carlist Wars in 1836–76, Spain in the 1880s reinterpreted the remnants of its colonial empire in the Caribbean (Cuba and Puerto Rico), Africa (strips of territory along the Moroccan coasts and the equatorial islands of Fernando Poo and Annobon plus the mainland foothold of Rio Muni) and the Pacific (Philippines and the chains of the Mariana and Karolina Islands) as a shared endeavor of Spanish nationalism. This not only clashed with irredentist movements fueled by homegrown nationalism in the periphery or colonial territories, most notably in Cuba and the Philippines (Schmidt-Nowara 2008), but also with the increasingly hegemonial view of the United States of its spheres of interest in the wake of President James Monroe's 1823 declaration of the 'Monroe Doctrine' to allow no future expansions of European states in the Americas but only to tolerate the status of existing European colonies.

With the meteoric economic rise of the United States throughout the nineteenth century, the appetite for picking up European colonial remnants in the Caribbean dramatically rose and the Spaniards were seen as the weakest regional adversary. While a pre-Civil War attempt to buy Cuba from Spain fell through in the U.S. Congress, by 1894 about 90% of that island's exports and 40% of its imports were traded with its northern neighbor, making it a de facto economic dependency of the United States (Perez 1995: 149). Naval opportunities which were becoming so vital in the great game of imperialism were lacking, however, and the United States became determined to build a canal between the Pacific and the Caribbean, either through Nicaragua or Panama. Clearly U.S. economic, military and public efforts were focused on Cuba, the richest prize in the imperialist game towards the end of the century, with very little if any attention given to an active engagement in Puerto Rico or across the Pacific in the Philippines and the myriad of tiny islands in the north-eastern Pacific.

With these vital and growing U.S. interests in Cuba, the second Cuban uprising for independence, following the so-called Ten Years' War of 1868–78, broke out in 1895 and was the combustive agent to put Washington and Madrid at ultimate loggerheads with each other. The increasingly successful Spanish efforts to suppress the insurgency were eagerly taken up by the radically anti-Spanish newspapers in the U.S., pushing the public and subsequently a rather reluctant government to take a firm stand against Spanish "oppression" of Cuba. While President William McKinley attempted a negotiated solution, the spark that set off the political powder keg was, befittingly, a massive explosion on board the USS Maine, docked in the harbor of Havana to protect U.S. citizens and interests, resulting in its sinking on 15 February 1898. While the incident was never definitively resolved, the jingoist U.S. press squarely blamed it on the Spaniards. The media pushed the reluctant political establishment in Congress and President McKinley over the line to declare a very lopsided war for revenge and to aid the cause of Cuban freedom on 21 April (Offner 1992). However, neither politicians nor military had any schemes, - not to mention maps! -, sitting in their drawers to deal with the upcoming additional theatres of war in the Philippines and on Puerto Rico.

## 3 Conquering the Philippines: Without Having a Map of Them

When the U.S. Navy's Asiatic Squadron under Admiral George Dewey defeated the Spanish fleet in Manila Bay on 1st May 1899, the outcome of the battle was immediately reported to President McKinley. He faced a complex decision over the fate of the archipelago of the Philippines and knew that the general public was torn between the part jingoism and part economically induced idea of following the European powers into a colonization adventure on the one hand and the radical rejection to rule foreign countries and people on the other—based on the United States' national identity as self-liberated former colony.

The President himself in a rather nostalgic way declared in hindsight: "I went to bed, and went to sleep, and slept soundly, and the next morning I sent for the chief engineer of the War Department (our map-maker), and I told him to put the Philippines on the map of the United States . . . and there they are, and there they will stay while I am President!" A more elaborate story of that evening, however, was later revealed by Henry S. Pritchett, in 1898 the newly appointed Superintendent of the Coast and Geodetic Survey. Pritchett wrote that on the evening of 2nd May, following the reports of Admiral Dewey's battle in Manila Bay, he "wondered, whether the President had any satisfactory map of the region" (Pritchett 1909: 398). So he prepared a map to be handed to the President and was surprised that he was asked to present the map himself in the oval office.

> He [President McKinley] was using a small map of the Philippine Islands about as long as one's hand, apparently taken from an atlas or school geography. On this small map, Manila Bay was no larger than a pea, and none of the details of the shore line and local topography could be made out (Pritchett 1909: 398).

Pritchett brought a British admiralty chart to the White House—by that time the most elaborate map of the Islands, though mainly emphasizing the coastal areas (Fig. 1). He was then asked to provide the President's staff with the necessary maps for all the theatres of war with the Spanish. Thus the Map Room at the White House came into existence. The story of the mapmaker whom the President asked to put the Philippines on the map—for the United States—leaves the question what type of map was meant—a survey map, a special operations map to be used by the White House or a rather metaphoric public one. Of course, this brings up the question, too, what maps of these islands already existed by 1898.

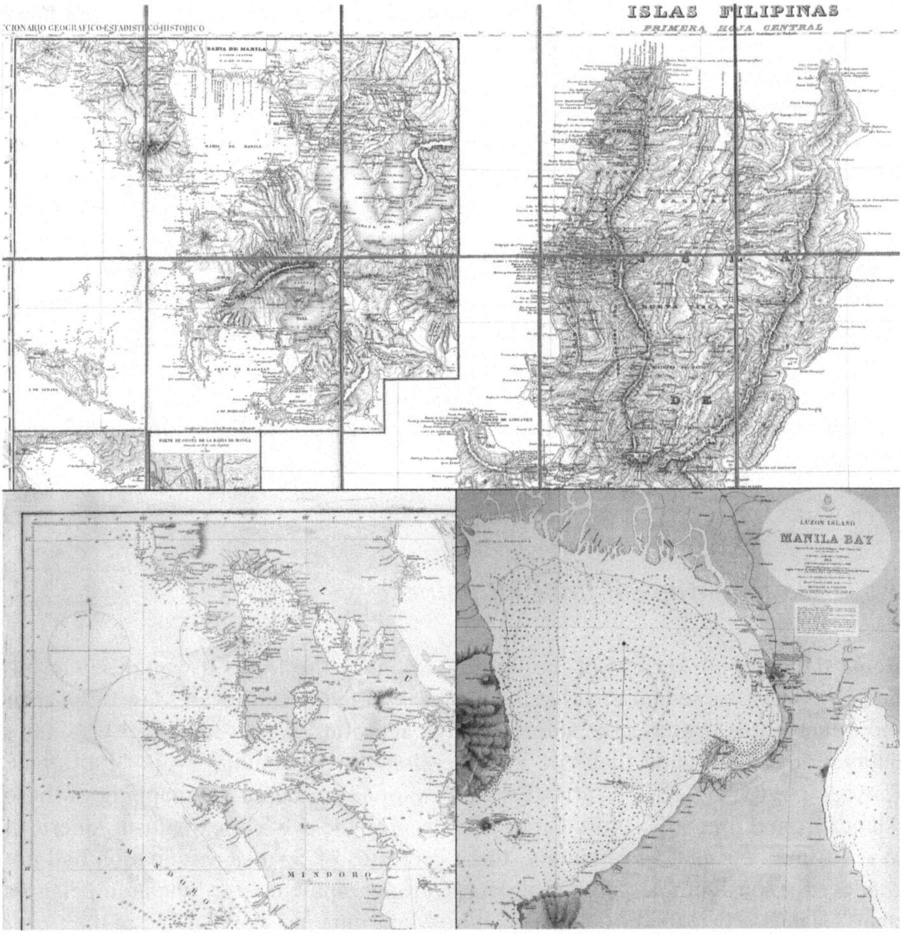


**Fig. 1** Hydrographical mapping of the Philippine Islands. This was the most up-to-date cartographic information available at the dawn of war in 1898. Detail enlargement from the British admiralty chart (*top*) and Carta General de Archipelago Filipinas, 1792 (*middle*) and from the Atlas de España y sus posesiones de Ultramar 1853 (*bottom*) (Private Possession, From Central Geographical Library Zentralbibliothek, Leipzig, HK 0064, From Biblioteca Universitaria, Universidad de Las Palmas De Gran Canaria, MDC 70775)

## 3.1 *American Maps of the Philippines up to the Spanish-American War*

It's hardly surprising that only a few single sheet maps of the Philippines appeared in pre-war U.S. atlases and geographic text books. This reflects that the Philippines had not been on the agenda for prospective territorial extension, whereas some South American states had become highly popular, like Venezuela where U.S. and British interests collided in the late 1870s. Furthermore, from the American point of view, the islands were situated at the fringes of traditionally covered geographic regions.

In analyzing the circulating atlases a characteristic pattern was the depiction of the Philippines as part of the East Indies (e.g. *Cram's Standard American Atlas of the World*, 1895). Rand McNally in the 1898 *New General Atlas of the World* put them in the Asian realm on a map covering China, Indochina and Malaysia. A year before, in their *Indexed Atlas of the World* (1897) the archipelago of the Philippines was part of the Oceania and Malaysia map. In Mitchell's New General Atlas they could be found unnamed on the eastern rim of Asia and as part of Malaysia named on a map of Oceania. And this marginal coverage was despite the fact that the Philippines had been part of the 1840s U.S. Exploring Expedition in the Pacific. But that survey had not produced detailed maps of the islands. Thus, the common practice to plagiarize maps of international provenance was how any of the very limited amount of more detailed pre-war coverage of the Philippines came into being.

### *3.2 Spanish Maps of the Philippines*

Being the colonial 'owner' of the Philippines since the 1560s it would be not too farfetched to assume that Spain at the outbreak of the hostilities in May 1899, after more than three centuries of administrative activities, would have had brilliantly detailed and up-to-date maps of it possession. In his seminal work "Philippine Cartography: 1320–1899" Carlos Quirino examines in-depth the mapping of these islands from thirteenth century Chinese maps until the eve of the Spanish-American War. It turns out that, surprisingly, three centuries of Spanish occupation had not produced a considerable number of detailed topographic maps of the archipelago's many islands. The expedition of Alejandro Malaspina in 1792 resulted in the *Carta General de Archipelago Filipinas*, with extensive coverage of the details of important coastal areas but leaving out the interiors completely (Fig. 1). After that it took another half-century to expand the mainly hydrographic charting under the supervision of Madrid based La Comision Hidrografica (Quirino 1969: 64). Between 1849 and 1853 Francisco Coello, a lieutenant colonel in the Spanish Corps of Engineers, produced three sheets on the Philippines on a scale of 1:1,000,000 as part of the *Atlas de España y sus posesiones de Ultramar* (Fig. 1) (see also Sect. 4), with insets of some areas in considerable larger scale (e. g. Puerto of Cavite in 1:50,000). These maps became the backbone of the new plans to create future living space for Spanish colonists. However, the resolution was still too coarse, and the critique arose that "Spain had not followed the example of other colonial powers in entrusting the study of the terrain and its vegetation to enlightened and competent officials" (Capel 1994: 64). None of the aforementioned maps could have been used to obtain strategic information on the interior of the islands. Considering that the Spanish population with only a few exceptions lived along the coastline and that the upcoming rebellion drew on indigenous people and their knowledge of the islands, the Spanish government set up a last plan for survey.

Between 1882 and 1898, Enrique D'Almonte y Muriel, a Spanish mapmaker working for the Ministerio de Ultramar, portrayed the whole archipelago in several maps scaled between 1:800,000 and 1:200,000. These became the base for a Spanish 'Atlas of the Philippines', a project that was never to be finished (Quirino 1969: 66). The maps he produced, of which the *Isla de Luzon et sus Adyacentes* (1887, 1:400,000) is the in the collection of the Biblioteca Digital Hispanica (bdh. bne.es) "have eclipsed both in accuracy and completeness all the other topographic maps of the period" (Quirino 1969: 66). Although the 1872 Cavite Mutiny, a precursor of the insurrection of 1896 under the Leadership of Andrés Bonifacio, often made the geological survey impossible, D'Almonte y Muriel was close to finish the atlas when in 1898 Manila was taken by American troops. He eventually left the Philippines to continue his work in the remaining Spanish colonies in Africa (Guinea and Spanish Sahara) (Quirós 1998). When the Americans occupied Manila on 13 August 1898, they found a manuscript atlas that contained 23 sheets, each 13 by 21 inches, presenting the entire archipelago on a scale of 1:800,000. Although this would have been a brilliant basis for an American atlas, the U.S. Coast and Geodetic Survey did not see it fit to publish and instead printed the *Atlas de Filipinas* by Father Jose Algué in 1900 (Quirino 1969: 65).

## 3.3 American War Maps

### 3.3.1 The (Yellow) Press

After President McKinley had requested the declaration of war with Spain from the U.S. Senate on 11 April, this was granted on 25 April 1898. The President had emphasized mounting public and political pressure in his request. And indeed, the media, then equivalent with newspapers and journals, had proactively stirred up the tensions to get into war with Spain. Thus it can be claimed that the jingoist Yellow Press partly caused the circumstances under which their products started to flourish. The first maps appeared in the newspapers just before the war operations began to roll. On 28 April the *New York Tribune* placed a detailed map of the northern half of Luzon on its front page. The "central scene of war" (Fig. 2) had been taken from a Spanish original, showing a detailed view of Manila's surroundings to the south, including the Cavite province on a clipped map on an approximate scale of 1:300,000. In numerous details it is reminiscent of the Luzon map produced by the Jesuit-Philippine cartographers of the Manila Observatory, later published in the *Atlas of the Philippines* by the US Coastal Survey. It also reflects the strong affiliation between the publisher and the U.S. consul in Manila, Oscar Williams, who in absence of correspondents also acted as a pivotal source of information on the islands.

**Fig. 2** The "central scene of war". Detailed map of the northern half of Luzon, published 28 April 1898 by the New York Tribune (from The Library of Congress, Chronicling America, Historic American Newspapers)

Besides this very elaborate map, three types of newspaper maps can be identified:

- Small-scale overview maps showing the Pacific from the American West Coast to the Asian shores, often including Java and Sumatra. Surprisingly, these maps appeared only at the dawn of the battle at Manila Bay (e.g. *The Times*, Washington, 28 April) along with the discussion of how to further treat the Philippine Islands, accompanying articles like "Our new colonies in the Philippines" in *The Times* (Richmond, VA) after the seizure of Manila.
- Medium-scale maps showing the Philippines, often with Hong Kong in the upper-left corner, were most common throughout the war, mostly published double/triple columned (110–180 mm width) and varying in scale between 1:10 and 1:12 Millions covering an area between 115° and 130° E and 20°–5° N. On most of these maps it had been difficult to place text east of the Philippine Archipelago, whereas the west was often spread out to include British North Borneo. It often reflects the base maps because 130°E was the edge for maps on South East Asia/Oceania. Some of the maps were simple hand-drawn sketches (Fig. 3) and some more meticulous copies (e.g. number and arrangement of the islands), obviously based on U.S. and up-to-date foreign atlases (Fig. 3).
- Large-scale but schematic plans and maps covering Manila Bay or the northern part of Luzon Island appeared *after* the victory of Admiral Dewey in Manila Bay on May 1. Surprisingly no copy of the admiralty chart which Pritchett had brought to President McKinley was used. Most of these maps were sketches,

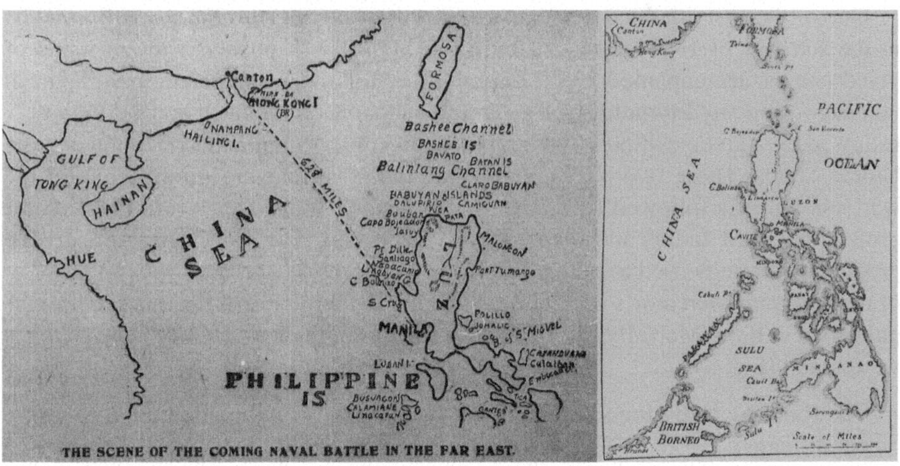


**Fig. 3** Medium scale newspaper maps of different qualities, both published end of April 1898. Sketch map (*left*) published 30 April 1898 by The Houston Daily Post, detailed map based on atlas maps (*right*) published 28 April 1898 by the New York Tribune (from The Library of Congress, Chronicling America, Historic American Newspapers)

often over-emphasizing the Cavite Peninsula, where most of the assault on the Spanish fleet had taken place.

Among the first newspapers to realize the public interest in maps directly attached to daily stories rather than appearing in costly and soon outdated supplements, were the *National Tribune* and the *Times* (both in Washington DC). Over the course of the war they published a mean of 0.2 maps per issue, spreading them over all pages instead of just putting them on the front page as other papers did.[3]

Following the outbreak of the Second Boer War (1899–1902) nearly all of the aforementioned newspapers widely used maps to visualize spatial information in their news coverage. Thus it is evident that the interest in the geographic implications of conflicts steadily grew from the Civil War (knowledge on the battlefields) via the Spanish-American War (knowledge on becoming U.S. overseas territories) to the Boer War and later the World War I (reflecting the involvement of the U.S. as an imperial power).

### 3.3.2 Co-operative Atlases: War Atlases, Mapping Companies and Newspapers

It took not long for atlas publishers to join in the competition for readers. As the production of new maps was a task that needed to be evaluated monetarily, most publishing houses looked for new ways to spread already existing maps. In this respect the war didn't primarily induce the production of new maps in the first place. It rather opened up new opportunities to extend the business.

One of the first atlases to appear was *The Washington Post Standard War Atlas* in the summer of 1898, produced by Rand McNally. It opened with an index of place names accompanied by "General rules for the pronunciation of Spanish language" and an "Etymology of geographical names in Spanish and Portuguese", followed by a juxtaposition of geographical and military statistics on Spain and the United States. These two pages constituted the most relevant information on the war, since it was followed by standard atlas maps (Cuba, The World in Mercator projection, West Indies and maps on all continents). The world map reflects the production strategy used to publish this kind of atlases. It is a true copy of the "Map of the world on Mercator's projection. Showing the direction of the ocean currents" published in Rand McNally & Co.'s. Indexed atlas of the world. Only three changes were made (Fig. 4):

- The Rand McNally logo was removed (where removed the graticule is missing, on all maps).
- Telegraph lines were added as overprint in red (sometimes not fitting the connected points).

[3]The *New York Times* published a mean of 0.05 maps per issue, all on the front page; the *New York Herald* 0.1; for a recent view on war maps in newspapers cf. De Wolf et al. (2005).

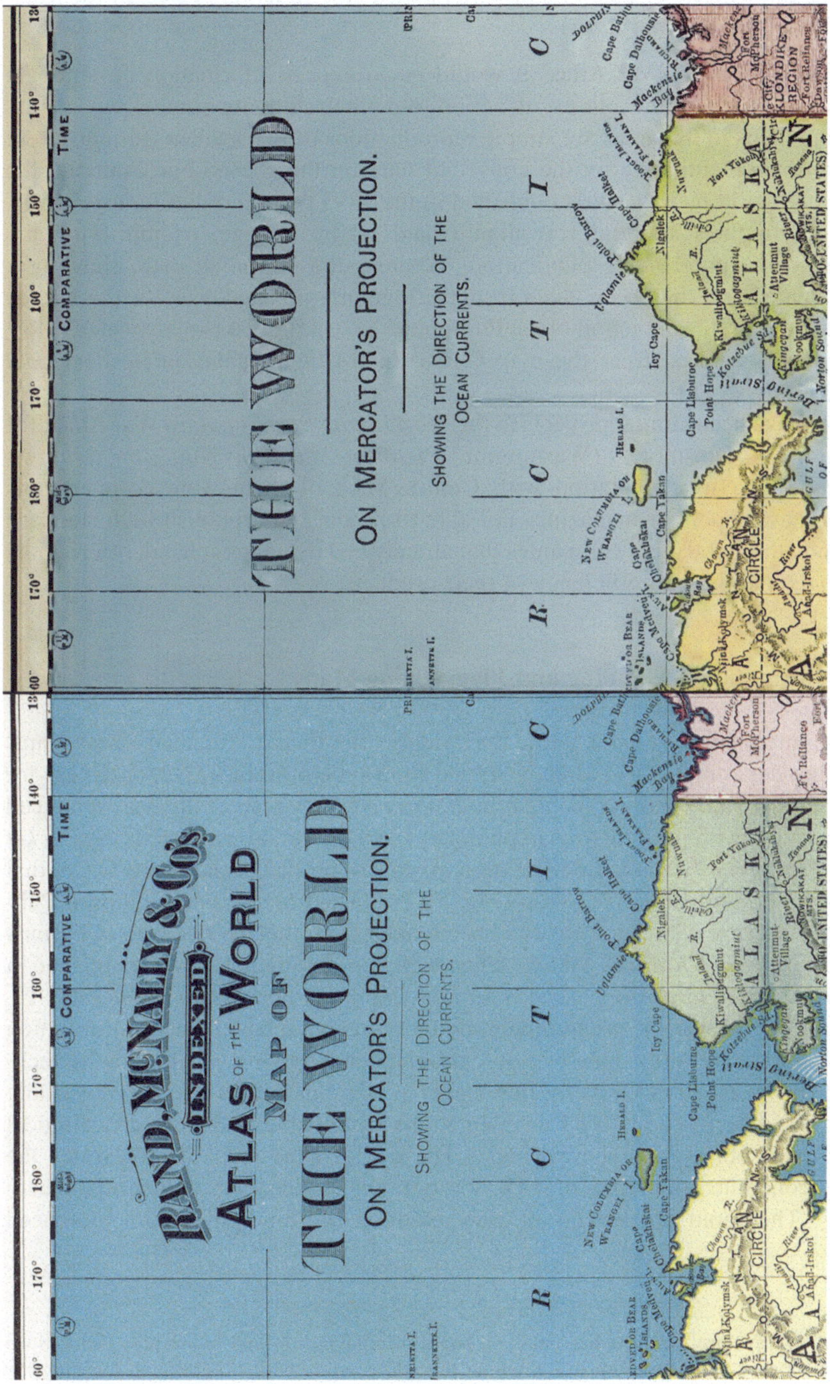


**Fig. 4** Details from the world maps in The Washington Post Standard War Atlas (1898) and Rand McNally & Co.'s. Indexed atlas of the world (1897) (from David Rumsey Map Collection, www.davidrumsey.com)

- A marginal index of place names was added (the index appeared in the 1899 edition of the *Indexed Atlas*).

Although from a 'War Atlas' it would be to expect information on strategic issues, like emphasizing places of action and vital shipping routes, but all the following maps in that atlas are simple reproductions of maps taken from different atlas editions, identifiable by the copyright date on the sheets. For example, the underlying world map was first published in the 1895 edition of the *Indexed Atlas*.

The Philippines and the Hawaiian Islands share a page, exemplifying the amalgamated nature of the publication. Whereas the Hawaiian map, showing a very delicate relief, had been taken from the regularly updated *New Business Atlas* of the same year, the depiction of the Philippines is strikingly unsophisticated. Most of the details derived from the map China, Indo-China, Malaysia in the *World Indexed Atlas*.

There have been similar projects to the *Washington Post Standard War Atlas*, for example R. A. Dinsmore's (Washington) *Pictorial Atlas illustrating the Spanish-American War* (in cooperation with Cram). Most of them were co-operations between newspapers or magazines not able to produce and publish high standard maps and atlas publishing companies that already had the maps at hand, offering the latter to extend the cost efficiency of their products.

### 3.3.3 Promotional Mapping and Plagiarizing Maps

A different business model were the widely distributed 'branded' war maps. Marketing departments had already started to use the prominent personalities[4] for advertisement when another idea got hold: the giveaway map of the war, produced in cooperation with prominent map publishers. A fairly good example is the *Pocket map of Philippine Islands*, produced as a giveaway for the Vermont MFG Co. that produced Vermont Butterine syrup (Fig. 5). Published by the Fort Dearborn Publication Company (Chicago), that had been taken from the 1899 edition of *Cram's Standard American Railway System Atlas of the World* which, too, was published in the *National Standard Atlas of the World* by Cram and Fort Dearborn in 1900.

As plagiarism was the predominant modality to obtain basic spatial information for reproduction (mainly done by non-cartographers applying wax engraving techniques) in the creation of many newspaper maps, this seems to be a rewarding subject of future research. Such investigations will enable to draw a genealogical tree of map sources and 'borrowings'. The use of atlas maps demonstrates the strong economic grounding of U.S. cartography at the end of the nineteenth century. The rapid succession of new editions of 'modern' atlases outpaced

[4]E.g. Dewey's picture, washing his hands on board of his flagship used to promote Pears' Soap under the slogan: "The first step to lightening the white man's burden is through teaching the virtues of cleanliness" surrounded by what seems to be the Philippines.

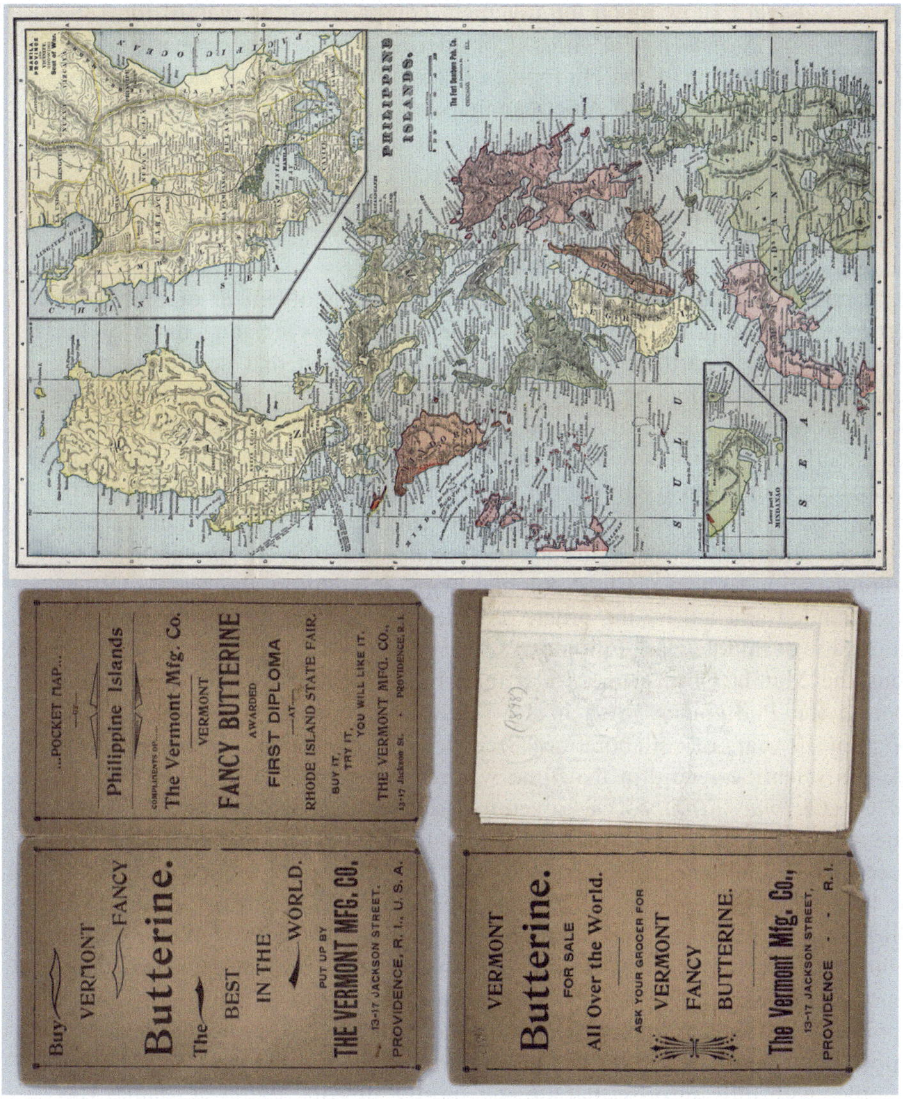


**Fig. 5** Pocket map of [sic!] Philippine Islands, published by the Fort Dearborn Publication Company (Chicago) 1899 (from David Rumsey Map Collection, www.davidrumsey.com)

traditional map producers like Mitchell and Armstrong and was based on cost effective printing techniques, new marketing concepts to extend the product life cycle of maps as well as new modes of use. The content of atlas maps, however, did not change significantly. A good case in point is the map of the Philippine Islands in Rand Mc Nally's seminal 1910 edition of the *Universal Atlas of the World* that, besides being printed as map No. 3 (side by side with a depiction of Puerto Rico as map No. 2), doesn't show much improvement compared to the map of the 1898 *Washington Post Standard War Atlas*. Considering the sketched scenario of the home front map coverage of the Philippines, it seems that McKinley's desire, to put the Philippines on the (official) American map, required a different approach.

## *3.4 The* Atlas of the Philippines

It might come as a surprise that there was a strong involvement of Philippine cartographers in the production of the, as mentioned, few existing maps of the Archipelago. This is partly based on a strong Jesuit input that here as in other corners of the (catholic) globe had gained scientific reputation on diverse research fields like geography, meteorology and seismology. Besides earlier unhappy relationships with the Spanish authorities (between 1772 and 1858 the Society of Jesus was banished from Spanish dominions), the international and educated members of the order were able to survey and draw maps of high standards that did not rely on the poor governmental surveys. In fact, the Jesuits were often being employed by the administration to help produce colonial maps. In addition, the Manila Observatory, founded and led by the order, provided an opportunity of education for ambitious lower middle class Filipinos.

Thus the Manila Observatory for a long time had already been a scientific hotspot in the Philippines when in 1897 the Jesuit meteorologist Father José Algué after 10 years of collaboration succeeded his predecessor, Father Féderico Faura. His scientific record at that time was already remarkable: the 1897 book *Baguíos y Ciclones Filipinos* was a seminal scholarly account of the Philippine climate and storms. Having been at Georgetown University in Washington DC for advanced studies in meteorology, seismology, and astronomy in the early 1890s, Algué then co-curated exhibitions such as the Columbian Exposition at the 1893 World Fair in Chicago and later the Philippine Village at the 1904 World Fair in St. Louis. His most important invention, however, the barociclonómetro (barocyclonometer), based on his collaboration with Faura, could foretell storms in the Philippines and Asia and immediately attracted wide interest, providing him with the opportunity to personally present his research in Europe and the United States.

It comes with little surprise that already shortly after the Battle of Manila Bay American military commanders recognized the importance of Algué, both as an internationally trained scientist and a fountain of knowledge on the Archipelago, even more so as he was about to finish a manuscript on the regional geography of the Philippines. Thus he was approached to function as intermediator and to help establish America's future success as a more enlightened colonial power than the driven out Spaniards. A meeting between Admiral Dewey and Father Algué on the flagship of the American fleet in November 1898 resulted in the immediate transfer of the Manila observatory from Spanish to American administration without the loss of a single day's pay for his employees (Warren 2009: 508). On the occasion the Jesuit presented the U.S. admiral with a comprehensive atlas manuscript, which turned out to literally be the answer to McKinley's desire to put the Philippines on the American map.

The U.S. government publication of the *Atlas of the Philippine Islands* in 1900 made no use of the very detailed topographic maps by D'Almonte Y Muriel. Rather, the U.S. Coast and Geodetic Survey under the auspices of H.S. Pritchett (McKinley's map maker) published the manuscript maps of Algué.

It was marked by an unusual bibliographic feature of two title pages (Fig. 6). "The first, in Spanish, and framed by a trellis of local flora, credits the work of unnamed Filipino draftsmen under the supervision of Father José Algué, Director of the Manila Observatory; the second one, less ornate, lists the volume as 'Special Publication No. 3' of the US Coast and Geodetic Survey" (Kirsch 2016: 2). The medium-sized atlas (300 by 360 mm) contains 30 full-page color maps ranging in topics from active volcanoes and earthquake frequency to ethnological distributions and political divisions. The latter became the basis for the organization of American administration of the newly acquired territories (Fig. 7). The scale of the maps ranges from 1:46 Million for a map of the Pacific preceding five thematic maps in 1:6 Million, with the rest of the maps showing the various islands adopting different scales appropriate for the depiction of the respective area. The provinces, or departments, are represented in different colors, the drainage, coastal waters and the sea in blue, and the relief by crayon. All maps are signed with the official logo of the Manila Observatory, mentioning the U.S. publisher in small capitals in the upper left corner, the printing company, Hoen & Partners of Baltimore, in the lower right corner, both outside the map frame. The maps reflect both the cartographic expertise of their makers and the harmonious print-out based on the lithographic expertise of the printing company.

The text part of the atlas, taking up the first 24 pages, opens with an introduction by Pritchett, describing the origins and the decision that lead to the publication of Philippine maps made by Filipinos, just mentioning that "some of the maps were redrawn at the Coast and Geodetic Survey Office" (Kirsch 2016: 3). Pritchett emphasized the "absence of accurate surveys of many of the islands" by the Spanish and the opportunity that opened up by cooperating with the scientists of the Manila observatory.

The following very brief geographic characterization, based on a manuscript by Algué, was the most up-to-date information available at that time, drawing very heavily on the extent and distribution of natural resources and the agricultural potential. The bilingual character of the title is also reflected by the general remarks and three different gazetteers (places, hydrographic features, mountains and volcanoes), the latter always accompanied by an instruction how to pronounce the Spanish place names (in two languages). Since none of the map contents had

**Fig. 6** Spanish (*left*) and English title page of the Atlas of the Philippine Islands/Atlas Filipinas (from David Rumsey Map Collection, www.davidrumsey.com)

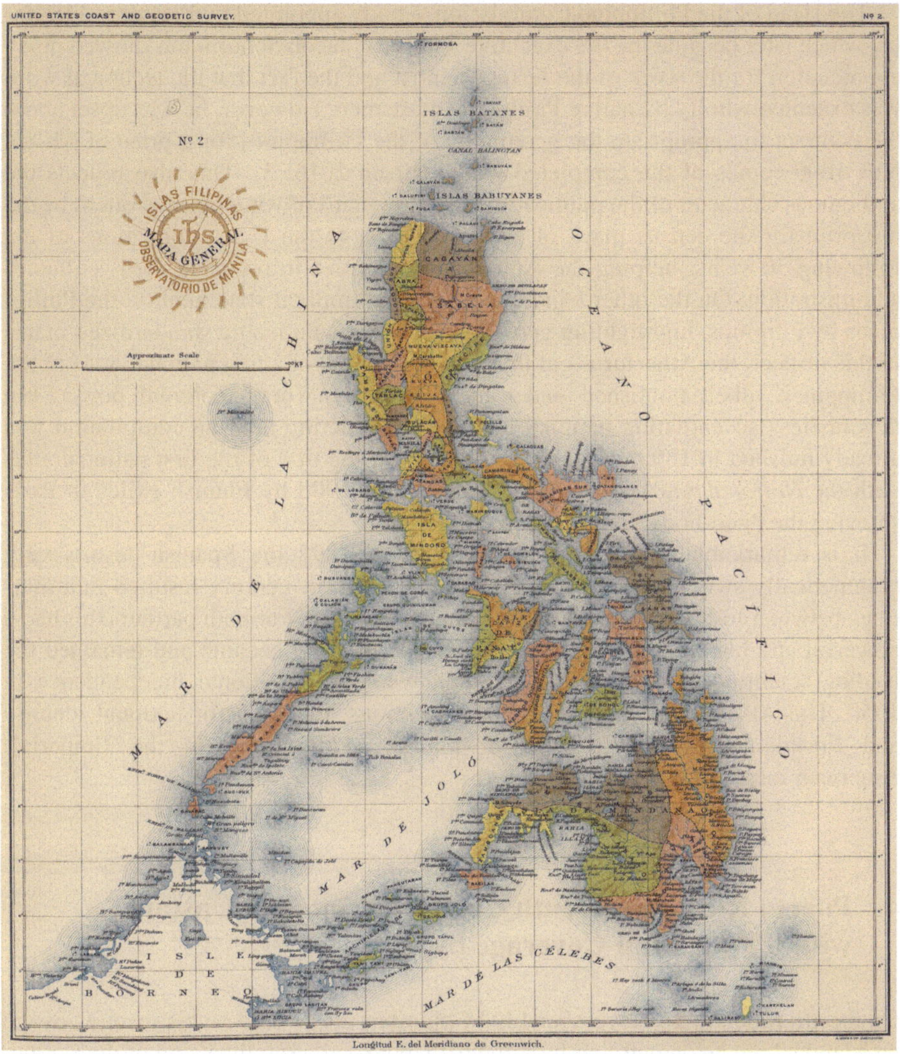


**Fig. 7** Mapa General (general map), Map No. 2, taken from the Atlas of the Philippine Islands/ Atlas Filipinas (from David Rumsey Map Collection, www.davidrumsey.com)

been translated, this was a good help for understanding the legends and remarks. This lingual dualism is a reflection of the cartographic bargain, "also hinting at tensions in the nature and ownership of geographical knowledge during a moment of territorial transition" (Kirsch 2016: 2). In a review in the *Bulletin of the American Geographical Society* Henry Gannett stated that "This atlas furnishes, in convenient form for reference, probably the best maps of the islands to be obtained" (Gannett 1901: 141).

The Head of the Philippine Commission, a body appointed by President McKinley which later became the US executive authority, Jacob Schurmann showed much appreciation for the work of the Jesuit scientist and the fact that the technical work was executed wholly by native Philippine draftsmen. However, he was upset about the contract to appropriate the geography of the Philippines for a price of $1085, plus 1000 copies of the completed work (Kirsch 2016: 2). This also reflects the ambiguous character of the publication, the base maps of which turned out to be the blueprint for the census maps of 1913 and most of the U.S. publications of the following 30 years, helping the American authorities to organize their 'colonial' administration. On the other hand, the atlas is an implicit statement on the Philippines by Filipinos, highlighting geographical aspects on iconic standardized maps. In this respect, the Atlas turns out to be the equivalent of a first national atlas of the Philippines, albeit published under the reigns of a (new) colonial power, but nevertheless accentuating national identity. Considering that the compilation was already finished in 1898, it can thus compete for the title of the first national atlas with the *National Atlas of Finland*, published in 1900 by Finnish exiles in Paris showing the "geo-body" of an occupied Finland.

It is remarkable that in the second half of 1898 the Spanish Jesuits very pragmatically switched allegiances, which for the new century resulted in a marriage between Jesuit science and American imperialism. For both partners this had a powerful and lasting legacy, as the Manila Observatory became and remained the leading scientific institution until the achievement of national independence in 1950. The Atlas perfectly reflects both intentions, the wish for national identity and President McKinley's wish to put the Philippines on the (imperial) American map.

## 4 Puerto Rico: Nineteenth Century Economic Upstart with a Poor Map Coverage by 1898

On the afternoon of 16 November 1494 Columbus discovered on his second voyage an island which he named San Juan Bautista, but soon after, the name of the capital, Ciudad de Puerto Rico or City of Rich Port, came into use for the whole island. Having an almost perfect rectangular shape, Puerto Rico's 9104 km$^2$—equaling about the size of Cyprus or the combined U.S. states of Delaware and Rhode Island—is about one-fifth of the size of the neighboring island of Hispaniola (with Haiti and the Dominican Republic) and only one-tenth of Cuba. Puerto Rico has fertile flat coastal areas in the north and south with marked valley systems especially to the north of the Cordillera Central, peaking at 1338 m, as the spine of an orography which certainly challenged all detailed surveying and topography from the Spanish conquista in 1508 to the advent of airborne surveys in the early twentieth century.

Though part of the Greater Antillean islands, Puerto Rico was throughout the almost four centuries it spent under Spanish rule in every respect in the shadow of the mainland Vice-Royalties of New Spain in Meso-America and New Granada in northern South America, until they fell from the Spanish crown in the 1810s. Along with Cuba it thereafter formed the remnant of Madrid's once colossal continental empire, but economically and strategically it was dwarfed by its sister island of Cuba until the end of the nineteenth century (Carrion and Morales 1983). It is therefore no surprise that, with respect to printed and publicly circulated cartography, Puerto Rico prior to 1898 was the least topographically mapped and hydrographically charted of all the Greater Antillean islands, sporting almost nothing better than general survey sheets.[5]

The large sheet on the *Isla de Puerto Rico* in the *Atlas de Espana y Sus Posesiones de Ultramar* (Madrid 1851) which contains two general orientation and eight topographical detail maps of the islands and some prominent tracts by Capitan de Ingeniero Francisco Coello (Fig. 8), summarizes the topographical knowledge which would not be surpassed significantly in the second half of the century. However, it is noteworthy that since 1873, based on an order by the Spanish Army General Staff, a series of Topographic Commissions carried out surveys and topographic reconnaissance sheets which subsequently covered the whole island. This project was decidedly a military undertaking to ensure a sustained upper hand against the repeated internal uprisings and increasingly feared foreign threats. After amassing over 4000 cartographic sheets and text pages, this project of the *Mapa Militar de Puerto Rico* unfortunately was aborted in December 1897, and less than a year later in October 1898 all the drawings and documents were shipped across the Atlantic along with the by then defeated Spanish troops and administrators. Although numerous large and mid-scale linear itineraries and areal croquis sheets plus several advanced topographical compilations were produced,[6] not a single sheet of the *Mapa Militar de Puerto Rico* was printed for public circulation or even for use within the Spanish administration until the U.S.-Spanish war. The *Mapa Militar* is thus to be disregarded in this chapter on *printed* maps of the island.

The American imperialist aspirations might never have set their sights on Puerto Rico if the notorious Spanish policy of shielding its American colonies had not been abolished by the Royal Decree of Graces in 1815, rewarding the loyalty of the island amidst the ultimately successful drive for independence in the mainland

---

[5]For a comprehensive overview of the islands cartographic history see Anibal Sepulveda Rivera: *Puerto Rico urbano: Atlas historico de la ciudad puertoriquena*. 4 Volumes. San Juan (Centro des Investigaciones CARIMAR y Departamento de Transportacion y Bras Publicas) 2004.

[6]Information on the *Mapa Militar de Puerto Rico* is by courtesy of Michael Deliz, who in 2013 researched the topic as a class project for the Transatlantic History PhD program at the University of Texas at Arlington under the supervision of Imre Demhardt. For a general orientation see Luis E. Gonzalez Vales / Luis A. Magallanes Pernas: *Documentación de Puerto Rico en el Archivo Cartográfico y de Estudios Geográficos del Centro Geográfico del Ejército*. San Juan (Historiador Oficial de Puerto Rico)/Madrid (Ministro de Defensa) 2007.

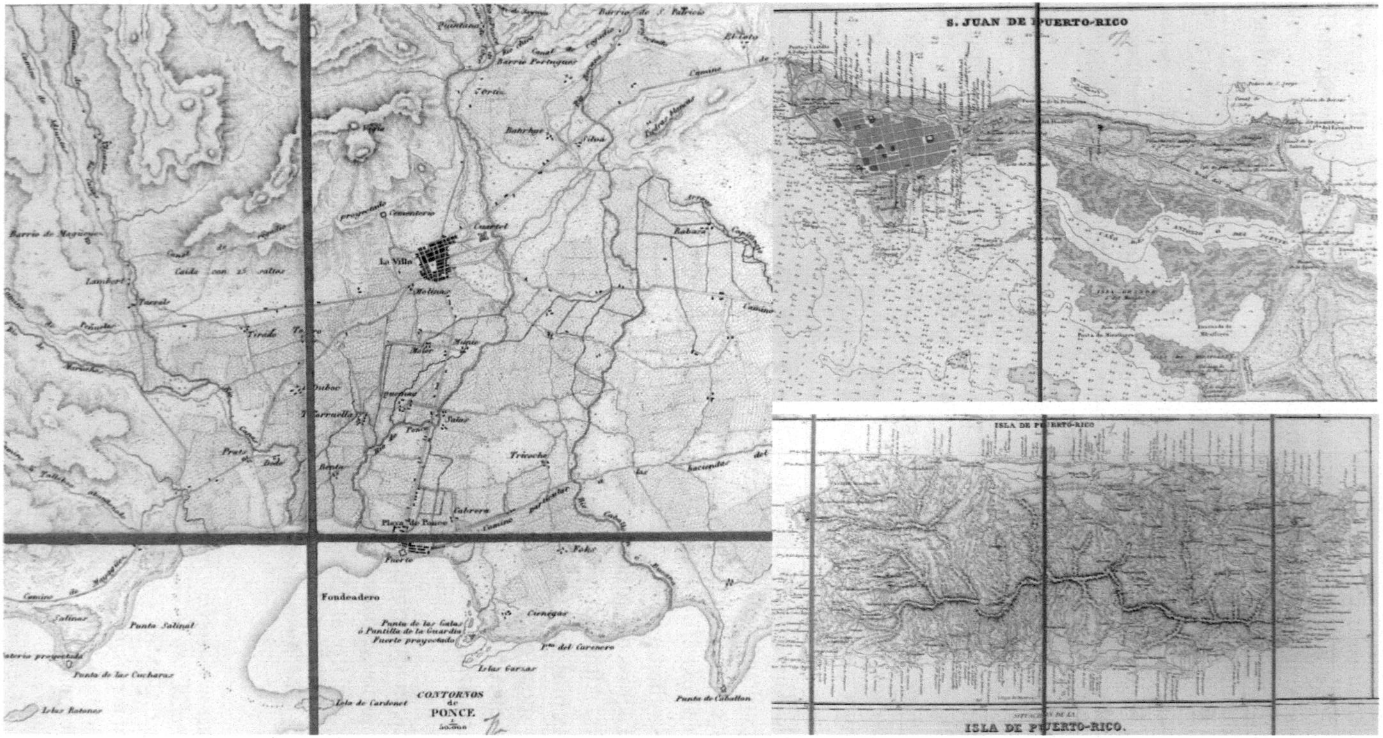


**Fig. 8** Details of *inset* maps on the large sheet by Francisco Coello's (cartography) and Pascual Madoz's (text) *Diccionario Geografico—Estatistico—Historico. Atlas de Espana y Sus Posesiones de Ultramar. Posesiones de America. Isla de Puerto Rico.* Madrid 1851 (from the American Geographical Society Library, University of Wisconsin-Milwaukee Libraries, Map Collection, 238 A-1851)

colonies, by opening up its ports and export-orientated agriculture to commerce beyond the colonial metropolis. The lifting of the *de facto* trade embargo triggered an exponential growth of import and export from \$269,000 in 1813, \$2,494,000 in 1823 and \$9,397,000 in 1843, to \$16,512,000 in 1863 with cane sugar as the principal cash crop, following the decades older role model of Cuba, and the runners-up coffee and tobacco. By 1897 Puerto Rico used about 14% of its surface for the production of more than 62,000 tons of sugar and syrup. This boom for the first time lured a significant immigration, not only from the Iberian Peninsula, but also a sizeable number of people from other European origins. The expanding plantations could import slaves more or less unhindered until the 1840s with slavery being abolished only in 1873. However, slaves never dominated the Puerto Rican economy as much as in other Caribbean islands.

Against this backdrop of prosperity in a comparatively small economy, the population grew almost 18-fold within a century from only 45,000 in 1765 to 785,000 in 1864, the time of the last Spanish census, and about 900,000 at the turn of the century. At that time about 6 in 10 Puerto Ricans were regarded as white, 3 as creole or mixed race and only 1 as black (former slaves). In spite of the economic boom within an overall growing significance of trade in the Caribbean, mainstream U.S. cartography until the end of the nineteenth century showed little interest for the perceived farraginous plethora of bigger and smaller islands lying to the south of the Union. In pre-1898 U.S. atlases the Caribbean was usually relegated to one sheet, or as in the case of the 1889 edition of *The Rand McNally Standard Atlas of the World*, all of it was crammed into the overview sheet on North and Meso-America (Fig. 9), with Puerto Rico literally marginalized and not even given the same political area color as Cuba.

Known for both its loyalty to Spain in the 1810s maelstrom of Latin American independence and persistent nineteenth century liberal tendencies and uprisings[7] of an increasingly diverse white population against the conservative regime in Madrid, the start of the Cuban War of Independence in 1895 prompted the crown to charter a Constitucion Autonomica in 1897. Early in 1898 municipal councils and an island parliament were elected with the latter constituting itself on 17 July 1898—3 months into the U.S.-Spanish War and only 8 days before the U.S. invasion to "liberate" the island.

For the Spaniards Puerto Rico was little more than an appendage of the larger and so much more valuable Cuba throughout the nineteenth century, whereas U.S. imperial interests in the economic boom on the hinge island of the Greater and the Lesser Antilles were an extension of the economic-political interests in Cuba, too. The only significant original strategic interest in Puerto Rico itself was seen along with the rise of American naval power, seeking a safe passage for navy and commercial vessels through a major strait separating the line of the Antillean islands for ships to and from the canal through the Isthmus of Panama, which in the

[7]Most notably the movements of Lares in 1868 and Yauco in 1897 with exiled leaders moving to the U.S. for their anti-Spanish machinations.

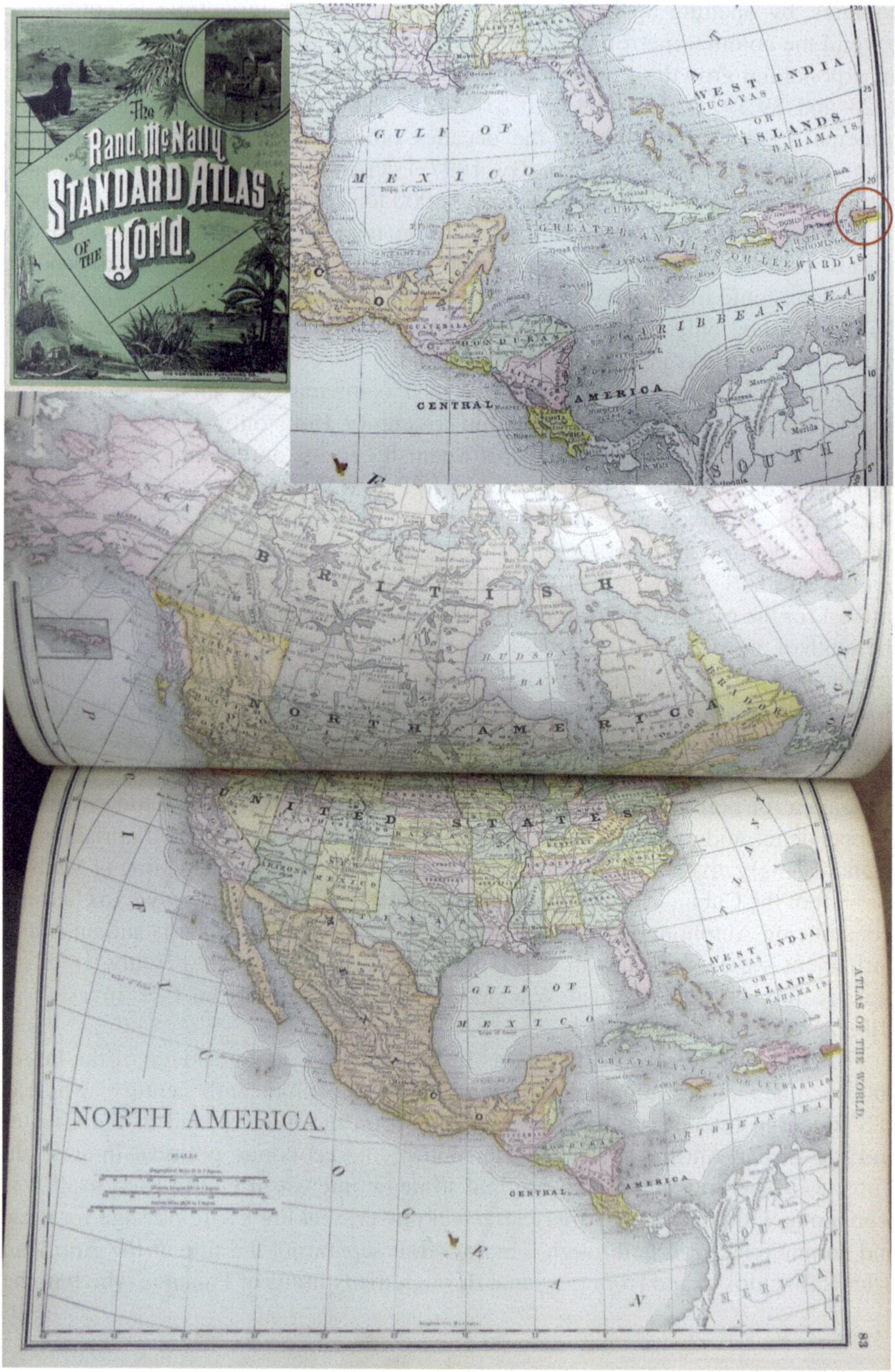


**Fig. 9** Title Page, Map "North America" (pp. 82–83) and detail of peripheral Puerto Rico in the 1889 edition of *The Rand McNally Standard Atlas of the World*. Chicago (Continental Publishing Group) 1889 (from the American Geographical Society Library, University of Wisconsin-Milwaukee Libraries, Map Collection, Atlases 1889)

1890s the Americans took over from the ill-fated French Compagnie Universelle du Canal Interocéanique. Sugar cane then was rather seen as a bycatch to securing the U.S. a naval station on Puerto Rico. Since 1894 the U.S. Navy drew up contingency plans in earnest for a war with Spain which included operations in Puerto Rican waters, only later to be dropped in favor of focusing on its prime U.S. target: Cuba (Beruff 2007: 7).

When the U.S.-Spanish War broke out on 21 April 1898, the military actions and coverage by jingoist U.S. newspapers (Fig. 10) focused—not surprisingly—on Cuba. The immediate navy blockade of that island was soon to be followed on 1 May by the U.S. Navy Asiatic Squadron's belligerent siding with insurgents in the Philippines by defeating a Spanish fleet in Manila Bay and taking the capital (also see Sect. 3). As so often before, Puerto Rico remained undisturbed in the backwaters of the military activities and public (newspaper map) attention. It was only on 12 May that the capital and main port of San Juan was bombarded and laid under blockade. It took a full 2 months of scratching together ships and men after the declaration of war before about 15,200 U.S. troops landed on 25 July, meeting only little defence in Guanica in the southwest of the island, far away from the capital San Juan on the north-eastern side where most of the about 7000 Spanish defenders were concentrated. Hampered more by the lack of detailed maps than the enemy, the U.S. invaders rapidly progressed from the landing bases along the south coast across the Cordillera Central towards the north coast. The Spanish collapse in Puerto Rico was swift: following the last major skirmish on 10 August, a truce took effect 2 days later and the Spaniards announced the cessation of the island to the United States on 29 September. The last vessels carrying Spanish administrators and troops—including the unfinished *Mappa Militar*—left the island on 18 October.

After a brief interlude of military administration, the Foraker Act of 1900 granted Puerto Rico (bearing the official anglicized name of the island *Porto Rico* until 1931 when by resolution of the U.S. Congress the original name was restored) limited internal autonomy, thereby ironically falling short(er) than the civic rights ceded by the Spaniards to the islanders in 1897/98 which were washed away in the U.S.-Spanish War of 1898. Not until 1917 were Puerto Ricans granted U.S. citizenship, but they still have no representation in the U.S. Congress because of their island's persistent status as Unincorporated Territory of the United States of America. However, the U.S.-Spanish War made the easily-won and handsome bounty of Puerto Rico an instant darling of American public interest, a fact which in turn prompted U.S. atlas producers to literally stop their presses in late 1898 in order to squeeze in a detailed coverage of that island. A good example is *The Century Atlas of the World* of late 1898 which added to the conventional map of the Caribbean—but other than Rand McNally in 1889 now showing Cuba and Puerto Rico in the same area coloring—a brand new sheet entitled *Porto Rico (Puerto Rico) and the Lesser Antilles* (Fig. 11). This was seemingly added shortly before going to press because the sheet was given the unique number of Map No. 68 ½.

U.S. atlas cartographers—at least initially—could satisfy the sudden interest in Puerto Rico rather easily by quickly salvaging the outdated survey-scale maps to

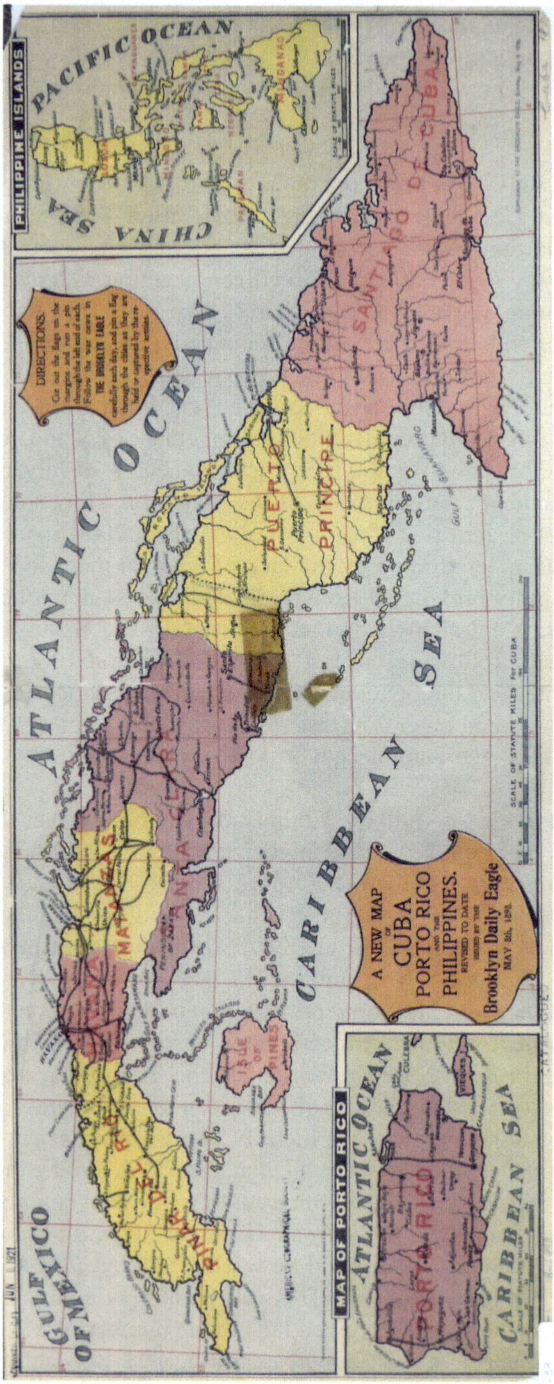


**Fig. 10** A newspaper (induced) war: *A New Map of Cuba, Porto Rico and the Philippines. Revised to the date, issued by the Brooklyn Daily Eagle. May 8th 1898* (from the American Geographical Society Library, University of Wisconsin-Milwaukee Libraries, Map Collections, 233 B-1898)

**Fig. 11** Suddenly atlas-worthy: Map No. 67 "West Indies" and Map No. 68 ½ (!) "Porto Rico (Puerto Rico) and the Lesser Antilles" in the (late) 1898 edition of Benjamin E. Smith (ed.): *The Century Atlas of the World*. New York (The Century Company) 1898 (from the American Geographical Society Library, University of Wisconsin-Milwaukee Libraries, Atlases 1898)

achieve a sufficient representation. The first U.S. troops landing in the Bay of Guanica on 25 July 1898, however, may not have encountered any substantial resistance, but when advancing into the Cordillera Central the most effective hindrance came in the form of a maze of bridle paths and the lack of decent maps. Unaware of the just-concluded quarter century long military survey and topographic reconnaissance by Spanish military engineers, who less than 3 months after the beginning of the U.S. invasion left the island along with the their cartographic treasure troves and without the Americans getting wind or even hold of the *Mapa Militar* draft maps and files, U.S. marines, cavalry and infantry battled more for orientation than against the enemy. Only a few outdated small-to-medium-scale maps of the island and even fewer large-scale maps were available for this high-relief terrain overgrown with tropical vegetation and with often nothing more than badly kept bridle paths to cross it with any wheeled mode of transport. The situation thus seemed to confirm the American prejudice that the Spaniards were lazy cartographers and, generally, unfit to manage an adequate development of such a resource-rich island.

Soon after Spain had ceded Puerto Rico to the United States on 29 September 1898, one of the first American larger-scale single sheet maps appeared. J.M. Lago's *Latest Map of Porto Rico. New Territory of U.S.A.* (Fig. 12) on a scale of 1:200,000 both compiled the content of the pre-invasion published maps but still depicted an at best tentative relief representation, coming with only a very basic overlay of infrastructural communication lines. Another noteworthy piece of 'instant cartography' is *A Map of the Island of Puerto Rico by J. Domingo Sulsana. Second Edition Revised and Corrected 1899*, published on a scale of 1:250,000 by A. Hoen & Company in Baltimore.

However, all these early single sheet maps were mere short-term band-aids without the surveys of the Spanish *Mapa Militar* at hand. In the long run, the U.S. military basically had to reinvent the wheel by preparing and undertaking a state-of-the-art survey of the mountainous island with a meticulous topographic filling by plane tabeling. A decade after conquering the island the recently formed Porto Rico Regiment of Infantry, which basically was the back-up of the island's police force, assigned several officers the task of producing a topo-tactical map. With the bulk of the work carried out by its task force leader, Lieutenant William Henry Armstrong, in 1908–09, the printing of the impressive result, the *Progressive Military Map of the United States eastern division. Porto Rico* on the very large scale of 1:62,500 in 12 sheets (Fig. 13), was delayed until 1914. The endeavor was part of an envisaged large-scale map series by the U.S. Army aiming at interconnected tactical maps of colonial dependencies. Consequently for the Puerto Rican leg, although accommodating a broad governmental scope of content, the emphasis was on enabling rapid military and police deployment and tactical support on the spot of action, including the now highly unlikely defence against foreign invasions like the one by the Americans 10 years earlier (Thompson 2014).

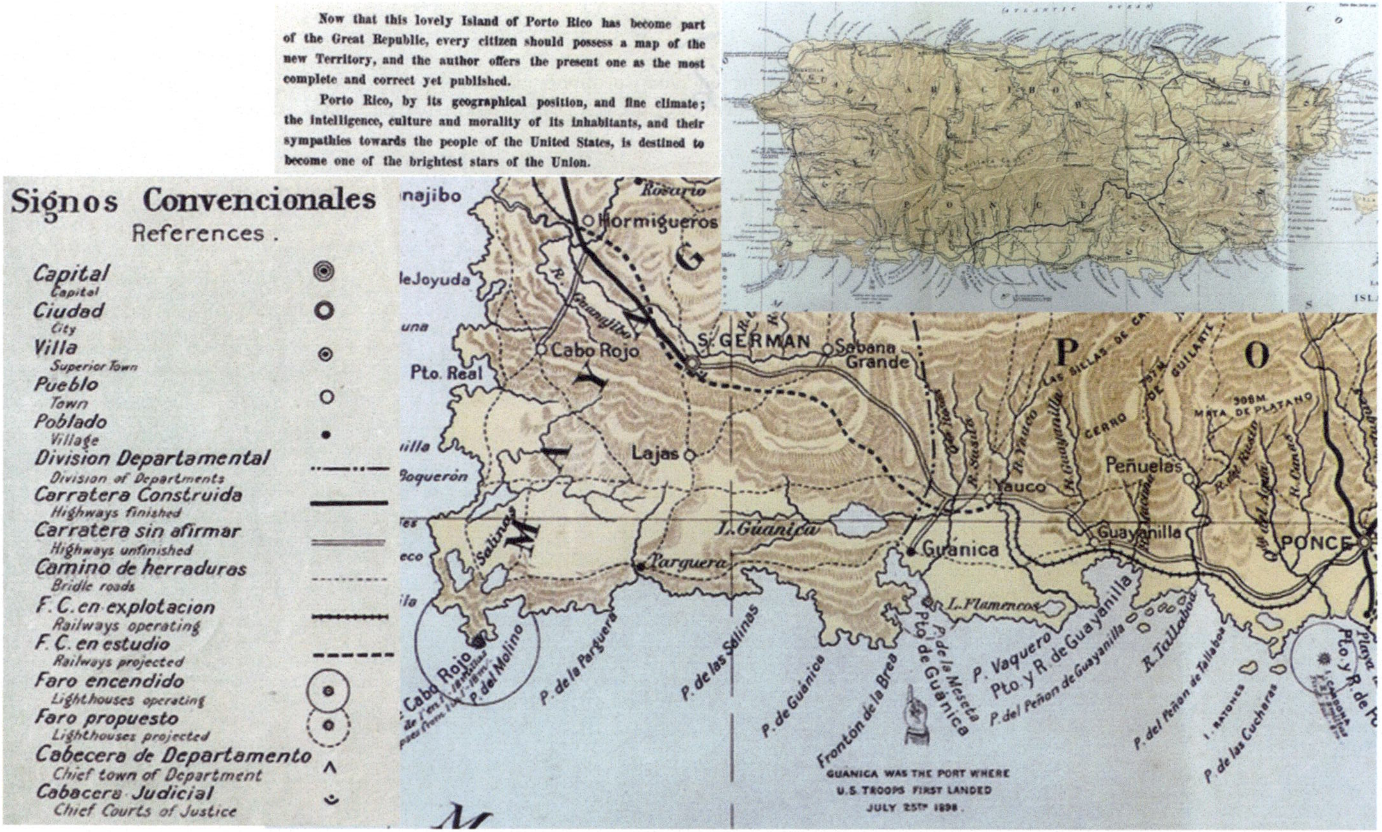


**Fig. 12** One of the first U.S. issued topographical maps of Puerto Rico: J.M. Lago's *Latest Map of Porto Rico. New Territory of U.S.A.* New York (Colton, Onman & Company) 1898 (from the American Geographical Society Library, University of Wisconsin-Milwaukee Libraries, 233 A-1898)

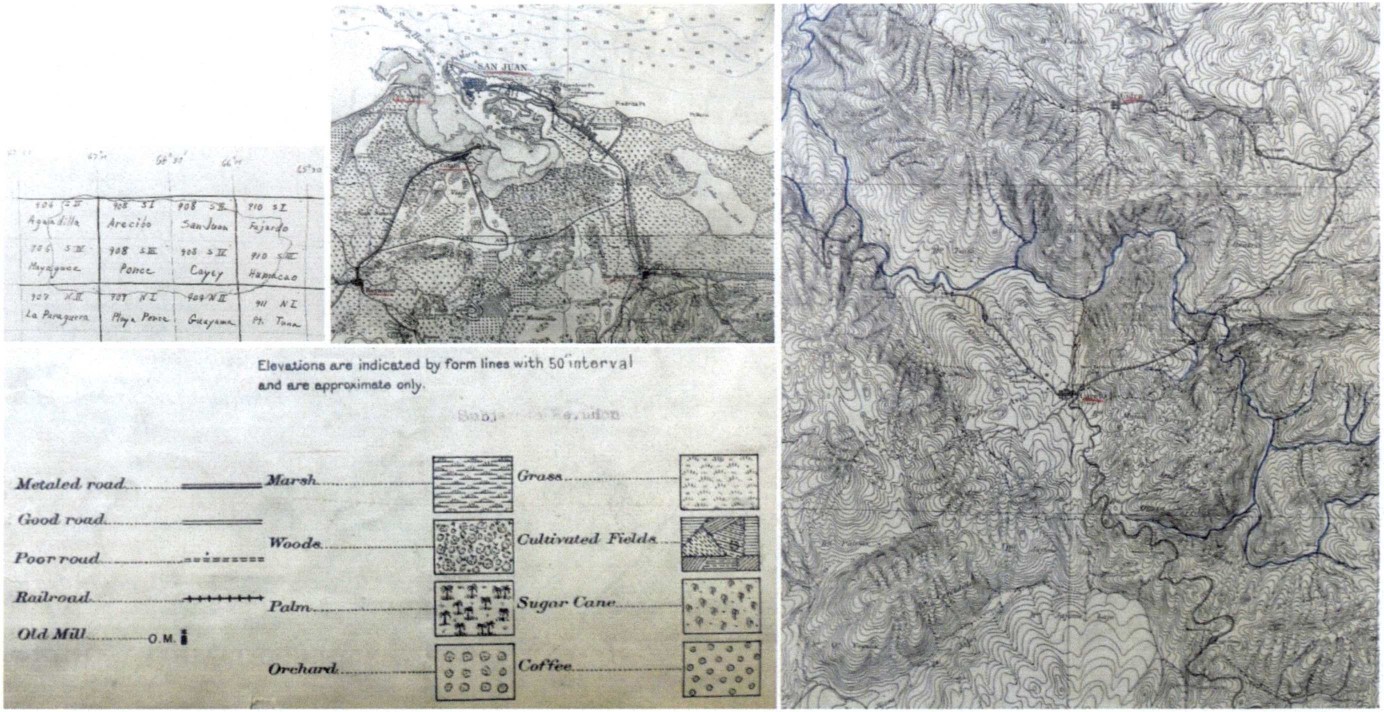


**Fig. 13** Manuscript index, printed legend and two details of Sheet 908 S II "San Juan" and Sheet S IV "Cayey" of the 12 sheet *Progressive Military Map of the United States eastern division. Porto Rico*. 1914 (from the American Geographical Society Library, University of Wisconsin-Milwaukee Libraries, Map Collection, 238b A-1:62,500)

## 5 Conclusion

The Spanish-American War of 1898 triggered a rapid and profound progress in the map coverage of hitherto only superficially mapped colonial territories because of their change of ownership. That change of sovereignty provided opportunities to tap into new resources of well-diversified official and private cartographic establishment which were the key factor in the almost overnight modernization of the map coverage of the Philippines and Puerto Rico.

In the case of the Philippines the U.S. administration embraced an already well advanced project of private cartography by Jesuit fathers and their indigenous team at the observatory of Manila. Profoundly different in Puerto Rico the U.S. administration was deprived of any significant official map series with no private cartography to fill the void. Thus Puerto Rico under U.S. administration required clean-cut surveying and topography from scratch, mirroring the approach of conquerors of virgin territory. The imperial aspirations of the United States became the ground on which soon after the acquisition a fusion of private and official cartography could flourish. And the lack of reliable maps of the newly acquired colonial territories strengthened the relevance of the U.S. Coast and Geodetic Survey (USC&GS) in the Philippines as the U.S. Army surveying in Puerto Rico.

## References

Algué PJ (1900) Atlas de Filipinas/Atlas of the Philippine Islands, Observatorio de Manila (1899) and US Coast and Geodetic Survey, Special Publication No. 3 (1900), Manila and Washington, DC

American Geographical Society Library (AGS Library), University of Wisconsin—Milwaukee, 2311 East Hartford Avenue, Milwaukee, WI 53211, USA. https://uwm.edu/libraries/agsl/cartographic/

Beruff JR (2007) Strategy as politics: Puerto Rico on the eve of the Second World War. Universidad de Puerto Rico, San Juan

Capel H (1994) Geography and the Spanish Empire. In: Godlewska A, Smith N (eds) Geography and empire, Institute of the British Geographers Special Publication Series, vol 30. Blackwell, Oxford, pp 58–73

Carrion A, Morales M (eds) (1983) Puerto Rico: a political and cultural history. W.W. Norton, New York

De Wolf N et al (2005) The quality of newspaper cartography in wartime: the United States of America against Iraq. In: Proceedings of the 22th international cartographic conference (ICA), La Coruña, Spain. Published on CD-ROM

Gannett H (1901) Map notices. Bull Am Geogr Soc 33(2):138–142

Kirsch S (2016) Insular territories: US colonial science, geopolitics, and the (re)mapping of the Philippines. Geogr J 182(1):2–14

Lewis MW, Kären KE (1997) The myth of continents. A critique of metageography. UC Press, Berkley

Offner JL (1992) An unwanted war: the diplomacy of the United States and Spain over Cuba, 1895-1898. University of North Carolina Press, Chapel Hill

Perez LA (1995) Cuba: between reform and revolution. Oxford University Press, Oxford
Pritchett HS (1909) Some Recollections of President McKinley and the Cuban Intervention. N Am Rev 189(640):397–403
Quirino C (1969) Philippine cartography: 1320-1899. N. Israel, Amsterdam
Quirós LF (1998) Dos geógrafos españoles en el 'noventa y ocho': Gonzalo de Reparaz y Enrique D'Almonte. Ería. Revista cuatrimestral de geografia 46:183–189
Schmidt-Nowara C (2008) The conquest of history: Spanish colonialism and national histories in the nineteenth century. University of Pittsburgh Press, Pittsburgh
Schulten S (2001) The geographical imagination in America 1880-1950. University of Chicago Press, Chicago
Thompson L (2014) Governmentality and cartographies of colonial spaces: the "Progressive Military Map of Porto Rico", 1908-1914. In: Goldstein A (ed) Frontiers of United States Colonies. Duke University Press, Durham, pp 289–315
Warren JF (2009) Scientific Superman. Father José Algué, Jesuit Meteorology, and the Philippines under American Rule, 1897–1924. In: Mc Coy AW, Scarano FA (eds) Colonial crucible empire in the making of the modern American State. University of Wisconsin Press, Madison, pp 508–521
Woodward D (1977) The all-American map. Wax engraving and its influence on cartography. University of Chicago Press, Chicago

**Eric Losang** is researcher at the Leibniz-Institute for Regional Geography in Leipzig, Germany. Having been responsible for the editing of the digital maps in the National Atlas of Germany project, his focus moved from mapmaking to into the fields of critical cartography and map production processes. His major fields of interest are the different aspects of atlas production and dissemination as well as their importance for the development of cartography throughout history. He is vice chair of the ICA Commission on Atlases.

**Imre Josef Demhardt** is interested in post-enlightenment cartography, colonialism and regional studies with a focus on Central Europe, Sub-Saharan Africa and the Americas. Besides authoring numerous articles and several books on these subjects he is co-editor of Vol. 5 (Nineteenth Century) in the encyclopedia project on the *History of Cartography*. He holds the Garrett Chair in the History of Cartography at the University of Texas at Arlington and currently serves as Chair of the ICA Commission on the History of Cartography.

# The History of Cartography of Western Border Areas of Russia in Eighteenth to Early Twentieth Centuries (Using the Example of the Smolensk Province)

**Natalia Mikhailovna Erman and Viacheslav Alekseevich Nizovtsev**

**Abstract** The article describes the history of mapping of the Smolensk region in eighteenth to early twentieth centuries, based on archival materials as well as published sources on the history of cartography in Russia. The article distinguishes three main periods of mapping and identifies activities that characterize the development of scientific knowledge and the understanding of the surrounding nature. The article makes a note of the role of scientific institutions, government agencies and individual researchers in conducting geographic research on the western border areas of Russia. It presents unique cartographic works (a number of which were previously unknown) of different periods, reflecting the transformation of the territory over the course of its exploration and economic use.

## 1 Introduction

The study of the history of geographical research in the Smolensk province allows to systematize the accumulated data and to expand the knowledge of its natural conditions and resources, population and economy. It enables to trace connections between the geographical study and the processes of territorial development and to determine the contribution of cartographic research and researchers to the study of geographical features of the region and the development of general geographic scientific ideas.

Smolensk land has at all times had an exceptional importance in the history of the formation and development of the Russian state. Its unique geographical position on the watershed of headwaters of several major rivers—the Volga,

N.M. Erman (✉)
S.I. Vavilov Institute for the History of Science and Technology of the Russian Academy of Sciences, Moscow, Russia
e-mail: erman.natalie@mail.ru

V.A. Nizovtsev
Lomonosov Moscow State University, Moscow, Russia
e-mail: nizov2118@yandex.ru

 
M. Altić et al. (eds.), *Dissemination of Cartographic Knowledge*, Lecture Notes in Geoinformation and Cartography, DOI 10.1007/978-3-319-61515-8_8

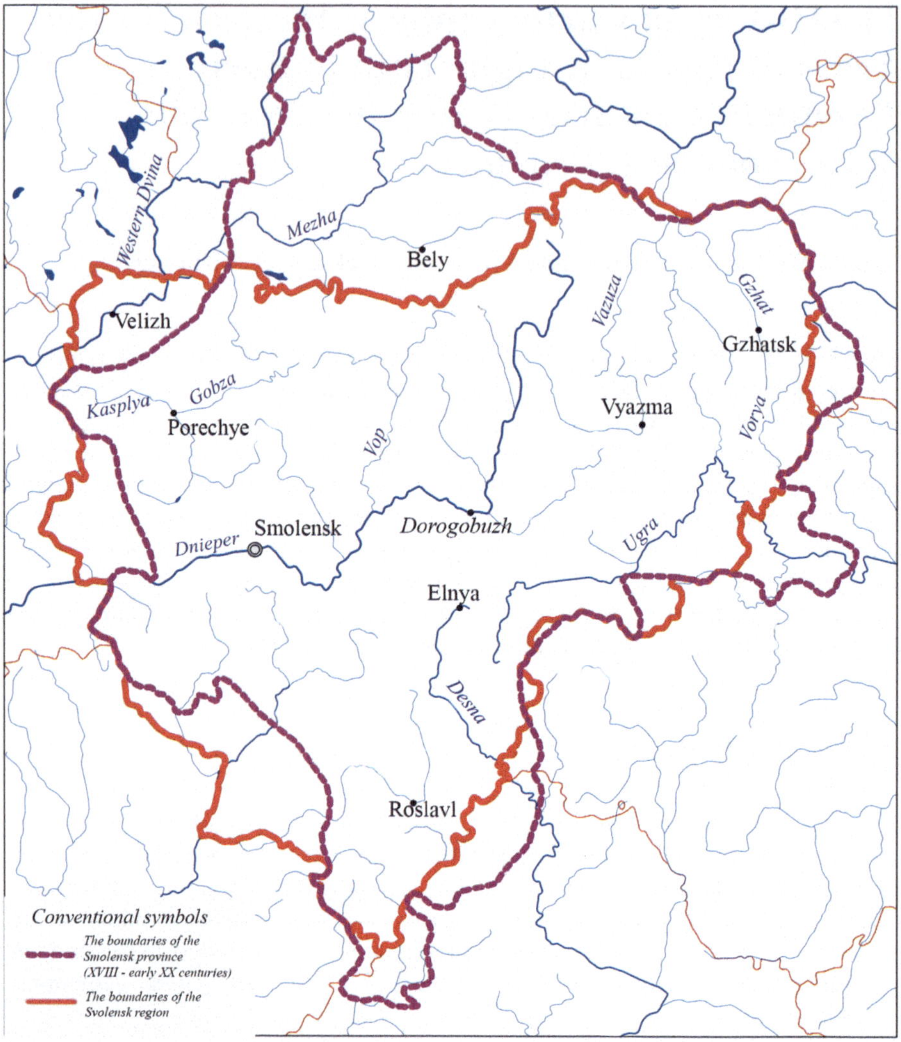


**Fig. 1** The geographical position of the Smolensk land. Scale 1:2,000,000. Drawn by N. Erman

Dnieper and Western Dvina and their inflows (Lovat, Sozh, Kasplya, Gzhat, Vorya) contributed to active settlement, economic development of this region and the emergence of many cities (Fig. 1). The city of Smolensk is one of the oldest cities in Russia, founded in 863. It was formed at the crossroads of transport waterways, the ancient trade route "from the Vikings to the Greeks" and "the Great Volga Route" and was the most important stronghold of trade routes connecting the Baltic, Black, Azov and Caspian seas.

The Smolensk principality was in the ancient period (eleventh to thirteenth century) one of the largest and most economically developed principalities. In the following centuries the boundaries of the Smolensk land and its administrative structure changed repeatedly. The land was the subject of a constant struggle with varying success between the Russian and Lithuanian states and Polish-Lithuanian Commonwealth. It finally became a part of the Russian state in 1654 as a result of the Russian victory in one of the Russian-Polish wars. The city of Smolensk then became an important outpost on the western borders of the Russian Empire, and the surrounding Smolensk province, occupying a unique geographical position and possessing rich natural resources, constantly attracted the attention of authorities and researchers.

For this study the following materials were used: the materials of the State Archive of the Smolensk region, Russian State Military Historical Archive, the Russian State Historical Archive, the Russian State Archive of Ancient Acts, the Archive of the Russian Academy of Sciences and stock materials of the cartography department of the State Historical Museum. A special place was given to the study of scientific papers which contain results of research on the Smolensk province by V.V. Dokuchaev (1879–1883), soil and geological studies by K.D. Glinka (1913), studies of rivers and lakes in the Smolensk region in the works by A.A. Tillo (1898), S.N. Nikitin (1896), and D.N. Anuchin (1891), as well as the statistical descriptions of the Smolensk land by J.A. Soloviev (1855), and historical and geographical descriptions of cities and trade in the Smolensk region by V.P. Semenov-Tian-Shansky (1900). A lot of useful information was extracted from the records of expeditions (travel route descriptions), in particular, "Topographical notes" (1780), "Journey of Her Imperial Majesty in the south side of Russia" (1787), "Notes of a trip to western provinces" by V.M. Severgin (1803), collections of scientific works "Reports" and "Works" by the Society for the Study of the Smolensk province (1912).

We distinguish three main periods in the geographic study and mapping of the territory of the Smolensk province, which differ in terms of volume and systematic nature of the research that was conducted, as determined by the peculiarities of the historical era and by the practical problems that required attention at a particular point in time.

## 2 Cartographic Works on the Territory of the Smolensk Province in the Eighteenth Century

In the eighteenth century Russia experienced a spike of geographical research related to the country as a whole and also its individual regions. The establishment of the Navy and the regular army, setting the new state borders, the reform of public administration and territorial division, trade development and education reform—all these events in the foreign and domestic politics at the time of Peter I were accompanied by mass mapping activities in Russia.

The unique geographical and strategic position of the Smolensk province contributed to the organization of a widespread scientific study of its natural

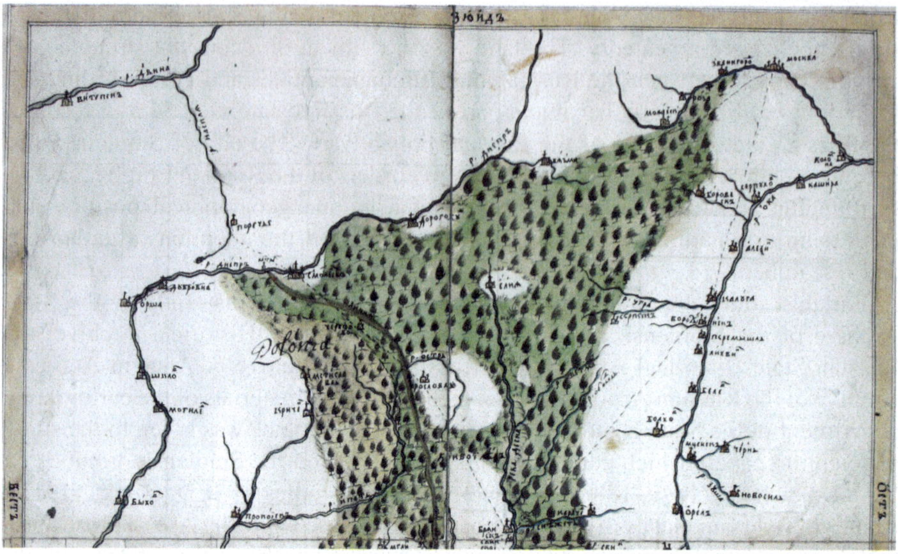

**Fig. 2** General map of the defense line from Smolensk to Chernigov of 1706 (detail). Russian Academy of Sciences Library. Manuscript division, the main collection No 658 (RASL, St. Petersburg)

environmental conditions and resources. A special role in these studies was played by the mapping efforts. The first scientific cartographic surveys in the Smolensk province began to be conducted in the eighteenth century, primarily under the influence of reforms and transformations of Peter I in the military, administrative and commercial spheres.

In preparation for the Northern War, aimed at protecting the western boundaries of the state, new fortifications were created, which were special lines of defense consisting of forest debris and fortifications (Postnikov 1996). In 1706, "the general map of the defense line" (Fig. 2) from Smolensk to Chernigov was drafted, which depicted the defense line together with a number of locations and cities (Smolensk, Vyazma, Dorogobuzh, Yelnya, Roslavl, Porechye), as well as forests, roads, and rivers (Dnieper, Western Dvina, Kasplya).[1]

The favorable administrative and physical-geographical position of the Smolensk province promoted the development of trade, especially along its numerous rivers with long-standing docks in Smolensk, Belsk and Porechye. Bread, flax and wood were transported from the Porechye pier along the rivers Kasplya and Western Dvina to Riga (Semenov-Tian-Shansky 1900).

Under Peter I the city harbour of Gzhatsk was founded in 1705 in order to provide the new capital St. Petersburg with grain products. It became an important trading center from which various goods and products were sent from the southern

[1]Russian Academy of Sciences Library. Manuscript division, the main collection No 658. General map of the defense line from Smolensk to Chernigov, 1706.

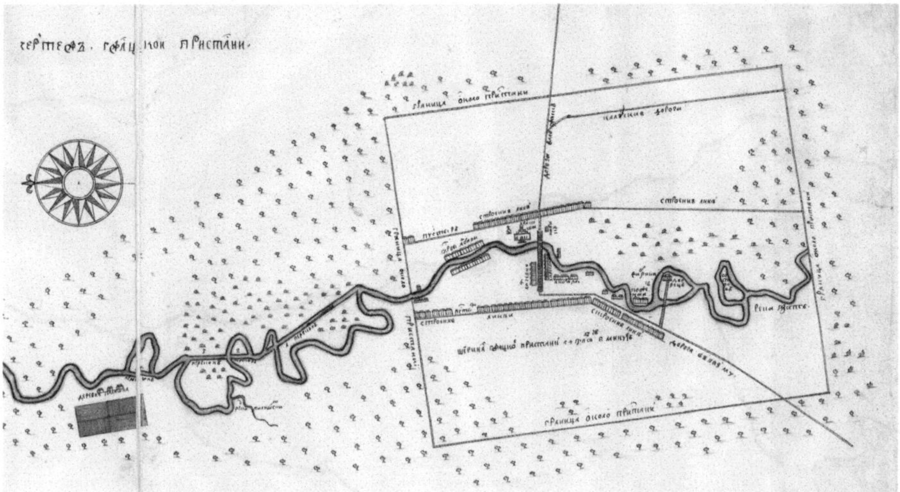

**Fig. 3** Drawing of the Gzhatsk harbour (detail). First half of the eighteenth century. Russian State Archives of Ancient Documents. Fund 192, list 1, No 2. (RGADA, Moscow)

provinces to St. Petersburg and Moscow. The Archive of Ancient Acts keeps the plan of the Gzhatsk harbour. The plan shows the river Gzhat and docks with commercial buildings[2] (Fig. 3). This plan deserves attention as it demonstrates the results of hydraulic engineering works such as straightening the curved river bed of the river Gzhat to improve navigation.

To improve navigation from St. Petersburg to Moscow for the development of trade, Peter I made an attempt to build a channel linking the rivers Gzhat and Vorya. The surveyors Vasily Leushinsky and Mikhail Yusupov were sent on site to conduct the hydrotechnical works on the river Gzhat (1717–1719). However, the attempt to construct the channel failed. The surveyors made a map of the project (Fig. 4). It shows the river beds and their inflows, with settlements, transport routes and mills on their banks, as well as the planned channel track, with an explanation.[3] The library of the Academy of Sciences preserves cartographic materials of this region.

Thanks to the reforms of Peter I and the organisation of practical mapping projects, Russian cartography and geography received new information about the territory of the Smolensk province. The cartographic work of the surveyors allowed to make a geographical representation of the territory of the Smolensk province of that period. Based on their maps the Chief Secretary of the Senate I.K. Kirillov published "The Atlas of the Russian Empire" and the general map of the Russian Empire of 1734. This map showed for the first time the outlines of the entire

[2]Russian State Archives of Ancient Documents. Fund 192, list 1, No 2. Drawing of the Gzhatsk harbour. First half of the eighteenth century.

[3]Russian Academy of Sciences Library. Manuscript division, the main collection No 154. The project to connect the rivers Vorya and Gzhat. First half of the eighteenth century.

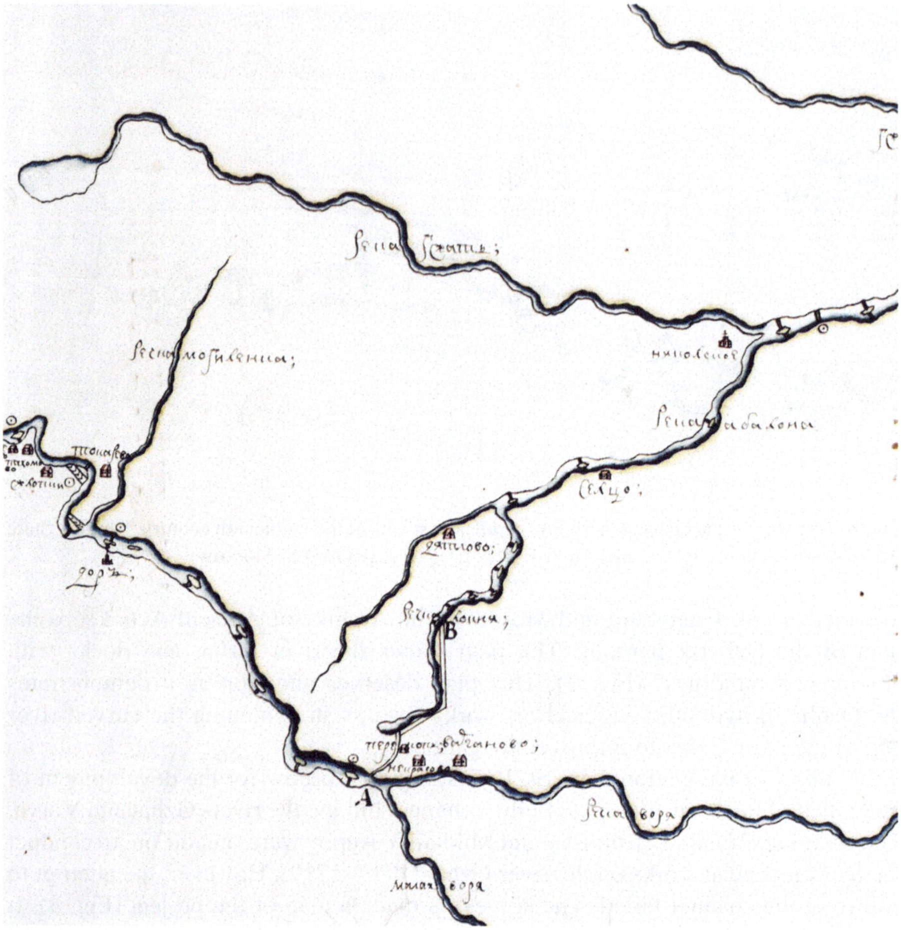


**Fig. 4** The project to connect the rivers Vorya and Gzhat (detail) dating from the first half of the eighteenth century. Russian Academy of Sciences Library. Manuscript division, the main collection No 154 (RASL, St. Petersburg)

territory of Russia that were close to reality. It displayed the territory of the Smolensk province, four of its settlements (Smolensk, Vyazma, Roslavl and Krasny), as well as the Dnieper and the Western Dvina rivers.[4]

One of the most important events for science in Russia was the creation of the Academy of Sciences in 1725. From that moment on, all geographic, expeditionary and geodetic works were conducted under the guidance of this academy. In 1745 it published "The Atlas of the Russian Empire", which contained, in addition to the general map of Russia, 19 maps of provinces with adjacent land, including the map

[4]Russian Academy of Sciences Library. Cartography department. The Atlas of the Russian Empire. Collection of maps I.K. Kirilov, 1734.

of the Smolensk province. The territory of the Smolensk province was displayed on a separate page with more detailed geographical information.

In the last third of the eighteenth century, extensive mapping works associated with the General Land Survey were conducted. Borders of parcels were plotted. The materials contain information about the economic value of land, fixed arable land, forests, hay meadows, factories, mills, and the elements of the natural landscape. Large-scale maps of landholdings of the largest part of the counties of the Smolensk province were created.

A large collection of such materials is kept in the Archive of the Smolensk region. Each large-scale plan shows an image of a separate domain or "dacha". It contains a stamp, a title cartouche with information about the name of the plot of land, its owner, the date and the author of the survey. Then the cartographic image itself has a scale of 100 fathoms to an inch (scale 1:8400), a legend revealing several graphic symbols and a list of adjacent possessions[5] (Fig. 5).

On the basis of such large-scale materials, combined maps of counties and provinces, as well as atlases were compiled. The archive of the Ancient Acts keeps two volumes of the atlases of the Smolensk governorship. The atlases were published by the Land Survey Bureau at the Land Survey of the Senate under the guidance of the collegiate adviser Waxell in 1781. They consist of county maps on a scale of five versts to an inch (1:210,000) of the counties of Smolensk, Porechsky, Duhovschinsky, Belsky, Elninsky, Sychevsky, Gzhatsky, Krasninsky, Yukhnovsky and Roslavlsky, as well as urban plans on a scale of 50 or 100 fathoms to an inch (1: 4200 to 1: 8400). As an example, Fig. 6 reproduces the map of the Viazemsky county, which shows all the settlements categorized as cities, villages or small villages, as well as graveyards and roads. It also shows the location of all factories and plants. Natural features indicated on the map are rivers, forests, marshes and—for the first time—relief, namely the main elevations (indicated as mountains), as well as some agricultural lands such as arable land and haylands.

The materials of the General Land Survey differ from the surveys of geodesists of the first half of the eighteenth century by the considerable thoroughness and reliability of the geographical image. They are exceptional source material for studying the physical and economic geography of the Smolensk province of the time, as well as for historical and geographical comparisons with subsequent periods.

The handwritten Forest Atlas of the Smolensk province in two parts, compiled by P. Trotskevich in 1797, which contains a description of the forests, was also of great importance. The maps show the cities, villages, settlements and rivers, while the location of the surveyed forests is coloured in. The work contains a geometric

[5]The State Archive of the Smolensk region. Fund 259, list 1, No 58. "Geometric special plan of village Ananyino Duhovschinskogo district of Smolensk province".

**Fig. 5** Geometric special plan of village Ananyino Duhovschinskogo, district of Smolensk province (State Archive of the Smolensk region. Fund 259, list 1, No 58)

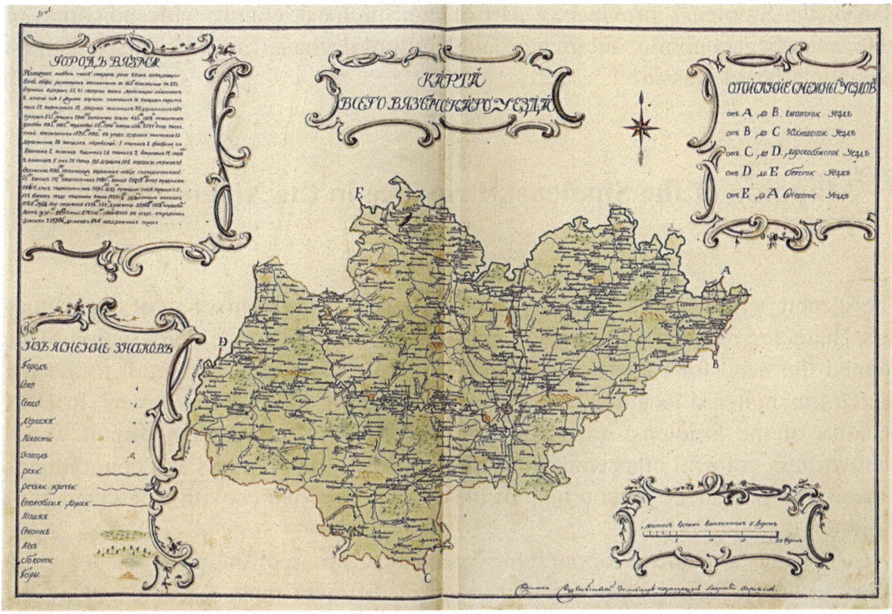


**Fig. 6** Map of Vyazma district. Atlas of General land surveying of Smolensk province of 1781 (Russian State Archives of Ancient Documents. Fund 1356, list 1, part II, No 5459)

**Fig. 7** Forest Atlas of the Smolensk province (detail). Russian State Historical Archives. Fund 380, list 28, No 32 (RGIA, St. Petersburg)

plan of the Smolensk province, a part of the Smolensk county with indication of state, specific, economic, one-manor and landlord dachas. (Fig. 7).[6] This is the first thematic atlas of Russia.

## 3 Mapping of the Smolensk Province in the Nineteenth Century

The second period covers the entire nineteenth century. The first half of the century was characterized by extensive mapping in preparation for military action. This created the need for topographic maps, depicting the surface in detail to aid the search for sheltered locations and to facilitate tactical military maneuvers. In 1811 an atlas of the Smolensk province was created by V.F. Blankengorn. It was a handwritten, colorful atlas consisting of 25 sheets of town plans and county maps as well as a one-sheet general map of the Smolensk province with a description of its rivers.

In 1812, as a result of topographical research, a map of the province depicting cities, towns, villages, churches and roads of different ranks was created by the Department of Maps. In addition the map gives important information about the traversability of the territory, with colours indicating swamps, forests and meadows (Fig. 8).[7]

After the war of 1812, plans of a number of cities of the Smolensk province were drawn up. To restore the destroyed towns of Porechye, Krasny, Vyazma, Yelnya, Gzhatsk and Smolensk, their new layout was designed (Fig. 9).[8] There were designated places for orchards, gardens, and the construction of mills, residential and industrial zones. These documents dating from 1817 to 1845 are stored in the Smolensk State Archive.

In the first half of the nineteenth century a variety of thematic cartographic work was also carried out in the province. In 1837 the first soil studies were undertaken under the direction of the Smolensk Governor N.I. Khmelnitsky. The map of Sychevsky county colours the soils to indicate the best, the mediocre and the worst, while the map of the Porechye county gives more qualitative characteristics of soils by assigning them letters: A-sandy, B-clay, C-podzol, D and F have no explanation (Fig. 10).

The most interesting local studies were mapping works from the 1850s. The "Statistical Atlas of the Smolensk province" by colonel K.I. Stiernskants (1850) is one of the most notable. The Atlas consists of ten sheets of thematic maps that show the territory of the Smolensk province. Among them a map serving a military-

---

[6] Russian State Historical Archives. Fund 380, list 28, No 32. Forest Atlas of the Smolensk province, 1797.

[7] Russian State Military Historical Archives. Fund. 848, list 16, No 21460. Semitopographical map of the Smolensk province, 1812.

[8] The State Archive of the Smolensk region. Plan of the city of Gzhatsk, 1833. Fund 1, list 1, No 41.

**Fig. 8** Fragment. Semitopographical map of the Smolensk province 1812. Russian State Military Historical Archives. Fund 848, list 16, No 21460 (RGVIA, Moscow)

topographical purpose, a map representing the water systems and the areas flooded by rivers, a map of the forests, a map showing all mills and factories, and a soil map. The forest map uses color to indicate different types of woods (pine, mixed, deciduous), and includes a table with statistical data on the surface area (in acres) of forested land in each county of the Smolensk province.[9]

The second half of the nineteenth century was characterized by scientific expeditions. Various ministries and departments initiated the systematic study of natural conditions and resources. A great deal of attention was paid to the study of soils as knowledge on this subject formed the basis for the economic development of Russia at the time. Particularly important in this regard were the studies of V.V. Dokuchaev. Dokuchaev conducted scientific expeditions throughout Russia: from his native Smolensk region to the south of European Russia. His comprehensive research laid the foundations of the knowledge of natural zonation, soil processes and podzolic soils. As a result of his work, N.M. Sibirtsev, G.I. Tanfilev and A.R. Ferhmin compiled and published a new soil map of

[9]Russian State Military Historical Archives. Fund 386, list 1, No 4823. The "Statistical Atlas of the Smolensk province" K.I. Stiernskants, 1850.

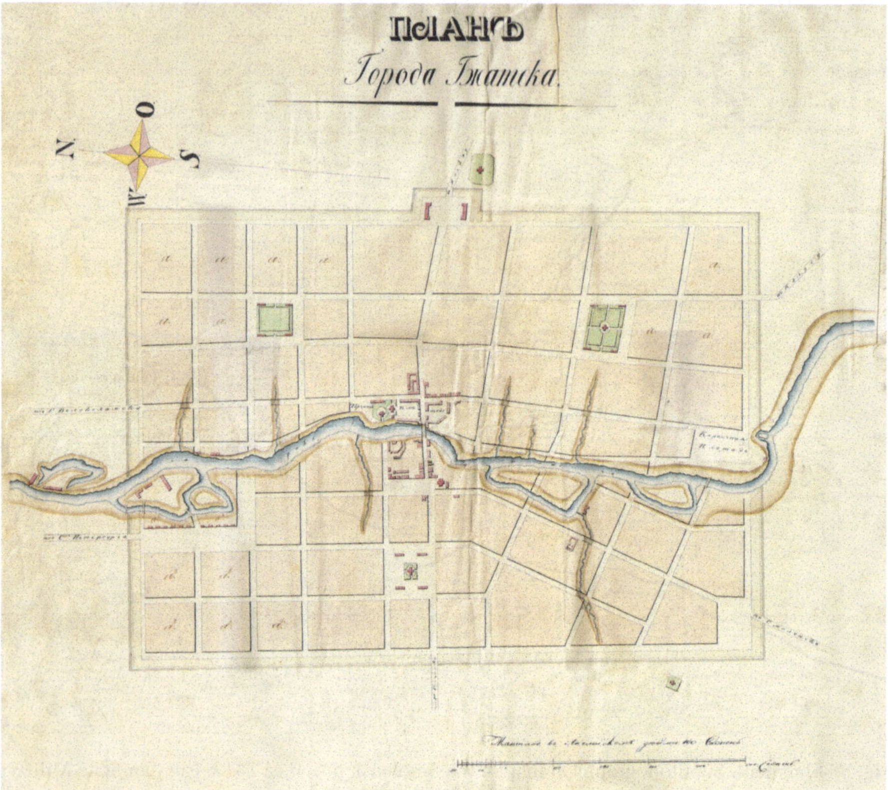


**Fig. 9** Plan of the city of Gzhatsk, 1833. (State Archive of the Smolensk region. Fund 1, list 1, No 41)

European Russia in 1900 “on the initiative and plan” of V.V. Dokuchaev. Soil types with zonal and azonal genetic types were first identified on this map (Tanfilyev and Ferkhmin 1902). For the territory of the Smolensk province, colours and symbols indicate sod-podzolic and podzolic loams—No 16; sod-podzolic and podzolic sandy loams and clay sands—No 17; podzols—No 18; sandy soils—No 26; and alluvial soils—No 27. Figure 11 shows a fragment of this soil map of the Smolensk province (Sibirtsev et al. 1900).

In the study of natural resources such as rivers and lakes, a special place is occupied by the “Expedition on the study of the sources of the main rivers of European Russia” under the leadership of A.A. Tillo (1894–1899). Scientists of different specialties participated: hydrogeologist S.N. Nikitin, soil scientist V.R. Williams, forester M.K. Tursky, botanist N.I. Kuznetsov, hydraulic engineer F.G. Zbrozhek, surveyor A.A. Fok, and geographer D.N. Anuchin. The purpose of the expedition was to identify the causes for the shallowing of European rivers. The expedition prepared a detailed physical and geographical description of the basins of the Volga, Dnieper, and Western Dvina rivers, determined the exact location of

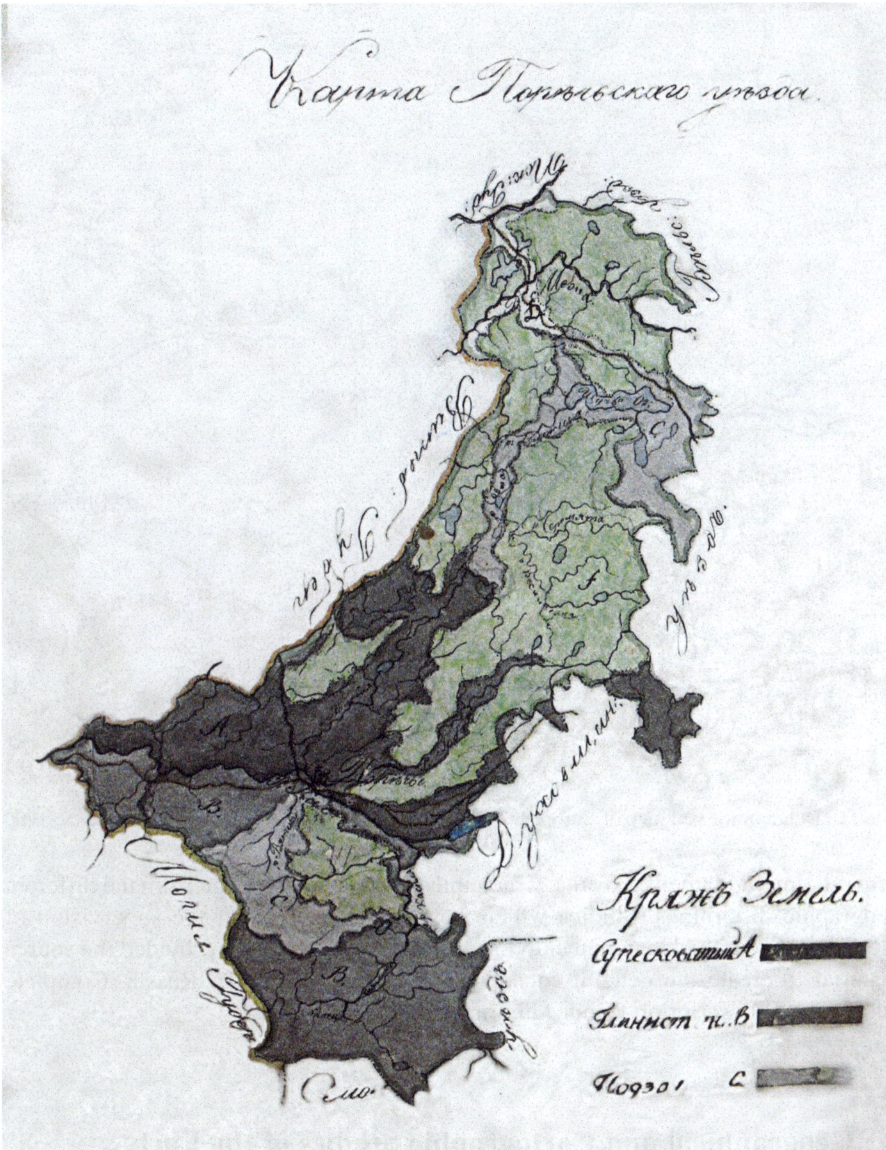


**Fig. 10** Map Porechsky County, 1837. (State Archive of the Smolensk region. Fund 1, list 1, No 3)

the sources of the main rivers of European Russia, and mapped these rivers for the first time on large-scales in a hypsometric map, a map of water availability and botanical maps (Nikitin 1896).

Large-scale complex studies and cartographic works devoted to the different components of nature, to population and to economy, became the basis for new

**Fig. 11** Detail of the soil map of European Russia "on the initiative and plan" of Dokuchaev, 1900

generalizing fundamental works, which integrated the knowledge from the different geographic disciplines. Studies which were carried out under the supervision of P.P. Semenov in the late nineteenth to early twentieth centuries provided the source material to create the general consolidating geographic work "Russia. Complete geographical description of our fatherland".

## 4 Geographical and Cartographic Studies of the Early Twentieth Century

The third period of geographic research and mapping covers the time from the beginning of the twentieth century to 1929. In 1929, instead of the Smolensk province, a new western region was formed with Smolensk as its administrative center. The external borders and internal structure of this new western region changed. The territory of the Smolensk province from the beginning of the eighteenth century to the beginning of the twentieth century did not undergo significant changes.

The multitude of cartographic works executed in the early twentieth century made the Smolensk lands the most studied Russian territory at the regional level. At the beginning of the twentieth century, the students of V.V. Dokuchaev and followers of his scientific ideas continued with comprehensive geographical studies of all components of nature. K.D. Glinka organized soil research in the Smolensk province following the program developed by V.V. Dokuchaev. The work was carried out by local scientists—agronomist A.A. Sondag, and soil scientists L.V. Abutkov and A.V. Kostyukevich. As a result of their studies, five volumes of "Materials for the assessment of the lands of the Smolensk province" were published in 1901–1912 containing large-scale soil maps.

One of the last major works of this period is "An outline of the natural historical conditions of the Smolensk province" published in 1925. This essay gives a detailed description of the geographical position, relief, hydrology, geological structure, minerals, soils, climate, vegetation and fauna of the Smolensk province, and contains geological, mineral, hypsometric and soil maps. Natural science studies by K.D. Glinka, A.V. Kostyukevich and Y.Y. Alekseev generated cartographic representations of all natural components of the Smolensk region.

In conclusion we can say that the organization of scientific geographic studies in Russia began in the eighteenth century. Due to the unique geographical position of the Smolensk province, all the important political and economic developments in Russia were reflected in its geographical studies. Its territory also became a sort of testing ground for numerous cartographic projects. Consequently, the geographic and cartographic works of eighteenth to early twentieth centuries on the Smolensk province are important sources to study the development of geographical research and Russian cartography. A number of these works were previously unknown but their discovery has now allowed to create a more comprehensive picture of the geographical study of the Smolensk province.

**Acknowledgements** The research presented here is performed under the Basic Research program of the Presidium of the Russian Academy of Sciences in 2017.

## References

Anuchin DN (1891) Iz poezdki k istokam Dnepra, Zapadnoy Dviny i Volgi. [From a trip to the sources of the Dnieper, the Western Dvina and the Volga rivers]. Severnyy vestnik 8:119–162

Dokuchaev VV (1879) Kartographiya russkikh pochv [Cartography of Russian soils]. Typography by V.F. Kirschbaum, St. Petersburg

Dokuchaev VV (1883) Russkiy Chernozem: Otchen Imperatorskomu Volnomu ekonomicheskomu obshchestvu [Russian Chernozem: a report to the Imperial Free Economic Society]. Free Economic Society, St. Petersburg

Glinka KD (1913) Predvaritelnyi otchet zemstvu o pochvenno-geologicheskikh issledovaniyakh Vyazemskogo i Sychevskogo uezdov Smolenskoy gubernii [A preliminary report to the country council on soil and geological studies of Vyazemsky and Sychevsky counties of the Smolensk province]. Provincial Typography, Smolensk

Nikitin SN (1896) Bassein Dnepra: Trudy ekspeditsii dlya issledovaniya istochnikov glavneishih rek Evropeiskoy Rossii [Dniper Basin: works of expedition for the study of the sources of the main rivers of European Russia]. Typography by K. Birkenfeld, St. Petersburg

Ocherk estesnvenno-istoricheskikh usloviy Smolenskoy gubernii [An outline of the natural historical conditions of the Smolensk province, with maps enclosed] (1925) Smolensk Statistical Provincial Bureau, Smolensk

Otchet obshchestva izucheniya Smolenskoi gubernii [A report of the Society for the study of the Smolensk province] (1912) Typography by Podzemsky, Smolensk

Postnikov AV (1996) Russia in maps: a history of the geographical study and cartography of country. Nash Dom-L'Age d'Homme, Moscow

Puteshestviya ee imperatorskogo velichestva v poludennyi kray Rossii predpriemlemoe v 1787 [Journey of her Imperial Majesty in the south of Russia, undertaken in 1787]. Printed at the Mining School, St. Petersburg

Semenov-Tyan-Shansky VP (1900) Torgovlya i promyshlennost Evropeiskoy Rossii po raionam [Trade and industry of European Russia by regions]. Typography by V.F. Kirschbaum, St. Petersburg

Severgin VM (1803) Zapiski puteshestviya po zapadnym provintsiyam Rossiiskogo gosudarstva [Notes of a trip to western provinces of Russian state]. Imperial Academy of Sciences, St. Petersburg

Sibirtsev NM, Tanfiliev GI, Ferkhmin AR (1900) Pochvennaya karta Evropeiskoy Rossii, sostavlennaya po pochinu i planu V.V. Dokuchaeva [Soil map of European Russia "on the initiative and plan" by V.V. Dokuchaev]. Edition of the Department of Agriculture, St. Petersburg

Solovev YA (1855) Selskokhozyaistvennaya statistika Smolenskoy gubernii [Agricultural statistics of the Smolensk province]. Typography by A. Semyon, Moscow

Tanfiliev GI, Ferhmin AR (1902) Kratkiy obyasnitelnyi tekst k pochvennoi karte Evropeiskoi Rossii [A brief explanatory text to the soil map of European Russia]. Typography "Gerolda", St. Petersburg

Tillo AA (1898) Ekspeditsiya po issledovaniyu istochnikov glavneishih rek Evropeiskoy Rossii: Kratkiy predvaritelnyi otchet po rabotam 1897 [Expedition on the study of the sources of the main rivers of European Russia: a short preliminary report on the works of 1897]. Typography by K. Birkenfeld, St. Petersburg

Topographicheskie primechaniya na znatneishie mesta puteshestviya ee imperatorskogo velichestva v Belorusskie namestnichestva [Topographical notes on the most notable places of travel of her Imperial Majesty to the Belorussian viceroyalties] (1780) The Imperial Academy of Sciences, St. Petersburg

**Natalia Mikhailovna Erman**, Ph.D. of Geographical Sciences, senior researcher at the Department of history of Earth Sciences of S.I. Vavilov Institute for the History of Science and Technology of the Russian Academy of Sciences. Research interests: history of geography, historic geography.

**Viacheslav Alekseevich Nizovtsev**, Ph.D. of Geographical Sciences, chief researcher at the Geographical faculty of Lomonosov Moscow State University, deputy head of the Department of Physical Geography and Landscape. Research interests—landscape studies, historic geography, cartography and pedagogics. He has authored and co-authored over 550 scientific, educational, methodical and popular works, including more than 20 monographs on geography and landscape studies and more than 25 textbooks and teaching aids for higher and secondary schools.

# Part III
# Map Production and Dissemination of Knowledge

# ‘Back to the Drawing Board’: Map-Making and the Royal Geographical Society (1830–1990)

**Francis Herbert**

**Abstract** Original holograph and manuscript cartographic materials, for potential or eventual publication by the Royal Geographical Society, survive in considerable variety, number and condition. The RGS’s first commissioned map (1830) was a facsimile of Hereford Cathedral’s medieval manuscript *mappamundi*. From 1854 Map Room staff became draughtsmen of large diagram maps for lectures; until 1877 external draughtsmen were generally employed to copy or produce the RGS’s published illustrations. Manuscript itineraries, astronomical observations for position-fixing, sketch maps (some derived from ‘native’ or indigenous sources), and printed items (Admiralty charts, War Office maps) were bases for compiling maps and large diagrams. Survey Department Egypt’s Capt. H.G. Lyons sent 6200 copies of the Nile basin orographical map for his *Journal* article (Lyons. Geogr J 32 (5):[449]–480, 1908). Compilation rules (data sources noted), toponymy (transliteration, Romanisation and orthography), lettering styles, and technical processes (steel, then quill, pens to scribing) evolved. To recoup its expenses the Society hired out or loaned its illustrations and artefacts to other events and institutions (Herbert. The Royal Geographical Society. In: Monmonier (ed) The history of cartography, vol 6. University of Chicago Press, Chicago, pp 1371–1375, 2015).

## 1 First Lecture Meeting, 22 November 1830: The Hereford *mappamundi*

The Royal Geographical Society of London (founded 1830, Royal Charter 1859) aimed to disseminate geographical information in the form of maps. This was demonstrated at its first official evening lecture meeting on Monday 22 November by transporting, from Hereford Cathedral to London, its large manuscript *mappamundi* of ca. 1290–1310 by Richard of Holdingham (Fig. 1). Antiquary, topographic historian and RGS founder member, J. Britton, opined it “as one of the earliest specimens of British Map-Drawing, (being supposed as early as the twelfth century), and consequently an interesting link in the history of what may be

F. Herbert (✉)
RGS-IBG (retired), London, UK
e-mail: Francis443herbert@btinternet.com

 
M. Altić et al. (eds.), *Dissemination of Cartographic Knowledge*, Lecture Notes in Geoinformation and Cartography, DOI 10.1007/978-3-319-61515-8_9

**Fig. 1** Hereford *mappamundi* (c. 1300): facsimile MS copy by T. Ballard, 1831; British Isles ('Hibernia' within outer border) and Europe section (courtesy of RGS-IBG, London)

called the written character of Geography" (Britton 1830; Herbert 1986). At December meetings the Council sought, from RGS map engraver members J. Gardner and J. Arrowsmith, advice and quotations for "the probable expence [!] of copying and engraving the Hereford map in different styles and sizes", and "of having an exact Fac-simile made . . . for the use of the Society, and for preservation by it." Capt. [F.] Beaufort's 18 December report to Council stated: "a Fac-simile of the same size could not be procured in London under from £25 to £50 . . . [and] that Mr [J.] Biddulph should be authorized to have it exactly copied on the spot, at an expence not exceeding £25, exclusive of the materials employed." This copy was commissioned from artist Thomas Ballard of Ledbury, during the winter of 1830–31, his £25 cheque dated 16 May 1831 (Herbert 1999–2001+).

## 2 Draughtsmen and Engravers/Printers for the RGS from 1831

The Society initially met in the Horticultural Society rooms, 21 Regent Street; from 1839 to 1854, at the publishing house of Rivington, 3 Waterloo Place, Pall Mall (Mill 1930). Meetings discussed international contemporary and historical subjects: geography, hydrography, meteorology, surveying, cartography, exploration and travel. Source materials for meetings were usually manuscript illustrations, maps, narrative descriptions and observation logs, such as resulted in the map 'Marocco, from Observations in 1830', listing 'Geographical Positions' and 'Authorities', by 'Mr. [John] Washington R.N.' (Washington 1831). ). On 15 December 1832 the RGS Council considered "Candidates for the situation of Librarian and Draughtsman to the Society, and unanimously appointed Mr Charles Bradbury . . ." (cf Worms and Baynton-Williams 2011: 105); this initial 'in-house' preparation of illustrations lasted until Bradbury's resignation. Also recorded was: "an application from the Editor of the Nautical Magazine to be allowed the use of one or two of the Society's plates for his publication, which was assented to on condition that they be returned in good condition . . . and without erasures being made . . ." (Council Minutes Book, RGS-IBG Archives). Maps for the *Journal of the Royal Geographical Society* and, from 1855, *Proceedings of the RGS*, might occasionally be sought or offered from elsewhere, lessening the RGS's initial expense of commissioning a new copper-plates or lithographic stone. An example is John Arrowsmith's copper-plate 'Map of the Kawára, or Quorra' engraved originally for *The Friend of Africa* (London: J.W. Parker, 25 February 1841, 4(1)): Capt. Becroft's exploratory information was added to the re-titled map as 'Africa, The lower course of the River Kawara 1841' (Becroft 1841). Maps, charts, topographical views, diagrams and tables in copper, lithography or wood for the *Journal* from Vol. 1 ([September] 1831), and the *Proceedings* to 1878, were entrusted to external draughtsmen-cartographers, engravers and printers. The Council discussed merits of using the relatively new, but cheaper, planographic (lithographic) *versus* centuries-old intaglio (copper- or steel-plate) techniques (Verner 1971:5–6): comparative speed, simplicity, ease of corrections, number of printed impressions, costs of optional additions (relief representation; hand or printed colour). Sheets larger than 8vo format required folding. Early *Journal* illustrations by London lithographic craftsmen were not always of good quality. Lithographic colour-printed registration improved and, later, lithographic transfers of original intaglio engravings appeared (Pearson [Cook] 1978/1980; Mumford 1999).

In approximate chronological order, the *Journal*, *Proceedings* (etc.) involved, besides those mentioned (Verner 1971), the external individuals or firms: J. & C. Walker (*via* RGS founder-member John Barrow, Admiralty), J. Shury (bankrupt, June 1831), C. Bradbury (RGS Librarian; lithographer, 1832–35), W. Day (lithographic printer), Day & [L.] Haghe, G. Scharf (lithographer), C. Hullmandel (lithographic printer), A. Findlay, A.G. Findlay, W. Hughes, A. Petermann (from Germany *via* Edinburgh to London; latter as lithographer/printer), Hullmandel &

Walton (lithographic printer), Stanford's Geographical Establishment, J.W. Lowry, J. Bartholomew (Edinburgh), E. Weller (succeeded by son, F.S. Weller), E.G. Ravenstein, Malby & Sons, W. & A.K. Johnston (Edinburgh, from 1869 also in London), G. Philip & Son (Liverpool & London) and Walker & Boutall. The RGS's publisher's imprint of John Murray, Albemarle Street, London, normally appears. Admiralty, Ordnance Survey (Royal Engineers), Survey of India, Survey of Egypt, War Office (Intelligence Department) officers, Colonial and Foreign Office officials were often RGS Members or Fellows, sitting on Committees or Council. Mutual exchanges of source material such as MS maps, printing plates or stones and impressions, featured regularly.

## 3 Map Room, 1854; Survey Instruction; Diagram Maps for Meetings

In 1854 the RGS moved to 15 Whitehall Place, receiving an annual government grant on condition that this 'Map Office of the Nation' should be freely accessible to non-members. A Map Room was created with geographer and established London map-seller, T.W. Saunders, as Map Curator until 1857. Commander C. George RN succeeded as Curator of Maps and Instruments. The RGS loaned scientific instruments (those from 27 March 1861 to July 1936 are recorded in the manuscript log-book in the Society's Archives' [*List of*] *Instruments lent to travellers*); and advised intending travellers from at least 1854 (RGS *Hints to travellers*). Also in the Society's Archives is a manuscript *Royal Geographical Society Catalogue of instruments*: items acquired from December 1869 (e.g. Cary [case of] 'Drawing Instruments'), with a section (pp. 40–118) recording 'History of Instruments'. Instruments and instruction were given in order that resulting observations might provide better data for map compilation—whether through external or 'in-house' cartographers—and thus enhance the Society's integrity. From the 1850s scientific instrument manufacturers regularly advertised in RGS publications, careful to mention their connections: such as L.P. Casella (addressed specifically to Fellows and naming Livingstone) and Cary & Porter. George's 1877 resignation letter mentioned that he had given survey instruction in his "spare time" and had gained a "first Class Medal from the Committee of the Paris Exhibition" (i.e. Société de géographie, Congrès international des Sciences géographiques, August 1875) for an artificial horizon and an improved barometer through his 1868 & 1873 British Patents respectively (George 1877); both sides of his medal were illustrated above Cary & Porter's advertisements in *J of RGS*, vols 48–50 (1878–80), and references to George in Cary's *Geographical J* advertisements to at least February 1902.

The RGS's 14 December 1854 Map Committee recorded that "Mr. [F.] Galton made some remarks to the effect that Maps might be hung in the Meeting room shewing the progress of discovery, and the advance of the several expeditions"; the Council authorised production of large diagram maps. The Society's 'Committee on Instruments' 25 April 1860 meeting resolved action on certain relevant items: '28. Case of wooden ruling cases. Retained for Map-Room'; '12. Pentograph [!]. Mr. George to report upon.'; '10. Brass triangular Chartometer. To be retained for use in Map Room'; and '4. [. . .] Map maker's rule. To be retained.' From 1858 to 1868 meetings were held in nearby Old Burlington House, from 1870 to 1920 in larger-capacity venues such as University of London's and Royal Society's shared new lecture theatre in Burlington Gardens, or in the St James' Hall in Piccadilly.

In 1871 the Society moved to a refurbished 1 Savile Row (Vigo Street and Burlington Gardens junction). George (1877) commented: "The present large sized Diagrams for the Evening Meeting were originally projected by me, and have been drawn under my direction to the present time." Prepared by Map Room staff or by external map-maker Weller they included: Asia, South America, Malay Archipelago and the world (Weller, 1860–67, 1867); and Australia and Africa (George, 1862, 1863–64). The RGS's 'Committee on Maps' record that Weller was "directed to furnish the Committee with a list of the authorities he intends using in the construction of the Map and a list of the names of places rivers mountains &[ca]. that he intends entering on the Map"; and that "The size of the letters and the quality of the paper was determined on." He was paid, on 12 March 1867, £60 for the 'Diagram of Asia', £40 for 'Diagram of S. America'. Large diagrams taken to outside lecture rooms sometimes added expense: "Three waterproof covers to protect Diagrams &ca. from rain during their conveyance to and from Burlington House . . . at a cost of fifteen shillings" (RGS Finance Committee, 5 February 1866). Regular payments for 'Steel pens', 'Cloth and drawing paper', 'Pins for Diagrams', 'Hoisting apparatus for Diag[ram]' (University of London), 'Colours', 'Putting up maps &c. at St. James's Hall', and 'Mounting cloth' from London-based suppliers are recorded in RGS Cash Books. From May 1858 to 1878 payments were made, for example, to: Stagg & Mantle ('Map-mounters'), [T.] Baker ('Cloth for mounting'), H. Gould (late Cary) ('Instruments for Map room'), S. & C. Harding ('Drawing paper'), both G. Myers & Sons and Turner & Co. ('Hoisting apparatus'), W.F. Stanley ('Pear wood curves'), Gould & Porter ('Brushes &c. for Map Room'), and A.E. Ball ('Touching up diagrams'). The 1877–78 Council report noted "The large Maps of the Society have been lent on several occasions during the past year for . . . illustrating geographical lectures in different parts of the kingdom . . . and the general public have made frequent use of the Society's collection of Maps and Charts, particular interest . . . shown in the series of Maps now exhibited on screens in the Map Room."

RGS Assistant Map Curator A. Keith Johnston jr, from April 1872 to October 1873, doubtless drew some of these; employed from 15 November 1873, W.J. Turner certainly did—as promised in a 11 November 1873 letter of recommendation from W.H. Bullock (afterwards: Hall):

> Mr. W.H. (Bullock) Hall has much pleasure in recommending Mr. W.J. Turner, as an intelligent worker on diagrams. Mr. Hall has confidence, from experience of his work, that Mr. Turner will prove a useful addition to the Staff of the Royal Geographical Society.

Succeeding George was another ex-Navy man, J. Coles FRAS (additionally, in 1879, as Instructor of Surveying). On his retirement in 1900 RGS Council, on 2 June, passed a 'Resolution' commenting: "But the chief claim of Mr. Coles to the gratitude of the Society is the system of instruction in map drawing and practical astronomy that he conducted . . . The object . . . was to enable travellers to lay down their routes correctly, to fix their positions, and to draw maps." From June 1878 Coles' assistant, E.A. Reeves, improved his map draughtsmanship for the "large hand-drawn wall-diagram maps, measuring at times from twelve to twenty feet . . . These . . . were, in the early days, also used to illustrate papers read at the Geographical Section of the British Association . . . They have now been superseded by lantern-slide maps. During the transition period we had hand-maps distributed at the meetings, with a large general diagram map hung up by the side of the lantern screen to show the locality dealt with" (Reeves [1933]). Map Curator from 1901 and, from 1902, also Instructor in Surveying, Reeves supervised the draughtsmen in their 'Map Drawing Room'. His skill became useful for the occasional MS map like that of the Caucasus, prepared ("in my spare time") and printed by Stanfords, for D. Freshfield's *Caucasus* book (Freshfield 1896). Reeves' genius for instrument improvement or invention is evident from those eleven relevant British Patents granted him from 1888 ('. . . for dividing a straight line into equal parts and measuring any portion of such line') to 1915 ('. . . relating to maps, charts, and the like and demonstration apparatus for use in connection therewith').

## 4 'In-House' Map Compilation and Draughtsmen from 1877

Articles, if recommended for publication with maps, had the costs for inclusion of an author's re-worked holograph or specially-commissioned map sought from external engravers and printers. It was not always insisted that the original ought to be returned to the Society. A May 1877 letter to RGS President Sir Rutherford Alcock from W.J. Turner included several positive suggestions, of which one was:

> That all map-drawing (besides the diagrams for the evening meetings) be done upon the Society's premises, under the superintendence of the Map-Curator. This is a matter which I have long considered very essential, from the following reasons—first, when the maps which illustrate the papers in the Society's transactions are drawn elsewhere, others have an opportunity of publishing the information before it can be issued by the Society—again, the original is away, and so, not available if required in the office—they are also often retained a considerable time after the map is completed & frequently returned in a dirty state—and, lastly, the work would be done much more expeditiously, indeed, I believe the delay in the issue of the present "Journal" has arisen solely from the maps not being ready.

Turner's application next month to become Map Curator was unsuccessful, although recorded as a London-based 'Map & Writing Engraver' (Census of England 1861), then 'Geographical Draughtsman' in the 1871 Census onwards.

Sir Henry M. Vavasour Bt's letter to RGS Assistant Secretary began "Having been asked by Mr. W.J. Turner to write a few lines in his favour as a Candidate for the Office of Map Curator, I have much pleasure in doing so . . ." (Vavasour 1877). The 'Duties of the Map Curator' were:

1. To have charge of the collection of Maps, Diagrams, and Atlases, in the Society's collection. To maintain them in classified order for ready reference, and keep a press marked catalogue of the same. To enter all accessions, and fill up, with the least possible delay, forms of acknowledgement of donations, for the signature of the President or Secretary.
2. To ascertain the desiderata of the collection; to report on the maps that it may be desirable to acquire, and afford all necessary information to the Map Committee with respect to them; such as whether the maps are to be obtained by purchase or donation, the names of publishers, price, &c., and to draft letters of application when they have to be solicited from any public department, or person.
3. To receive all visitors to the Map-Room, and shew them whatever maps they require, or afford them information.
4. To furnish the Council with information respecting materials and cost of maps required for the Journal and Proceedings.
5. To draw up for each Evening Meeting a list of the donations received since the previous Meeting, and a report on [signed: Rutherford Alcock] || the business of the Map Department for each Monthly Meeting of the Map Committee; also a General Report for each Anniversary Meeting.
6. To compile and edit a catalogue for printing, of the Society's collection if required.
7. To have charge of the instruments belonging to the Society, and see that they are kept in good repair. Also to give advice to travellers regarding their outfit of instruments, and instruction in their use if required.
8. The hours of attendance are from 10.30 a.m. to 5 p.m.; on Saturdays to 2 p.m., and on days of Council and Committee Meetings, to the hour when such Meetings are terminated. Three quarters of an hour are allowed for lunch, except on Saturdays; to be arranged so that he is not absent at the same time as the Assistant Curator.
9. To superintend the hanging of the Diagrams at Evening Meetings, and to attend afterwards the Meetings.
10. He is to consider himself under the immediate direction of the Assistant Secretary, through whom his Reports to the Council and Committees are to be presented.

On 29 June Turner accepted instead the post of 'Map Draughtsman and Assistant Curator'. The RGS's 'Adjourned Map Committee' Minutes of 12 November 1877 record correspondence, and consequent actions taken, regarding his appointment:

> The Committee having before them, practically only one candidate for the appointment of Assistant curator, in the Maproom [!]; hesitate to make any recommendation for a permanent appointment to the Council; but, recommend that further enquiries should be made, and that meanwhile, Mr. Turner be appointed for three months, at the rate of £100 per annum. As assistance is urgently required, the Committee have empowered Mr. Bates, to employ Mr Turner in the Map room, on the terms above stated, subject to confirmation by the Council.

The conjunct leaf by the 8-item 'Duties of the Map Draughtsman and Assistant Curator', but here transcribed from Turner's signed copy, reads:

1. To draw all diagrams, required for the Evening Meetings, and General purposes of the Society.
2. To compile and draw Maps for the Journal and Proceedings, when considered necessary by the President and Secretary.
3. To aid the Assistant Secretary in reading proof sheets, and correcting proofs of Maps, of the Journal and Proceedings.
4. To afford special geographical information whenever required by the Council or by travellers.
5. The hours of attendance are from 10.30 A.M. to 5 P.M, on Saturdays to 2 P.M., and on days of Council and Committee Meetings to the hour when such Meetings are terminated. Three quarters of an hour are allowed for lunch, except on Saturdays; to be arranged so that he is not absent at the same time as the Curator.
6. To be in attendance in the Map-Room, and perform the duties of the Curator, at all times when the latter is absent or when such attendance is considered necessary by the Assistant Secretary.
7. To aid the Curator in hanging the || diagrams for the Evening Meetings, and to act as pointer for the papers read at the Meetings.
8. He is to consider himself under the general direction and supervision of the Assistant Secretary.

[signed:] William John Turner | 29th June 1877.

A copy, similar except for occasional punctuation and capitalisation differences, in Committee Minutes Book (*supra*) is signed: Rutherford Alcock | Chairman.

From May 1878, in a similar letter to RGS Council repeating his 1877 advice on in-house map drawing, Turner was re-designated as the Society's first official in-house 'chief draughtsman'. With a professionally-trained assistant, H.A. Milne, these appointments were undoubtedly reflected in the increasing number of maps issued in the *Proceedings of the RGS* in the immediately ensuing years: 26 in 1878 to 39 in 1881. It is interesting to note that at least one of Turner's maps printed for the *Proceedings* may have gained his employment with the Edinburgh firm of W. & A.K. Johnston. In February 1881, the year when †A.K. Johnston sr's younger brother, Thomas Brumby Johnston, proposed Turner for RGS Fellowship, Turner resigned from the RGS and went to work for Johnston (Fig. 2).

Fortunately, assistant draughtsman Milne's 7 March 1881 application letter notes his professional connection since 1879:

I beg to tender my services for the position of Draughtsman to the Royal Geographical Society, which will become vacant consequent on the resignation of Mr. Turner, . . . I have since May 1879 to the present time had the pleasure of assisting Mr. Turner, and to the best of my knowledge have carried out the work, entrusted to my care in a satisfactory manner; during that period, I have constructed upwards of Twenty diagrams for the Society and [the] Map of King William Land, for Proceedings illustrating Mr. Clements Markham's paper. I also have acquired a perfect knowledge of the routine of the work appertaining to the Society, and trust that my previous experience under a practical map maker (Mr. J. Bartholomew) [1] and Map Publishers' (Messrs. Letts) [2] whose Testimonials I enclose, will be deemed by you of some value in the position I seek. [. . .] I enclose specimens of the class of work I have been engaged upon for your inspection.

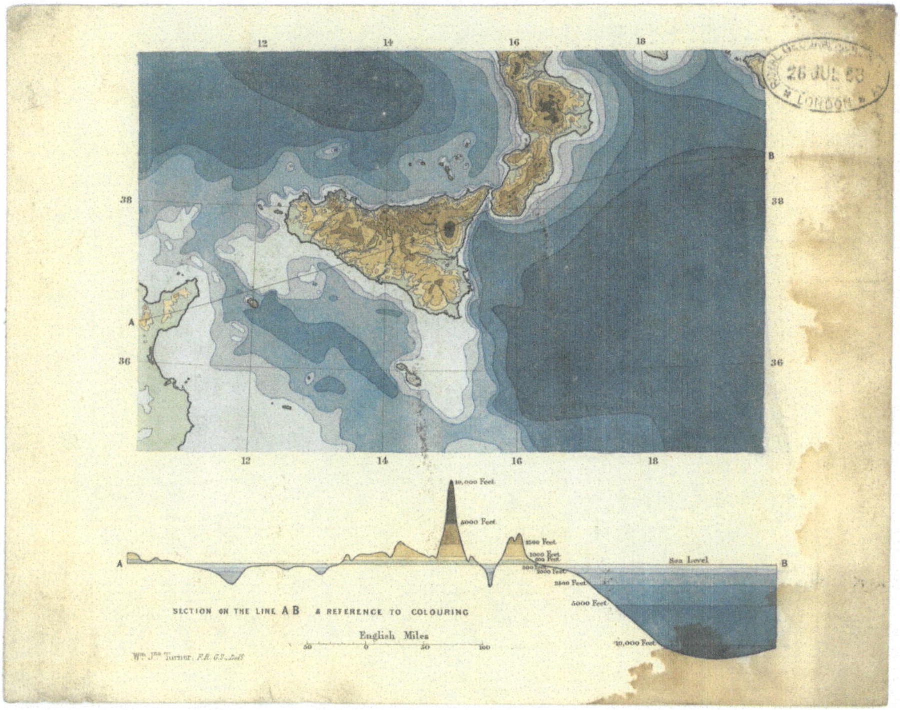


**Fig. 2** Bathy-orographical map of Sicily region: untitled holograph by W.J. Turner FRGS (c. 1881–88) (courtesy of RGS-IBG, London)

[1] I have employed Mr. H.A. Milne as Geographical Draughtsman for some years and have been quite satisfied with his work. He is a very neat and careful draughtsman and I have always found his work correct. I have pleasure in recommending him for the situation he is applying for | [signed] John Bartholomew (from 31 Chambers Street, Edinburgh, 7 March 1881; the letter is not in his hand)
[2] Mr Milne | Dear Sir, I have very much pleasure in recording my testimony for what it may be worth as to the excellence of your map drawing, punctuality and attention, and shall be pleased to learn you have been successful in your application | I am | Yours faithfully | for Letts Son & Co Lim[ited] | T. Letts (from 33 King William Street, London EC, 5 March 1881; the letter is not in his hand) (Milne 1881)

Milne was accepted as temporary assistant, but Turner's position was given to H. Scharbau. Sharbau (a naturalised British subject with occasional anglicised spelling) continued as draughtsman for RGS publications and for diagram maps; even his son, Frederick, was paid on 21 December 1887 for "(Assist[an]ce prepar [in]g diag[ram]s &c.)". The 'Map drawing Account' of 3 February 1886 calls for payment to T. Culleton for 'Stencil plates'—tools aiding a faster and consistent manner of filling in blocks of hand-colouring ('Culleton's Heraldic Office . . . Engravers to the Royal Geographical Society' appears in *Geographical Journal* 'Advertisements' pages from November 1893, 2(5) onwards). The Society had a

**Fig. 3** 'Tibet and the surrounding regions. Compiled from the latest information': holograph by H. Sharbau (1894, with revisions to 1906), for *Geogr J*, July 1894, 4(1) and for Chandra Das' 1902 & 1904 books (courtesy of RGS-IBG, London)

policy problem in 1898 with the folding map for inclusion in the second volume of Savage Landor's *In the forbidden land* published that year by Heinemann. Writing to RGS Secretary J. Scott Keltie on 9 August, Sharbau informs: "I expect to finish Landor's map tomorrow (Wednesday). After seeing Sir Thos. H[oldich']s report with which I quite agree, the shortest way to get rid of this unpleasant business appears to be by your presenting him [Landor] with the map under the condition that he does neither mention nor publish that it was done at the R.G.S. But I very much question we can trust him or his publisher. The temptation to create a controversy between the R.G.S. and the publisher is great & would cause a greater sale of the book. I shall be glad to receive your direction." (Sharbau 1898). This subsequently anonymous map—'South-western Tibet reduced from the survey of A.H. Savage Landor, 1897'—showing his route and return journey was 'Lithographed by W. & A.K. Johnston, Edinburgh & London.' (Fig. 3).

## 5 Hand Maps 1885–1914[1]

On 2 December 1885 a Special Meeting of the RGS Library and Map Committee was held. Its minutes record: "Proposed by Mr. Delmar Morgan, that small maps be prepared for distribution at the [Monday] Meetings [. . .] it was resolved to recommend that small maps of the kind proposed . . . should be supplied during the present Session, for the use of Fellows at the Evening Meetings. The maps to be on stiff white paper. The estimated cost is thirty shillings [£1.50] a Meeting, or £20 for the Session. The stamp or name of the Royal Geographical Society to be inserted on the face of the map." Engraving and printing costs of these hand maps, in addition to the regular maps for the Society's *Journal*, *Proceedings*, and *Supplementary Papers*, are recorded in Ledgers in the Society's Archives.

But the first recorded use of a hand map—produced before the 2 December official proposal—seems to be that for the 30 November 1885 Evening Meeting, to accompany W. Montagu Kerr's paper 'A journey from Cape Town overland to Lake Nyasa'; its published version in *Proceedings of the Royal Geographical Society and monthly record of geography, New Series*, vol. 8 no. 2, February 1886, carries the name of W. & A.K. Johnston (see footnote 1).

Compiled in the Map Drawing Room, the hand maps were then customarily 'finished' and printed externally, being treated as preliminary or proof copies; corrected, revised, or—in rare cases—redrawn completely, they were next published in the Society's *Proceedings NS*, *Geographical Journal*, or *Supplementary Papers*. Not officially on Map Drawing Room staff, but contributing to making hand maps, was a Sharbau contemporary, German-born B.V. Darbishire, who drew several between 1892 and 1895. A typical example of differences between an ephemeral hand map and its published state is explained in [Herbert] 2005; another hand map (the first of three states) is reproduced here (Fig. 4).

The Finance Committee Meeting of 3 May 1886 records a cheque authorised to Edward Weller in payment for his "Hand maps for Meetings [£]10.6.3" [ca £10.31)]—about half the estimated sum permitted by the 2 December 1885 Library and Map Committee Meeting. Glass lantern slides (3 × 3″) maps added to draughtsmen's duties. In Reeves' 'Preface' to his lectures on *Maps and map-making* (Reeves 1910) he remarks: "Most of the illustrations, diagrams, and maps . . . were shown as lantern slides when the lectures were given"; and "In addition to the lantern slides, the interest . . . was considerably increased by the exhibition of old and new instruments, and examples of British and foreign cartographical production of various dates and styles, which were shown on screens round the Hall." Reeves thanks "Messrs. Cary & Porter, and Messrs. Casella & Co."—familiar to the RGS as instrument suppliers and advertisers in its publications. Of the 134 figures and five plates in Reeves' book 91 and two respectively survive today (2017) as his

[1]The Johnston's printed catalogues of their products in the 1890s, used the term 'hand map' in a different sense: for single-sheet maps ranging from ca 14 × 17 in. (ca 35 × 43 cm) in size, often of a simple nature and/or for educational purposes, for sale to the general public or schools.

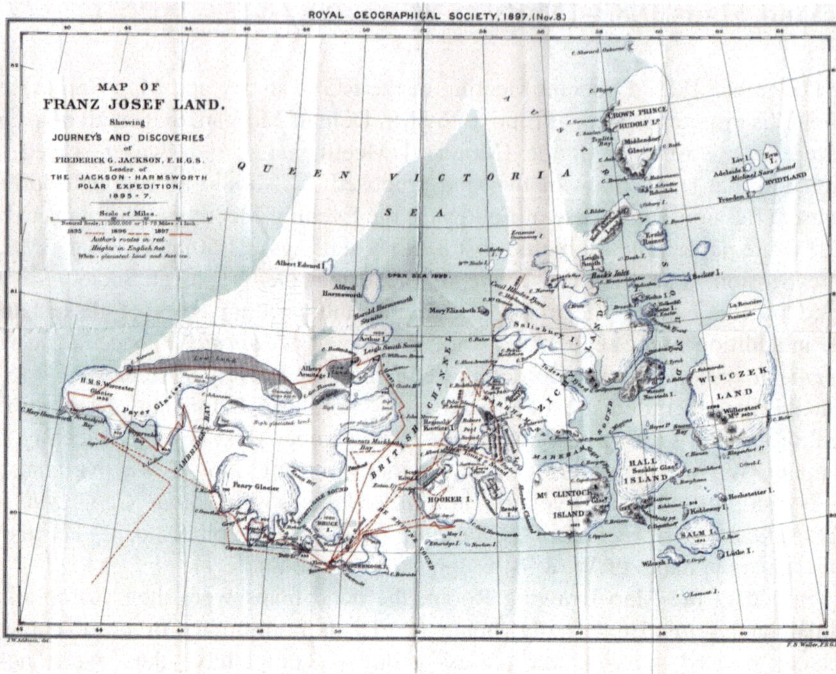


**Fig. 4** 'Map of Franz Josef Land. Showing journey's [!] and discoveries . . . 1895–7' by J.W. Addison: printed hand map for RGS Meeting, 8 Nov. 1897; revised (omitting Addison's name and Meeting date) for *Geogr J*, February 1898, 11(2), and for Jackson's 1899 book (author's collection, London)

1909 lecture lantern slides. The Drawing Office manuscript Log-Book's entry no. 751, under April 1914, lists 'handmap' of 'Nigeria—The Eket District' to accompany P.A. Talbot's 27 April talk (Talbot 1913); a lantern slide version (LS27) is in the Society's Picture Library. The MS for Talbot's 'handmap' was passed from the Secretary to draughtsman [H.F.] Milne, whose compilation was sent to the lithographers Martin, Hood for processing; in May their proof was returned to RGS, checked, approved and sent to printers Clowes as 'Southern provinces of Nigeria. The Eket District from a survey by P. Amaury Talbot . . . 1913' (Talbot 1914: 286)

Resigning from Johnstons in December 1886 Turner returned to London and advertised first in *Proceedings RGS*, *NS*, February 1887, 9(2), p. [ii] as: "Formerly Geographical Draughtsman at the Royal Geographical Society, And for the last five years in charge of the Geographical Department of Messrs. W. & A.K. Johnston's establishment, 8, Warwick Street, Regent Street, London, W . . ." He was paid by RGS for making 'diagrams' (e.g. Australia, British North America), 'constructing wall maps, &c.' and 'Contoured maps' from May 1887 to May 1889. Sharbau's assistant, H.A. Milne, from May 1887 suffered increasingly from a stroke and had to resign that autumn. RGS staff, some officers and Fellows, acknowledged his work by contributions for him and his family. A manuscript 'Subscription List' begins with J. Scott Keltie's invitation:

> Assistance is sorely needed by Mr. H.A. Milne, late Assistant Draughtsman to the Royal Geographical Society, who was stricken down by paralysis in the summer of 1887, and has no prospect of recovery. He has a wife and young family and is now in great distress, having come to the end of his savings. He was in the Society's service from January 1885 to July 1887. Temporarily employed from 1880.

The list (Keltie 1887) is headed by Coles in his own hand, followed—mostly in Keltie's hand—by, amongst others, Keltie, E. Delmar Morgan, Sir F.J. Goldsmid, John Bolton [Stanford's], W.J. Turner, H. Scharbau, Admiral John Dalrymple Hay (own hand), F.S. Weller, E.G. Ravenstein (own hand, giving 1 Guinea), Douglas W. Freshfield and W.J.L. Wharton (own hand): 16 individuals each contributed £1, with the RGS's inked institutional stamp against £10, brings the total relief sent to Milne to £26.1.0.

## 6 Map Compilation: Dalton's 1892 Proposals; Disposal of RGS's Plates, 1894

Another upheaval in the RGS's publishing programme was ceasing the *Proceedings* (*New Series*) in December 1892, and deciding on the format and contents of a newly-designed, but still monthly, periodical, from January 1893 to be called *The Geographical Journal*. Lt-Col. J.C. Dalton drew up a critical *Memorandum* suggesting the Society's maps should note sources used in their compilation:

> I should like to be allowed to draw the attention of the Council of the Society to a subject which I have long felt to be an important one, and of which I have recently been reminded by those officers of the Intelligence Department whose duty it is to compile maps and deal with geographical questions. The subject I allude to is that of the maps which appear from time to time in the 'Proceedings' of the R.G.S., generally to illustrate the journeys of travellers and the geographical papers which are published in the 'Proceedings' . . . These maps, which are doubtless carefully compiled, often fail, I would submit, to carry the weight which they ought to carry with those whose duty it is to compile maps, because they do not, as a general rule, bear upon the face of them, nor are they usually accompanied by, a sufficiently full explanation of the authority for the map, and of the sources from which the compiler has taken his details; also his reasons for either accepting or rejecting certain positions. The system adopted in Petermann's 'Mitteilungen' is, I venture to think, a sound and correct one, and I would refer as examples to the following numbers of the 38th Band of 1892, viz.:- [. . .]

Dalton's models are Dr Ruge, the Justus Perthes firm, Ravenstein's 'Eastern Africa', Kiepert, Habenicht, and the "brilliant example of . . . Mr.Curzon's map of Persia" (Bederman 1992; Ala'i 2005). Dalton recommended, too, that "Special rules for the style of writing should be drawn up by a small Committee and adopted as the R.G.S. rules [to] define the different styles of type which would denote Countries, Provinces, Tribal divisions . . . towns (according to importance and population), rivers, mountains, . . . for showing boundaries and other conventional signs . . ." He opines on colours for printed maps ('Details and heights' = black; 'Water' = blue; 'Hills' = brown; 'Travellers' routes' = red), and questions "as to making the R.G.S. Maps copyright, placing them on sale through an agent, &c.," (Dalton 1892 ; Herbert 1983). The 'Report of Meeting of Special Sub-Committee delegated to the R.G.S. Library and Map Committee to enquire into and report on the proposals [of Dalton's *Memorandum*]' of 13 December (RGS Report 1892) records reactions. Present were Gen. J.T. Walker CB (Chairman) E.G. Ravenstein, Dalton and "Mr. Darbishire, the Cartographer of the R.G.S." (Capt. W.J.L. Wharton RN was "Unavoidably absent"). Amongst the recommendations "unanimously agreed" and submitted to the Library and Map Committee were (here abbreviated): a map compiled for the *Journal* should have "an explanatory leaflet or paragraph containing a list of the documents, observations, &c., on which the map is based"; "travellers' routes . . . should . . . adhere as closely as possible to the original work"; "the Society should . . . occasionally publish standard maps of interesting regions"; "Geographical details for which there is not satisfactory authority should as a rule be shown by broken . . . lines"; "Colour printing should be resorted to as a general rule . . ."; and "That as far as is possible a uniform style be adopted in the writing on the maps." Was this the incentive for Sharbau's 'Tibet and the surrounding regions' map with Gen. Walker's accompanying 'Note' in the *Geographical Journal* (Walker 1894)?

The majority of the RGS's maps from external craftsmen were commissioned; the resultant copper plates and lithographic stones considered as its property. On 10 January 1871 H.W. Bates (RGS Secretary) wrote similar letters to Arrowsmith, Stanford and Weller asking to know "the number of map-plates & stones belonging to us in your possession & mention the years of the 'Journal' to which they belong". The number accumulated in Savile Row by November 1893 prompted Weller's son, Francis Sidney, to advise selling the older plates. During summer 1894 a specially-formed 'Copper Plate Committee' was formed of Lt-Col. J C. Dalton, E.G. Ravenstein, Col. J.K. Trotter, and J.W. Malby. Its minute of 8 October ran:

> The Committee appointed for the purpose of examining and reporting on the Copper plates in the Map Room of the Society have carefully gone through the plates in accordance with a list drawn up by Mr. Coles. They have marked off some fifty plates which they think should be retained by the Society—at all events for the present. The remainder they recommend should be sold as old Copper. [. . .] The Committee were greatly assisted in their work by the preliminary careful examination of the plates and the maps corresponding to them which Mr. Coles & Mr. Reeves carried out, and the Committee were much pleased at the admirable manner in which the inventory of the maps & plates is kept.

An 'addendum' in the hand of, and signed by, Coles reads: "I concur in the proposals regarding the disposal of the Plates". The list numbered at least 359 plates, this being the highest number amongst the fifty (covering 46 maps) for retention. These included works from 1837 to 1894 of Arrowsmith, Findlay, E. Weller, Saunders, Stanford, F.S. Weller, Turner, Turner & Shawe, and Johnston. The successful bidders on 5 November 1894 for the rejected map copper plates were Messrs Malby & Sons (engravers and printers to the Admiralty's Hydrographic Office et al.).

As indicated earlier, original RGS maps were used ('re-cycled') by other academic or commercial societies or publishers. They usually retained the original RGS imprint with 'By permission of . . .', or substituted that of Longmans; Macmillan; Sampson Low, Marston; *et al.* An example is 'West Africa. Map illustrating a journey to Bontuku from Cape Coast made in 1888 by R. Austin Freeman' drawn by W. Shawe FRGS (although dated '1891'), lithographically printed by Turner & Shawe, for a similarly-titled article in *Supplementary Papers Royal Geographical Society*, 1893, vol. 3. Omitting the original RGS credits it was re-used as 'Map showing the distribution of vegetation on the Gold Coast and life in Ashanti and Jaman' for Freeman's *Travels and life in Ashanti and Jaman* (London: Constable, 1898; reprinted Cass, 1967). Reeves did similarly (Reeves 1910).

Around this period both the Weller firm and Ravenstein were regular map-makers for the volumes published by The Hakluyt Society. Ravenstein (appointed in 1888 as George Philip & Son's Geographer) was on The Hakluyt Society's Council, and also edited volumes from the 1890s onwards. Weller was a regular engraver for The Alpine Club's publications by the Longman firm—like The Hakluyt Society, another close 'partner' with often overlapping memberships with the RGS. The Hakluyt Society commissioned RGS map draughtsmen, too: Sharbau (1899/1900); J.W. Addison and G. Mackay (1904); H.F. Milne (1908–28); W.R. Rand (1931); and G.S. Holland (1931–55).

A case of an external (Admiralty Hydrographic Department) draughtsman specially employed was that of J.E. Tuttle. The RGS paid him from February 1908 to May 1910 or (for corrections) to October 1911 for compiling and drawing, at 1:1,000,000 scale for printed reduction to 1:2,000,000, the map 'South Peru and North Bolivia: including the rubber yielding Montaña'. Sir C.R. Markham initiated this through his three typescript leaves 'Proposal to the Council for the construction of a special Map' (June 1906), followed by a similar leaf, 'Proposed South American Map. Memorandum by Sir Clements Markham', from his home address on 30 April 1907. His paper's published form 'The Land of the Incas' states: "It will have been seen that the compilation and construction of the map has been a very difficult task, and reflects the highest credit on our map curator, Mr. Reeves, and on Mr. [F.] Batchelor, the draughtsman [*sic*], who has acquired a firm grasp of the various intricate points which have constantly arisen." (Markham 1910: 391). True to Dalton (1892)'s 'Memorandum' recommendations, although not the sole good example, is Reeves' contribution 'Note on map of South Peru and North Bolivia' on pp. 398–404; this includes sections on 'Material used in the construction of the map' and 'List of positions' (but cf Washington's 'Marocco' of 1830, etc.).

## 7 Technical Manuals Owned by H.A. and H.F. Milne

Donated to the RGS by family members are some text-books owned or used by Milne father & son: *An explanation of the gnomonic projection of the sphere* . . . by Augustus de Morgan (London: Baldwin & Cradock for SDUK, 1836); *Map and*

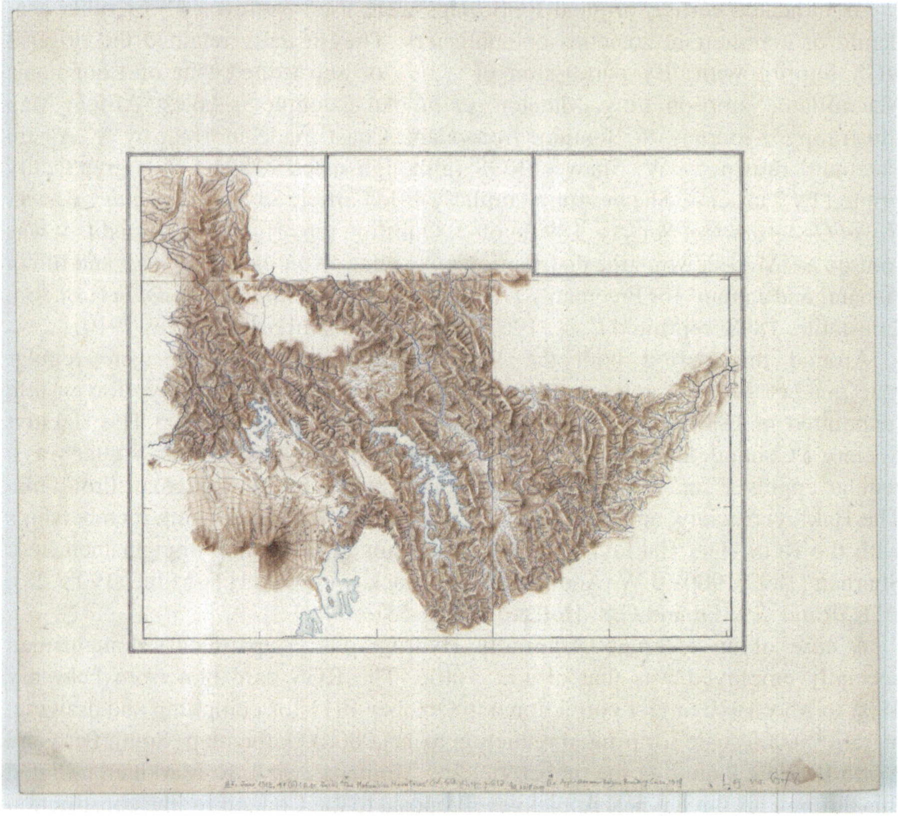

**Fig. 5** 'Survey of the Anglo-German-Belgian boundary region by the British Commission under the command of Capt. E.M. Jack R.E. 1911': untitled holograph by H.F. Milne of January 1913 onwards, for *Geogr J*, June 1913, 41(6) (courtesy of RGS-IBG, London)

*plan drawing* by C. Cooper King (London [etc.]: Cassell, Petter & Galpin, [1873]); *A complete epitome of practical navigation, and nautical astronomy* . . . by J.W. Norie . . . new edition . . . by W.H. Rosser (London : Norie & Wilson, 1889); *Text book military topography, Part I & II* (London : HMSO for War Office, 1898); *Military sketching made easy, and military maps explained* by Maj.-Gen. H.D. Hutchinson, 6th edition, revised . . . by Captain R.F. Pearson (London [etc.]: Gale & Polden, [1906]); and *Graphical methods of plotting from air photographs* by Lt.-Col. L.N.F.I. King (London [etc.]: HMSO for War Office, 1925) (Fig. 5).

## 8 External Finishers or Processors of RGS Maps

The original drawings of maps made by the compilers were copied by the lithographic draughtsmen, and were not always complete when the copying began and usually lacked

> nicety of finish, which was left to the litho-artist. The essentials for the compiler were to letter tidily (but without finish), indicating the style, size and positioning of names, and the quality of line and nature of symbols to be used to depict the features. A layout of the title, scales, key and notes was sometimes made independently, to be set in the most convenient corner when the border or frame was drawn. Carefully painted and highly imaginative hill-shading models were prepared in sepia washes on Whatman paper, or were drawn in lithographic crayon on grained paper, which the litho-draughtsman would chalk, or directly transfer, on to the grained stone or plate. The compiler governed the map's scale, limits, content, arrangement and general style, but the details and the appearance of the printed map were the work of, and had the stylistic characteristics of, the litho-artist. As photolithography became general, the balance of the work shifted, and the basic drawings were completed in every detail by the Society's compiler-draughtsmen. This added considerably to the work by the more careful finish required, but this was eased by the advantage of being able to prepare the drawings at larger than reproduction scale—usually about one-third. (Holland 1980)

Predominantly London-based lithographic draughtsmen, printers, photo-etchers, photo-engravers and block-makers used by the RGS from about October 1901 to WW2 are recorded in the former Map Drawing Office's MS log-book (Royal Geographical Society, Map Drawing Office 1900). These were (earliest mentions recorded): Manly; Danielsson (November 1901); F.S. Weller (March 1902 [but "work too bad to accept" so passed to Manly]); W. & A.K. Johnston (May 1902); Bale and Danielsson (February 1902 [this firm also prepared images for The Geographical Association]); Stanford (July 1902); Weller & Graham (February 1903); Stanford [i.e. William Stanford, Oxford Geographical Institute] (December 1903); Philip & Son (May 1904); Justus Perthes [Gotha] (February 1905); Martin, Hood & Larkin (September 1905); Cattell (January 1907); Leonardson (February 1909); Bartholomew & Co. (July 1910); Indian Govt Svy Office (August 1914); Menpes Press (July 1916); Vitty (January 1919); Haycock, Cradle & Graham (November 1920); Vitty & Seaborne (December 1924); Siviter, Smith & Co. (August 1925); West & Partners (March 1927); Swain & Waterlow (December 1928); Waterlow & Sons (January 1929); Swain (April 1929); J. Swain & Sons (July 1929); Hudson & Kearns (August 1929);Strand Engraving Co. (December 1933); and Harrison & Co. [or: & Son(s)] (April 1939) (Fig. 6).

## 9 Hinks as Secretary: Drawing Office Re-organisation, 1913; The Great War, 1914

After appointment as RGS Assistant Secretary in 1913 A.R. Hinks began to reorganise and update the RGS Drawing Office. This procedure included the purchase of manuals, text books and reference books "on Surveying and cognate subjects" as "many of them are much worn and defaced by notes." His 23 October 1913 memorandum to RGS Honorary Secretaries is complemented by his holograph list of ca 42 desiderata (some items in multiple copies) to cost, over 2 years, about £35. The list has pricing and availability MS comments from Sifton, Praed &

**Fig. 6** 'Average annual number of mornings with snow-cover 1912–1940': holograph for G. Manley's article in *Geogr J*, June 1944, 103(6) (courtesy of RGS-IBG, London)

Co Ltd ('Mapsellers, Geographical and Military Booksellers & Stationers', 'Royal Geographical Society's publications') of The Map House, 67 St James's Street (Hinks 1913). In 1914 (when RGS moved to Lowther Lodge, 1 Kensington Gore), he was also busy arranging accommodation for War Office and Naval

Intelligence Division personnel, and into employing extra and temporary map- and gazetteer-compiling staff for the Great War effort (Heffernan 1995).

## 10 Calligraphy: Change to Quill Pens by Draughtsmen, 1926

Hinks, Secretary from 1915, improved RGS Map Drawing Office's style, for both production speed and aesthetic reasons, by sending, in 1926, three teenage recruits (C.E. Denny, G.S. Holland and W.R. Rand) for tuition in italic calligraphy and use of the quill pen. Having a good italic hand, perhaps he was aware of hydrographic publisher D.W. Kettle's *Pens, Ink, and Paper: a discourse upon The Caligraphic [!] Art*... (Kettle 1885/1886). Hinks's 'Note on style . . . required in sketch maps to illustrate books . . . and for blocks in text of Geographical Journal' begins:

> "Scarcely any books of travel are published nowadays with adequate maps, on account of the great expense of map draughtsmanship and reproduction. Such maps as are drawn . . . have the names typed, which is always unsatisfactory . . . and the old style . . . with names in Roman elaborately drawn and finished is impossibly expensive . . . and compels us to look for an italic or script that can be drawn sufficiently well by single strokes of the pen. I suggest that such a style might be based upon the best Dutch engraved maps of the 17th century, especially those of Hondius". He ends: "The style should be funded on the best results of the past, but should not slavishly imitate them . . . It is probably wrong . . . to aim at too great mechanical perfection of lettering, which takes an immense time . . . We should aim . . . [to] leave a certain amount of freedom to the draughtsman to adapt himself to the special qualities of each map . . . I should like to encourage junior draughtsmen to experiment in developing such a style from the best 17th and 18th century models." (Hinks 1925/1926?).

Hinks reiterated his thoughts on style and "the need of a machine for writing names on maps a good deal better than the dreadful typed names that economy too often demands" in his presidential address, 'The science and art of map-making', to Section E 'Geography' at the British Association on 27 August 1925 (Hinks 1926; Herbert 2010). The Second World War inevitably halted regular Drawing Office operations; members left for service. The last 'traditional' products were two 'stand-alone' maps for the British Council on which Denny and K.C. Jordan worked: sheet 1 'Europe and the Middle East' (in separate English language and Arabic script printings in 1941 and 1945 respectively), available as flat or folded in covers; sheets 2 'Far East' and 3 'North America to Britain' remain in outline (Hinks 1944/1947; Crone 1964). The RGS Map Drawing Office Archive is the repository of such materials. Developments are also recorded in a continuation of Mill (1930) by Crone (1955)—*Royal Geographical Society: a record 1931–55*.

From 1967 to 1984 the RGS Drawing Office operated—with men and women students from (then) Luton College of Technology and Oxford Polytechnic—annual periods of in-house training. Holland's successor, Anne Stanton from London School of Economics, introduced techniques and materials practised there. Students would have used Rotring and Graphos pens; then innovations of

**Fig. 7** Four selected pear-wood splines (or: railway curves, or radius curves) by '[William] Cary London'; 'Elliott Bro[ther]s 30 Strand [London]'; 'Casella London'; and 'Hudelo a [!] Paris F. Pomet Succ[esseu]r.' Two ebony parallel rulers with brass hinges: laid horizontally = 24″ (graduated in 10ths of 1″), by 'Cary London' with 'Royal Geographical Society No 4' (purchased in December 1901); laid vertically = 14″ (ungraduated), anonymous with 'R.G.S.' Another pear-wood curve (broken, like the Stanley example above) is by 'A.G. THORN[TON] MANCHES [TER]', with 'A R H[INKS]'. For more information on drawing instruments see illustrated book by Riches (2012)

Letraset, and of scribing—"using orange plastic film and working in negative with special tripods fitted with sapphire cutters" (Stanton 2017). The last was for production mainly of the RGS's 'stand-alone' (separate) maps: 'The highest part of Rockall Bank' and 'Socotra' (printed by Cook, Hammond & Kell Ltd, 1975 and 1978 respectively). For lettering, either Letraset was used or type was bought in from a film type-setter. Size changes of Drawing Office compilations to reproduction size for the *Geographical Journal* continued: although a pantograph was still available, a purchased 2nd-hand enlarging/reducing machine was used. Over 100 years of changes in style and techniques are evident from Turner's 'Explorers of Africa in modern times' ("Published by the African Exploration Committee of the Royal Geographical Society", printed by W. & A.K. Johnston, 1877), *via* Holland's 'The Mount Everest region' (1961 edition printed by Waterlow & Sons Ltd; [1973] edition by Cook, Hammond & Kell—also printers of the 1975 'Mount Everest Region' edition), N. St. J. Groom's compiled 'A sketch map of South West Arabia showing pre-Islamic archaeological sites' (printed by Cook, Hammond & Kell Ltd, 1976), the illustrated map 'Royal Geographical Society: history of exploration 1830–1980' researched and drawn by E.J. Hatch, J. Humpston & P. Hewitt-Taylor (printed by J. Bartholomew & Son Ltd, 1980), to 'The mountains of Central Asia' by RGS & Mount Everest Foundation (printed by Mapping & Charting Establishment RE, 1986) (Figs. 7, 8, and 9).

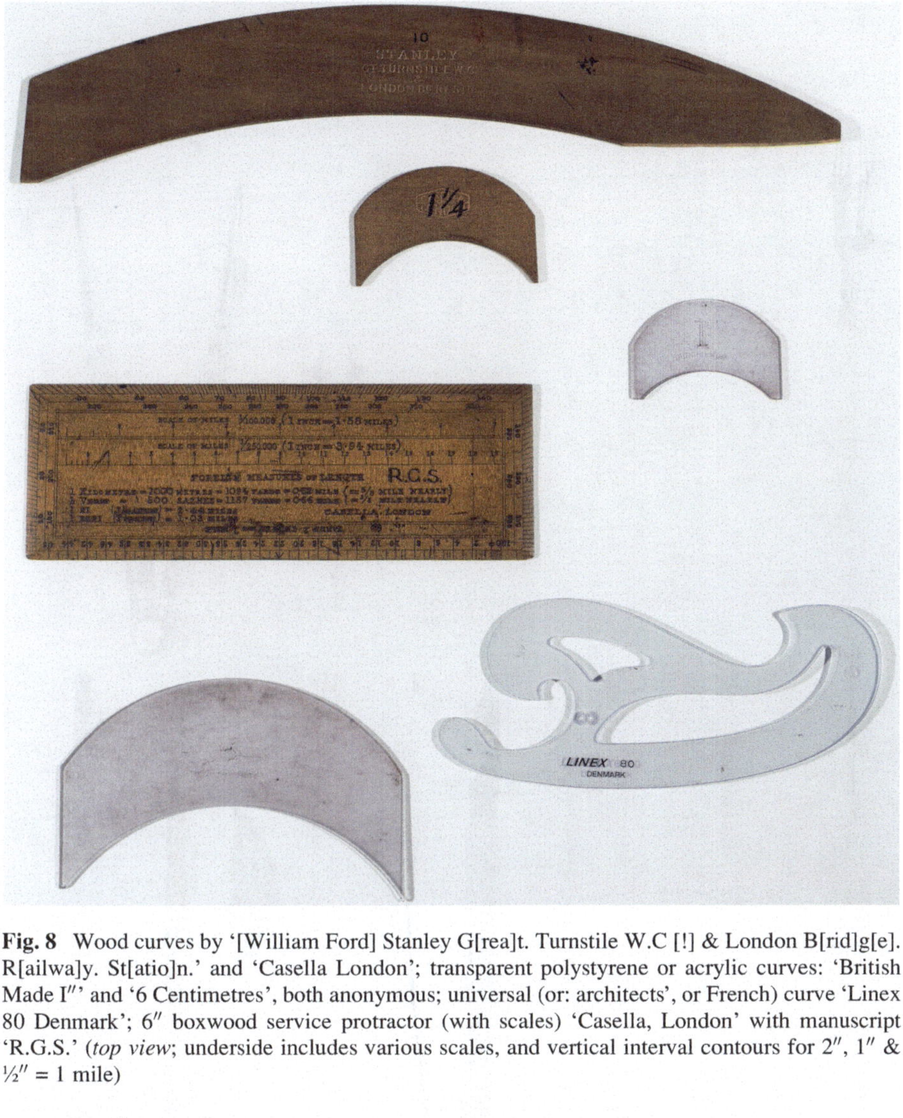


**Fig. 8** Wood curves by '[William Ford] Stanley G[rea]t. Turnstile W.C [!] & London B[rid]g[e]. R[ailwa]y. St[atio]n.' and 'Casella London'; transparent polystyrene or acrylic curves: 'British Made I″' and '6 Centimetres', both anonymous; universal (or: architects', or French) curve 'Linex 80 Denmark'; 6″ boxwood service protractor (with scales) 'Casella, London' with manuscript 'R.G.S.' (*top view*; underside includes various scales, and vertical interval contours for 2″, 1″ & ½″ = 1 mile)

## 11 Publicity and (Inter)national Exhibitions, 1904–99

The RGS exhibited examples of its explorers' holograph maps (or reproductions), recommended survey instruments, its Drawing Office manuscript compilations and published maps at national and international events. At the Geographical Association's 1904 travelling exhibitions (Geographical Association 1904) the Society displayed bathymetrical charts by J.G. Bartholomew and Sir J. Murray (from 1892 *Proceedings NS* and from 1899 *Geographical Journal* respectively), H.R. Mill's

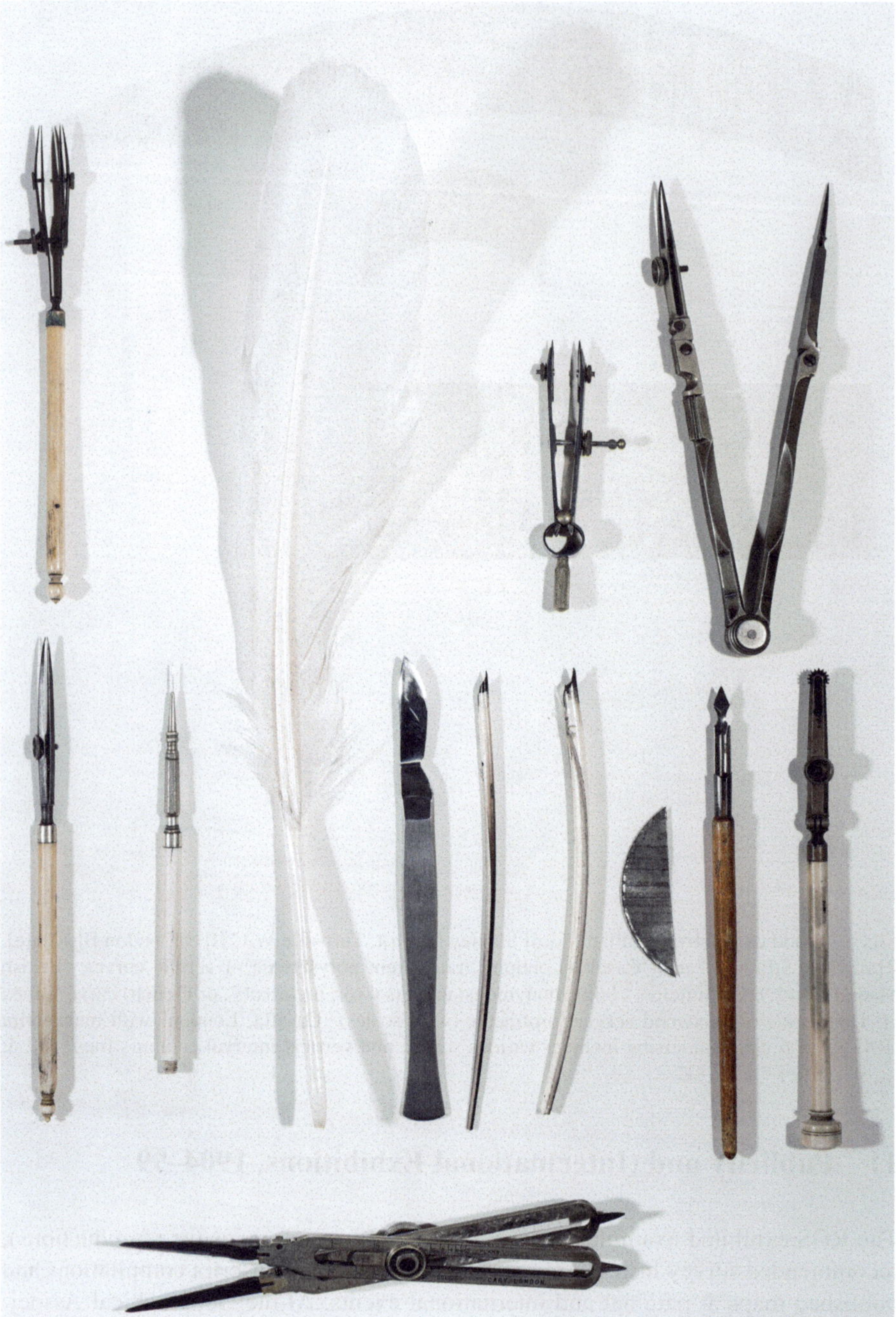


**Fig. 9** Proportional dividers by 'Cary, London.' with 'Royal Geographical Society No 2' (*bottom*); cog-wheel for marking pecked lines, with ivory handle (*bottom right*); 'Joseph Gillot maker accommodating pen holder England 403A' (*bottom*, 2nd from *right*); dividers 'R.G.S. 19' (*top*

and Sir J. Murray's & L.F. Pullar's bathymetric surveys of English & Scottish lakes/ lochs (to 1903), and—in Chelsea Public Library—sets of the RGS *Journal* (1858 [!]–80), *Proceedings* (both series: 1857 [!]–92), and *Geographical Journal* (1893–1903). Perhaps the biggest and most prestigious twentieth-century international display was that for the 1904 St Louis International Exposition: its *British official catalogue*, for the RGS, covered pp. 8–18 (pp. 10–13: 'Selection of maps published by the Royal Geographical Society in recent years'; pp. 14–15 'Selection of reproductions of photographs in the *Geographical Journal* in recent years, exhibited with maps they accompany'). The New Zealand International Exhibition (1906–7)'s *Official catalogue* (New Zealand International Exhibition 1906: British government exhibit), RGS section on pp. 145–154 include 'Selection of maps published . . . in recent years'; 'Selection of works [periodicals and monographs] published . . .', and 'Instruments for explorers and geographical surveyors'. In London: the 1908 Franco-British, 1910 Japan-British, and 1911 Coronation exhibitions, and the *Official guide* to the 'Anglo-American exposition 1914 of art, science and industries' all displayed smaller, but still representative, selections. The1910's *Fine arts [official] catalogue* devoted sections in 'Division 7. Geography. Organiser—Dr. J. Scott Keltie . . .' to 'Specimens of maps published by the Royal Geographical Society' (pp. 60–61); 'Photo-reproductions of original MS. maps of early British explorers in Africa' (p. 62); and 'Latest pattern instruments for geographical surveyors' (pp. 62–63). The 1911 Coronation exhibition's *Official guide & catalogue* (RGS section, pp. 188–191) included 'Latest pattern instruments for geographical surveyors' ("Made by Cary Porter Ltd.") where Reeves' name appears against sextants, his 'Folding Telescopic Alidade' [etc.], 'Patent Micrometers', and 'Proportional Dividers'. His 'Astronomical Compass and Time-Indicator'—"Made by Edward Stanford . . ." (Reeves 1910: 39–40, with two ill.)—ends the RGS exhibits list. Hosted by the RGS in its garden in the winter of 1945–46 was an army lorry with printing facilities to demonstrate similar civilian map production methods; an accompanying eight-page booklet (Lewis [1945]) was a result. An 'in-house' exhibition was mounted of RGS-produced maps selected from 1831 to ca. 1980. Offered by this author, two of the four 'quarters' of the RGS's 1830–31 MS copy by Ballard of the Hereford *mappamundi* were displayed, alongside the original, in 'Medieval world maps: an exhibition at Hereford Cathedral from 29th June–1st October 1999' (see its *Exhibition catalogue*).

**Fig. 9** (continued) right); spring bow [compass] by 'WHH' [i.e. W.H. Harling (drawing instrument maker): see advert in *Geogr J*, 4(5), p. xxiii] (*top*, 2nd from *right*); double ruling pen with ivory handle, anonymous (*top left*); ruling pen by 'Stanley' with ivory handle & 'EJH' [i.e. Edward ('Ted') J. Hatch, RGS draughtsman] (*bottom left*); scribing pen with bone (?) handle, anonymous with 'R.G.S.' (*bottom*, 2nd from *left*); swan feather, 'Stainless England' scalpel, and transformations by RGS Drawing Office staff of crow & turkey (?) feathers into quill pens using home-made semi-circular metal 'cutter' (all *bottom centre*)

**Acknowledgments** The author thanks Tim Bryars for inviting him to be guest lecturer at the annual London Antique Map Fair at the Royal Geographical Society-IBG, on 6 and 7 June 2009; for the same occasion, at the author's request, retired RGS draughtsman E.J. ('Ted') Hatch kindly displayed an accompanying small selection of RGS Map Drawing Office instruments, some of which are amongst those illustrated here. Former RGS-IBG Collections colleagues have made both the source materials and PowerPoint presentations possible. To his younger nephew, Robert Johnston, he is grateful for technical help in preparing a revised PowerPoint version of the 2009 lectures for delivery at the ICA History of Cartography Commission's Symposium 'The dissemination of cartographic knowledge: production, trade, consumption, preservation' in Dubrovnik, 13–15 October 2016.

## References

Ala'i C (2005) General maps of Persia 1477-1925: 249-52. Brill, Leiden

Becroft J (1841) On Benin and the upper course of the River Quorra, or Niger. J RGS 11:184–192. Map opposite p 184

Bederman SH (1992) The Royal Geographical Society, E.G. Ravenstein, and *A map of eastern equatorial Africa*—1877-1883. Imago Mundi 44:106–119

Britton J (1830) Mr Britton's Remarks on Hereford Map (endorsement [Capt. A. Maconochie, RGS Sec.]). RGS-IBG Archives, JMS/19/1 (holograph [3] pp)

Crone GR (1955) Royal Geographical Society: a record 1931-55. Murray for RGS, London

Crone GR (1964) Publications of the Royal Geographical Society: a chronological list of periodicals, special publications and maps, 1832-1964. RGS, London (Library Ser., 7)

Dalton JC (ca November 1892) Memorandum on a proposed system of map compilation for the Royal Geographical Society. 'For Members of Council only'

Freshfield D (1896) The exploration of the Caucasus, vol 2. Arnold, London [Holograph map without toponyms 'The peaks, passes & glaciers of the central Caucasus' in RGS-IBG Map Collection]

Geographical Association (1904) Catalogue of the geographical exhibition arranged by the Geographical Association, and shown in London and elsewhere in 1904: a Supplement to *The Geographical Teacher*, vol II. George Philip & Son Ltd for the Geographical Association, London; Philip, Son & Nephew, Liverpool

George C (1877) 26 April letter of resignation from RGS. RGS-IBG Archives (holograph) CB6 1871-80

Heffernan M (1995) Geography, cartography and military intelligence: the Royal Geographical Society and the First World War. Trans Inst Br Geogr NS 21(3):504–533

Herbert F (1983) The Royal Geographical Society, the map trade and geographical publishing in Britain ca 1830-1930: an introduction with a listing of some 250 Fellows. Imago Mundi 35: [90]–125

Herbert F (1986) The Map is a very pretty one. Map Collector 35:22–27 [**NB (to editors): the title is a quotation from a 19th-century MS comment on a map presented to RGS**]

Herbert F (1999–2001+) The Hereford *mappamundi*: manuscript, and selected printed, references 1802-49 collected . . . 1999-2001 as basis for article(s) on the Royal Geographical Society's interest in the original . . . in Hereford Cathedral and the Society's commissioning from Thomas Ballard of its 1831 manuscript facsimile, upon which that made in 1841 for Edme Jomard of the Bibliothèque royale [=Bibliothèque Nationale de France] in Paris was based [Database since extended to 1891]

[Herbert F] (2005) Cartography of convenience. Hand-map of a portion of the interior of Spitsbergen, Svalbard (1897). Geographical 77(5):22 [Martin Conway's 1896 Spitsbergen travels]

Herbert F (2010) The RGS/OS 1927 plan of Cambridge for 1928 IGC: cartography's and calligraphy's contribution... Cartographiti: The Newsletter of the Map Curators' Group of the British Cartographic Society 84:13–19

Herbert F (2015) The Royal Geographical Society. In: Monmonier M (ed) The history of cartography, vol 6. University of Chicago Press, Chicago, pp 1371–1375

Hinks AR (1913) Memorandum on need of small Technical Library for special use of Instruction Room and Drawing Office. [3] pp holograph and typescript 2 leaves + and carbon copy: top copy with MS note "Please put this on Agenda of next Library Committee. H.J. Lyons hon. Sec.", "Recommended to Council". 'Books required for use in Map Drawing Department' (MS list of 15+ items, ca 1909-14 inclusive); letters of 8 and 22 Oct. 1913 between Hinks and Sifton & Praed's A.J. Sifton FRGS, and one of 23 May 1914 between Sifton & RGS Chief Clerk regarding the order. RGS-IBG Archives, CB8 1911–20

Hinks AR (1925/1926?) Note on style of drawing required in sketch maps to illustrate books of travel, and for blocks in text of Geographical Journal ([4] pp holograph + 3 pp typescript)

Hinks AR (1926) Report of the ninety-third meeting of the British Association for the Advancement of Science, Southampton: [87]–100. London

Hinks AR (1944/1947) Maps and survey, 5th edn. Cambridge University Press, Cambridge

Holland GS (1980) The centenary of the Society's Drawing Office. Geogr J 146:[210]–217

Keltie JS (1887) Subscription List [for assistance to H.A. Milne] RGS-IBG Archives, CB7 1881–90 (holograph)

Kettle DW (1885/1886) Pens, Ink, and Paper: a discourse upon The Caligraphic [!] Art; with Curiosa, And an Appendix of Some Famous English Penmen. "by Bro: Daniel Walter Kettle, F.R.G.S., Cosmographer. Read at a Meeting of 'The Sette', at Freemasons' Tavern, on Friday, November 6th., 1885. . . . Presented to the Sette on January 8th, 1886 . . . Imprynted by ye COSMOGRAPHER, At ye Sign of 'Ye Golden Buck', over against Fetter Lane end, 53, in Fleet Street, London. MDCCCLXXXV." (No. X of '*Privately Printed Opuscula Issued to the Members of the Sette of* ODD VOLUMES.'; 'Edition limited to 233 copies')

Lewis CL (1945) The making of a map: a brief summary. Estates Gazette, London (RGS Pamphlet Ser., 1)

Lyons HG (1908) Some geographical aspects of the Nile. Geogr J 32(5):[449]–480. Folded map opposite p 548

Markham CR (1910) The land of the Incas. Geogr J 36(4):[381]–404. Folded map opposite p 512

Mill HR (1930) The record of the Royal Geographical Society 1830-1930. RGS, London

Milne HA (1881) Letter of application. RGS-IBG Archives, CB7 1881-90 (holograph)

Mumford I (1999) Milestones in lithographed cartography from 1800. PhD thesis, Dept. of Typography & Graphic Communication, University of Reading, 11.12.1999

New Zealand International Exhibition (1906) British government exhibit: Official catalogue . . . Geography and exploration. British Government Committee, London. 'Introduction' dated 30 July 1906

Pearson [Cook] KS (1978/1980) Lithographic maps in nineteenth century [!] geographical journals. PhD thesis, University of Wisconsin-Madison, 1978; University Microfilms International 1980

Porter & Cary (1878) Geographical Journal Advertiser 6 (top)

Reeves EA (1910) Maps and map-making: three lectures delivered under the auspices of the Royal Geographical Society ['Special lectures on maps and map-making': 5, 12 & 19 March 1909]. RGS, London. In back pocket: special printing of 'Central Equatorial Africa. Survey of the Uganda—Congo Boundary on the western frontier of the Uganda Protectorate by the British Commission under the command of Lieut.-Col. R.G.T. Bright, C.M.G. 1907-08.', scale 1:750 000, headed above border "FROM THE GEOGRAPHICAL JOURNAL 1909" and "TO ACCOMPANY REEVES' "MAPS AND MAP MAKING [!]" (see 'Survey and exploration in the Ruwenzori and lake region, Central Africa' by Major R.G.T. Bright (1909) in GJ 34(2): 128–156, folded map opp. p 240)

Reeves EA (1933) The recollections of a geographer. Seeley Service, London

Riches D (2012) Collecting drawing instruments. Hexagon Press (2017: D&M Heritage Press, Huddersfield, West Yorkshire)

Royal Geographical Society (1854) Hints to travellers. Murray for RGS, London

Royal Geographical Society (1892) Report of meeting of Special Sub-Committee delegated to the R.G.S. Library and Map Committee to enquire into and report on the proposals of Lieut.-Col. J.C. Dalton, R.A., as to the Maps published by the R.G.S. Held at 18 Queen Anne's Gate, S.W. 'For the use of Members of Council only'

Royal Geographical Society, Map Drawing Office (ca 1900–1978) Manuscript log-book of Map Drawing Office. Earliest entry 1899, continued to July 1943; entries from March–December 1969, and Easter 1976–March 1978

Sharbau H (1898) RGS-IBG Archives, CB7 1881-90

Stanton A (2017) RGS Drawing Office 1970s [personal communications]

Talbot PA (1913) 'Southern Nigeria Sketch Map of the Eket District to illustrate the paper by P. Amaury Talbot (1913)': hand map lacking scales

Talbot PA (1914) The land of the Ibibios, southern Nigeria. Geogr J 44(3):286–306. Folded map opposite p 340

Vavasour HM (1877) Letter of recommendation for W.J. Turner. RGS-IBG Archives, CB6 1871–80 (holograph)

Verner C (1971) Maps by John Arrowsmith in the publications of the Royal Geographical Society. Map Collectors' Circle, London (MCC Ser., 76). Page 8 [layout details] explains "*State:* It has not been possible to make a complete identification of states"—hence nothing concerning map No. 29 'Sketch of the Delta of the Indus by Lieut. T.G. Carless I.N. 1837.' (*J RGS*, 1838). There exist at least three states, identical in title and marginalia (one with mis-spelling '*Longitude E. from Grenwich*' [*sic*]), with orthographical variations from author's holograph (with toponyms edited in pencil by J. Washington) and its printed *Journal* text

Walker JT (1894) Note on the R.G.S. map of Tibet. Geogr J 4(1):52–54. Folded map opposite p 96

Washington J (1831) Geographical notice of the Empire of Marocco . . . Read 11th and 25th April, 1831. J RGS 1:123–155. Folded map opposite p 123

Worms L, Baynton-Williams A (2011) British map engravers: a dictionary of engravers, lithographers and their principal employers to 1850. Rare Book Society, London

**Francis Herbert** worked in RGS Map Room for 35 years, retiring as Curator of Maps, RGS-IBG, in 2006. 1981: founder member, Charles Close Society for the study of Ordnance Survey Maps; 1995: IMCoS-Tooley Award recipient; 1997: Hon. FRGS; 1998: Fellow, British Cartographic Society; 2013: Fellow, Society for the History of Discoveries. 1976–2005: compiler, 'Imago Mundi Bibliography', standardizing style and increasing information content. 1987: Research Editor, *The Map Collector*; 1996: Editorial Advisory Board member, *Mercator's World*; 2001: Editorial Advisory Board member, *Terrae Incognitae*; 2004: Honorary Advisory Board member, *MapForum*. Articles/reviews in *Brussels International Map Collectors' Circle Newsletter* (from January 2016: Brussels Map Circle's *Maps in History*), *The Cartographic Journal* (BCS), *Cartographica Helvetica*, *Cartographiti* (BCS Map Curators' Group Newsletter), *Geographical Journal*, *Geographical Magazine*, *Imago Mundi*, *International Map Collectors' Society [IMCoS] Journal*, *LIBER quarterly*, *MapForum*, *Scottish Geographical Journal* (RSGS), *Society of* [University] *Cartographers Bulletin*, and *World magazine*. Contributor to *Lexikon zur Geschichte der Kartographie* (1986), *Cartographical innovations: an international handbook of mapping terms to 1900* (ed. H.M. Wallis & A.H. Robinson, 1987), *Information sources in cartography* (ed. C.R. Perkins & R.B. Parry, 1990), *Tooley's dictionary of mapmakers, rev. edn* (1999–2004), *British map engravers: a dictionary . . . to 1850* (2011), *The history of cartography, Vol. 6*, entry 'Royal Geographical Society' (2015), and to *History of military cartography: 5th International Symposium of the ICA Commission on the History of Cartography* (2016)

# Directory Maps: A Neglected Cartographic Resource

**John Moore**

**Abstract** Research in historical cartography has frequently focused on several specific fields, namely, the major products of an individual cartographer, state agency or nation or the study of a particular collection or genre of mapping. The advent of the digital era has seen this focus reinforced by the wider availability of high resolution digital images of the most important or significant maps in the world's leading map collections. While the publishing of trade and street directories was a relatively short-lived phenomenon in the history of printing, the information included in the individual volumes has been recognized as a valuable resource for students of urban and social history. However, although the directories themselves have proved a fertile field for researchers, little if any work has been carried out on the maps which frequently accompanied the published volumes. This paper introduces the initial findings of research on those maps accompanying the variety of Scottish directories published in the period up to the outbreak of the Great War in 1914.

## 1 Introduction

Recent discussions on the problems in locating cartographic material which was formerly widespread but is now increasingly difficult to trace has been highlighted by the work of Demhardt on the cartographic resources found in nineteenth-century German geographical journals. Following his paper presented at the 26th International Conference on the History of Cartography in Antwerp (2015), the ephemeral nature of such material engendered considerable discussion. This paper should be seen as a progress report on current research supported by a Harley Fellowship which is investigating a style of cartography which I believe is equally 'at risk', namely those maps associated with the various Scottish and Welsh trade, street and, in particular, Post Office directories published up to 1914. In recent research on the urban cartography of Scotland, I have grown increasingly aware of the need to investigate what seems a much neglected area of map production and I wish to

J. Moore (✉)
University of Glasgow Library, Glasgow, Scotland
e-mail: john.moore@glasgow.ac.uk

 
M. Altić et al. (eds.), *Dissemination of Cartographic Knowledge*, Lecture Notes in Geoinformation and Cartography, DOI 10.1007/978-3-319-61515-8_10

record my thanks to the Trustees for this opportunity for such study. It also provides an opportunity for some reflection on the various paths taken in the manner in which cartographic heritage is being presented to the wider world.

Although the research considers two constituent parts of Great Britain, this paper will focus on my home country in presenting some features of what is proving to be an intriguing (if frustrating) piece of detective work. Why Scotland? Parochialism apart, town expansion probably grew faster here than anywhere else in Western Europe between 1750 and 1850 (Devine 1988: 29) and the period under consideration was to see that drift continue. By the 1911 census, the Scottish urban population had reached more than three million people (66% of the national total). Another key factor was that two of the firms responsible for producing many of the directory maps, John Bartholomew and W. & A.K. Johnston were Edinburgh-based and the Bartholomew Archive, now held in the National Library of Scotland (NLS) provides particularly unique information on specific plans and production figures.

## 2 Context

This research has thrown into perspective several critical issues faced by libraries in the digital age. Since the early work of Project Pont, Scotland, and especially the NLS, has built up an enviably impressive record in the digitization of documents. The Library's web-site now provides users with online access to over 160,000 high-resolution images. The Library has also collaborated with local authority library services to digitize and host access to nearly 700 directories published up to 1911, covering both towns and counties.

With such a wealth of historical sources, the need to research directory maps as a specific topic might be questioned. Indeed, it is worthwhile discussing whether or not directory maps themselves can be seen as a particularly unique genre of cartographic production. They form only part of the range of map-associated material which was characteristic of publishing in Victorian and Edwardian Britain. Fueled by developments in lithographic printing, the transfer technique and the introduction of the steam press, maps appeared in increased numbers in a bewildering array of works including newspapers, official documents, guide books, gazetteers, encyclopedias and, of course, scientific journals.

Considerable growth in commercial activity undoubtedly added to an increased demand for a style of map which met specific needs, whether this was the identification of premises or the new streets of the rapidly expanding urban areas of industrial Britain. Railways, the telegraph and the development of a postal system all created a call for a more comprehensive provision, which led to the appearance of specialist publishers competing to offer annual directories. In particular, the extraordinary growth in the volume of letters delivered following the establishment of a uniform postal service and the introduction of the Penny Black postage stamp led to an increase in what can be termed 'Post Office' maps. In 1839, there were

76 million letters posted in Britain but the next year saw that figure escalate to 168 million. Along with the introduction of pillar boxes, such developments necessitated the production of both directories and maps showing the location of post offices and sorting houses. It is also clear that immediacy in the availability of information had an impact on production and survival rates were affected by the rapidity of updated versions.

Regardless of such riches and background, the research proved to be remarkably apposite to a Harley Fellowship. In his guide to British map sources for local history Brian Harley (1972: 14) demonstrated the importance and availability of town plans, specifically mentioning the value of directories as a resource. Despite his call to action, the extensive literature on directories themselves, and their apparently widespread availability, little research has been conducted on any aspect of the range, regional variation, content and sources of such maps. In fact, while interest in directories themselves as valuable sources has helped preserve them in libraries, it has done nothing to safeguard the survival of their maps.

In many respects this is not really surprising. Functional in nature and showing both a limited use of color and lack of depiction of the underlying topography, the plans have attracted little attention from either collector or historian (Fig. 1).

Nonetheless, it can be argued that they provide more regularly up-to-date information than many contemporary maps and, of equal importance, it is possible

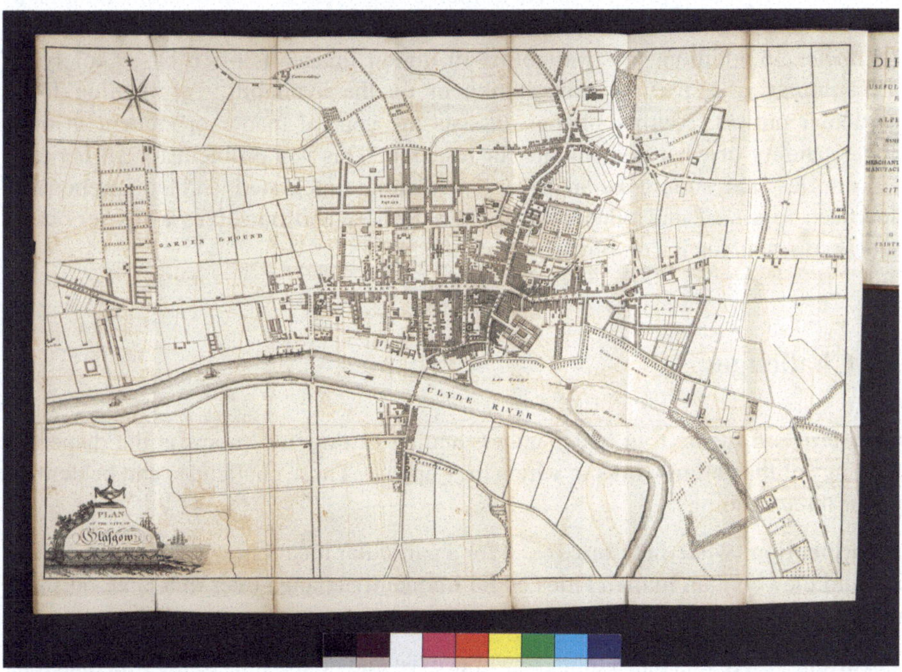


**Fig. 1** Plan of the City of Glasgow from *Jones's Directory*, 1790 (Reproduced by permission of the Librarian, University of Glasgow Library)

that they reached a wider audience through the circulation of their parent directories than the work of many official or professional surveys. Nevertheless, it is important not to ignore the fact that the driver of immediacy over comprehensive coverage could well have had an impact on the accuracy and quality of the mapping. Frequent examples from some of the smaller Scottish burghs which were mapped indicate that revision was often limited to crude additions made to a map without consideration for its appearance. These maps met a demand from an audience moving away from the luxury end of the market and, in their day, were produced in their thousands. Theirs is a history which is not particularly old and rarely attracts the interest of scanning projects. This is largely due to their size and inclusion within smaller volumes. This was very much the case in Scotland where the digitization of Directories failed to include most of the accompanying plans and it is only recently that these have been added to the NLS Maps website.

Unfortunately, this results in a separation of the map from its parent work and this brings into sharp focus the difficulties faced in assessing what is 'heritage' in a time of seriously reduced financial support for libraries in general. Within the wider area of UK research libraries, much emphasis is being placed on 'unique and distinctive collections' and while it provides a focus for establishing priorities to understand, promote and safeguard that significant material, particularly at a time when subject knowledge is in decline, I fear that we may be slowly moving away from a more balanced view of the totality of what has survived from the past. Maps can too readily be digitized because of their attractiveness, leading, for example, to an excessive number of images of Blaeu maps from the atlas *Theatrum Orbis Terrarum*. In addition, the availability of digital images is often based solely on the holdings of a particular institution or on the availability of funding for a particular project. Neither of these assists the general reader or student in their understanding of what is available for research. This is a problem for all heritage repositories and should be a driver for a more integrated and comprehensive approach to visual and other images. We do user and ourselves a dis-service if access to resources is complicated or confusing.

## 3 Challenges

One serious obstacle to a better understanding of directory maps is the dispersed holdings of the parent works in which they appear. These challenges can be defined under three headings:

**Identification** what directories exist for a particular location, what were their titles and did they contain maps? Titles could frequently change over time and, on some occasions, there were competitor publications.

**Location** where do copies of these works exist and do the runs of the title match its publication history? Research suggests that libraries frequently have markedly broken runs but also duplicate copies of titles.

**Survival** are the maps still within the volumes and what condition are they in? There were frequent cases of earlier maps being inserted in later volumes or loose maps being incorrectly placed in the wrong issue.

While the standard guide to British directories (Shaw and Tipper 1996) includes most of the relevant Scottish publications to 1950, a scrutiny of their list underlines just how broken are the runs for most titles in the major libraries. Nor is there any clear indication of frequency of publication and certainly little in the way of recording which volumes contained maps. Although the title pages or prefaces of many specifically note their inclusion, this is not an assured guide—as is the case for Aberdeen, where none of the volumes published after 1841 have any mention of a plan. Given the comparatively late date of the first Aberdeen 'Post Office' plan (1872), earlier depictions might have existed but have not survived. One additional difficulty in creating what could be described as a 'hit list' is the lack of a union catalogue for the holdings of relevant public and special libraries. While access to research collections is well provided through COPAC, the online catalogue of about 90 major UK and Irish research libraries, there are many significant holdings lying outside this facility. As an aside, as a librarian myself, I found it extremely embarrassing to discover that the cataloguing of both directories and their maps is, to say the least, somewhat cursory in many libraries.

Yet another drawback is the limited survival of the maps themselves. Being often produced on lighter and flimsier paper, they were prone to either theft, deliberate removal to ensure their safe-keeping or to serious damage as a result of inexpert folding along the creases designed to reduce them to a portable size—something exacerbated by Bartholomew's strict adherence to a six-inch scale for many of their town plans (Fig. 2).

As shown in this table, the growth in size of Scottish burghs led to larger plans, all needing to conform to a directory's dimensions. One factor highlighted in an early review of Edinburgh maps (Cowan 1932: 102) suggested that excess printings were often re-issued in later editions but investigation proved this to be exceptional, largely because of the astute approach which Bartholomew, in particular, took in calculating publication figures. During the research it became increasingly obvious that the directory plans for any burgh were more likely to be damaged or missing in that area's local libraries and that better copies were to be found where less heavy usage occurred. It also became apparent that there is a wide network of specialist libraries, local history societies and archives which can provide alternative sources of material. One good example of this was the library of the Royal College of Physicians and Surgeons of Glasgow which has a largely undamaged run of nineteenth-century Glasgow directories.

**Fig. 2** The growth in size of plans for the six major Scottish urban areas

| | | |
|---|---|---|
| Aberdeen: | 1840 (170x255 mm) | 1914 (730x580 mm) |
| Dundee: | 1829 (215x220 mm) | 1914 (525x840 mm) |
| Edinburgh: | 1793 (155x250 mm) | 1914 (810x1190 mm) |
| Glasgow: | 1790 (365x410 mm) | 1914 (1020x1130 mm) |
| Greenock: | 1861 (215x400 mm) | 1914 (340x780 mm) |
| Perth: | 1837 (495x365 mm) | 1913 (675x680 mm) |

## 4 Findings

What has the research revealed? To date, I have traced 475 individual maps or states covering either all of Scotland, particular districts or counties (such as the Borders or Fife) or individual towns. These range from the major cities, through significant industrial centers (such as Paisley) or ports (such as Greenock) to county towns and smaller commuting settlements or resorts (such as Portobello). The example of Hamilton, county town of Lanarkshire, is a particularly interesting one as this directory map is, in effect, a cut-down version of Thomas Boyd's 1846 plan of the burgh which does not appear to have survived in any original copy (Fig. 3).

Unfortunately, 27 (or 5.58%) of the plans traced appear to have no surviving copy, emphasizing the point of this genre becoming a disappearing heritage. Two burghs in particular have been badly affected by this loss, namely Greenock, where 8 (21%) of the 38 recorded plans cannot be traced and Kilmarnock where 4 (44%) of the 9 plans are missing. Over the period, John Bartholomew came to dominate much of this map output, producing 214 (or a little over 43%) of those found, particularly in the period after 1865. W. & A.K. Johnston, their nearest rivals, accounted for only 39 (or 7.88%).

The pattern is, however, a little more complex than a gradual takeover. It is also true that several burghs were mapped entirely by local practitioners—most notably Aberdeen. Certain towns were not of a sufficient size to make it economically viable for any outside involvement. While Bartholomew provided the plan of Dundee itself for the city's directory, plans covering such nearby locations as Broughty Ferry and Carnoustie were all prepared by the local burgh surveyors for subsequent volumes. One significant fact is that, of the 179 plans with a specific title reference to the Post Office, only the eighteen Aberdeen depictions between 1873 and 1895 actually mark local post offices or pillar boxes (Fig. 4).

Overall, the places where such plans were available has as much to do with the existence of an enterprising local publisher as with a town's relative importance in the Scottish economy, as seen in the numbers of maps covering individual towns. For example, Paisley, the sixth most populous town in Scotland by 1884, has 14 plans for this period whereas Helensburgh, with a population of less than 8000 in 1881, has 17 within the same time-frame. Twelve of these latter depictions were engraved by Bartholomew but the local publishers, Macneur and Bryden were equally important in the promotion of both directories and maps. This is particularly significant for pre-1860 depictions as it indicates a little known network of individuals and firms producing maps well before Bartholomew's appearance on the scene and it adds to the limited knowledge we have on such nineteenth-century Scottish printers and engravers as Brodie & Salmond of Arbroath and W. & J. Gardner of Perth.

**Fig. 3** Plan of Hamilton from *Brown's Directory* (1855) (Courtesy of the National Library of Scotland)

**Fig. 4** Detail of the Plan of the City of Aberdeen by Andrew Gibb (1888) showing location of pillar boxes (Courtesy of the National Library of Scotland)

# 5 John Bartholomew and W. & A.K. Johnston

Turning to the major producers of directory maps, it is instructive to consider the commercial rivalry between the two major Edinburgh map publishers of the period. Naturally, this competition extended beyond the production of directory maps alone but certain unique features in the resultant maps help indicate the strengths and weaknesses of the differing approaches to recording detail.

While John Bartholomew founded his engraving business in 1830, it was only under the direction of his son that the company expanded into producing its own maps in the 1860s. Part of the firm's success was based on a wide network of contacts in the immediate area providing local details to update their maps. Although renowned for introducing layer coloring to indicate topography, their directory plans have little in the way of color, restricted mostly to identifying boundaries and wards (Fig. 5).

In an area where its use was less essential, Bartholomew maps faced competition from other producers, most notably W. & A.K. Johnston. This rival firm, established in 1826, had a similar background in the Edinburgh engraving trade. Royal recognition came early to their endeavors, as they were appointed engravers to both King William IV and Queen Victoria. They established their name with a series of atlases, in particular, their Royal Atlas of 1861. In 1866, Keith Johnston

**Fig. 5** Bartholomew's domination of the Scottish directory plan market, 1864–1895

| | | |
|---|---|---|
| Dundee: | 35 | first appearing 1874 |
| Edinburgh: | 32 | 1864-1895 |
| Glasgow: | 50 | first appearing 1865 |
| Greenock: | 36 | first appearing 1871 |
| Helensburgh: | 12 | first appearing 1885 |
| Paisley: | 11 | first appearing 1883 |
| Perth: | 14 | first appearing 1885 |

became Edinburgh agent for Ordnance Survey maps, one of only two outside London, and the company used every opportunity to promote sales. Members of the family were subsequently to be appointed Geographers to the Queen in 1877 and 1897.

One particular rivalry with Bartholomew appears to be the case with plans for the Edinburgh directories. From 1832 the Johnstons produced a short series of plans for *Gray's Annual Directory* in direct competition to those engraved by W.H. Lizars for the more established *Post-Office Directory*. This latter work had been published by James Ballantyne since 1826. Gray's publication only lasted 5 years but Ballantyne was to introduce a new, improved map by the Johnstons into the 1860 issue of the directory, followed the next year by the innovative Time-Gun map of the city showing the calculation of the correct time based on distance from the One o' Clock Gun fired from Edinburgh Castle (Figs. 6 and 7).

Again, this was a short-lived association, possibly due to Johnston's failure to update the plan and in 1864 Bartholomew started a 32 year association with the Edinburgh publication. Interestingly, when the publishers Murray & Gibb took over production of the *Post-Office Directory* in 1867, Ballantyne, that same year, issued the rival, but equally short-lived *New Edinburgh, Leith & County Directory* with plans again prepared by W. & A.K. Johnston. Only two plans seem to have been produced and the company turned elsewhere for business. However, when George Harvey Johnston became Geographer Royal for Scotland in 1897, the firm resumed the position of cartographer for the plans for the *Post-Office Directory*.

Both companies used the volumes to publizise their own business. Bartholomew was to promote recently opened venues as well as the location of their own premises and products (Figs. 8 and 9).

In 1889, the company moved to new premises, described as 'The Edinburgh Geographical Institute' and in that same year W. & A.K. Johnston started to describe their Edina Works as the 'Royal Geographical Establishment', drawing attention to their position as Geographers to the Queen—although they had opened the building 10 years earlier! (Figs. 10, 11, and 12)

When Johnston resumed production of the Edinburgh directory plans in 1896, there was an immediate indication of their premises on the plan but no similar identification of the Bartholomew works in Newington.

**Fig. 6** W. & A.K. Johnston's Time-Gun Map (1861) and explanation (Courtesy of the National Library of Scotland)

Elsewhere, it was much more the case that where Bartholomew had the contract to produce such plans, they retained it throughout the period, largely because of the improved quality of the mapping. In truth, the competition between the firms was for the more lucrative market in touring maps which had a far wider audience. To put things in perspective, cycling maps frequently had print runs of over 60,000 whereas the records show numbers of directory plans at a much lower figure—between 2000–2500 for Dundee and 3000–6500 for Edinburgh.

# HISLOP'S TIME-GUN MAP OF EDINBURGH AND LEITH,

***Shewing the distance to which the Audible Report of the Daily Time-gun's Fire travels in every second, up to twelve seconds, according to the best ascertained velocity of sound, viz., 1,116 feet in one second, when the air is at a temperature of 54° Fahrenheit.***

THE EDINBURGH DAILY TIME-GUN is a large cannon, mounted on the "Half-Moon Battery" of the Castle of Edinburgh, and after having been duly loaded with four lbs. of gunpowder by the Master-Gunner and men of the Royal Artillery, is fired off at the precise instant of One o'clock P.M., (Greenwich mean time, now also the City, and, generally, Railway time,) through the agency of electric influence from the Royal Observatory on the Calton Hill, whereby the accuracy of the time of the actual flash and fire is guaranteed to the fifth part of a second.

The Map, as above described, is adjusted to the temperature of 54° Fahrenheit, because that is the annual mean temperature of the air for One o'clock P.M. in the neighbourhood of Edinburgh; and if any one particular day should chance to be so much as ten degrees warmer than what is here assumed, it will only increase the velocity of sound by 1-100th of the whole, a quantity which it is seldom in practice worth while to apply.

The effect of *wind* on the velocity of sound is even smaller still, though influencing much both its intensity and volume; that is, according to the strength and direction of the wind at the instant, an observer at the same place may hear the gun exceedingly loud one day, and exceedingly weak the next; but he will hear it on each occasion after sensibly the same interval of time from the fire, measured in seconds and fractions of a second, according to his position and the circles printed on the Map.

---

*Examples of the Mode of Using the Time-Gun Map.*

*1st Example.*—What correction must the captain of a ship in the new dry dock at Leith apply to the observed time of the sound, or report of the gun as heard *there*, in order to obtain the time of the *fire*, or of One o'clock true?

*Answer.*—The dry dock at Leith being placed on the map half-way between the eleventh and twelfth circles, the answer is, he will have to subtract eleven and a-half seconds.

*2d Example.*—A gentleman who had been to the top of Arthur's Seat, and heard the Time-gun fire when he was there, asseverated amongst his friends on returning to them, that his watch had pointed to One o'clock *exactly* at the very instant when the gun sounded in his ear. Required the error of the gentleman's watch?

*Answer.*—By looking at the map it will be seen that the *seventh* circle from the Castle passes over the summit of Arthur's Seat. Seven seconds, therefore, from the fire must have elapsed before the sound could reach that spot, and the gentleman's watch must have been consequently, and without any doubt, seven seconds slow.

**Fig. 7** W. & A.K. Johnston's Time-Gun Map (1861) and explanation (Courtesy of the National Library of Scotland)

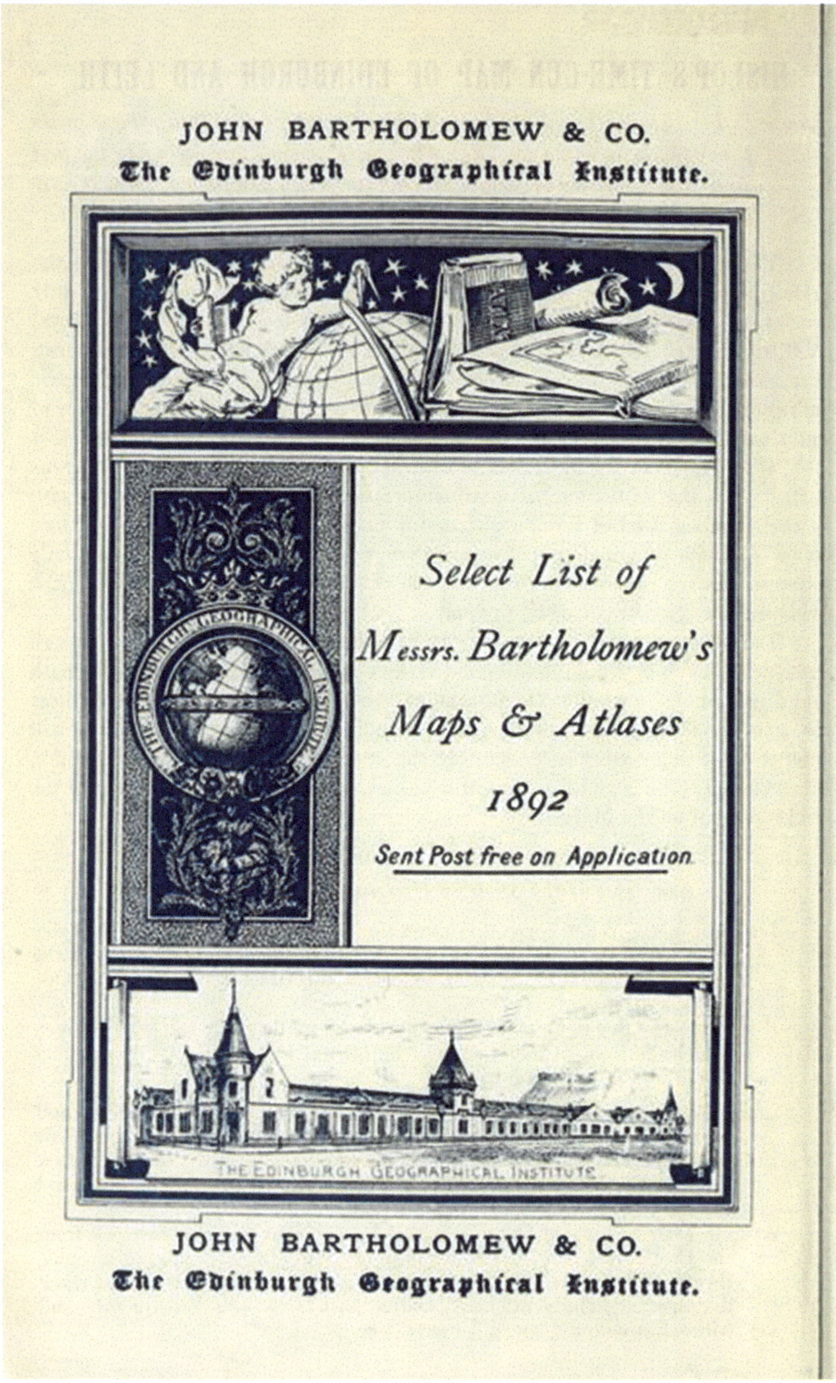


**Fig. 8** Bartholomew's advertisements in the *Post-Office Edinburgh & Leith Directory* (1888) and *Post Office Glasgow Directory* (1889) (Courtesy of the National Library of Scotland)

NOTICE OF REMOVAL.

PARK ROAD, EDINBURGH, *1889*.

MESSRS JOHN BARTHOLOMEW & CO. beg to intimate the Removal of their Business from 31 CHAMBERS STREET to their new Premises, "THE EDINBURGH GEOGRAPHICAL INSTITUTE," at Park Road.

The Institute has been specially planned for the production of Maps, Atlases, and all kinds of Geographical Work, and is fitted with the finest new machinery, and all the latest appliances calculated to aid in perfecting the various processes of Engraving, Lithographing, and Printing.

The staff of skilled Geographical Draughtsmen, Engravers, and Artists has been considerably augmented, and all work is carried out under the personal supervision of Mr J. G. BARTHOLOMEW, F.R.S.E., F.R.G.S.

The arrangements for the production of Geographical work of the highest class are now so complete that the execution of Contracts for Governments, Public Companies, and Societies, can be confidently undertaken with the full assurance of giving satisfaction.

---

JOHN BARTHOLOMEW & CO.,

Geographers, Engravers, Printers, and Publishers,

"THE EDINBURGH GEOGRAPHICAL INSTITUTE,"

PARK ROAD, EDINBURGH.

**Fig. 9** Bartholomew's advertisements in the *Post-Office Edinburgh & Leith Directory* (1888) and *Post Office Glasgow Directory* (1889) (Courtesy of the National Library of Scotland)

**Fig. 10** W. & A.K. Johnston advertisements in the *Post-Office Edinburgh & Leith Directories* (1898, 1900, 1905) (Courtesy of the National Library of Scotland)

**Fig. 11** W. & A.K. Johnston advertisements in the *Post-Office Edinburgh & Leith Directories* (1898, 1900, 1905) (Courtesy of the National Library of Scotland)

**Fig. 12** W. & A.K. Johnston advertisements in the *Post-Office Edinburgh & Leith Directories* (1898, 1900, 1905) (Courtesy of the National Library of Scotland)

## 6 Content

Investigating the contents of individual plans provides an incomparable time-series insight into changes in Scottish urban society during the period. Apart from the indication of the growing number of streets, the location of particular industries and works, docks and shipyards, parks, prisons, water and sewage works, and cemeteries, there is a valuable record of the incursion of the railway system into the urban fabric and the introduction of tramways in the major cities, as was the case, for example, with Edinburgh beginning in 1873 (Fig. 13).

Such features as the construction of stations, the development of railway workshops and the establishment of suburban lines (e.g. the Cathcart District Railway in Glasgow in 1886) give a ready image of the complex impact of the competing railway companies on the structure of cities which survives to the present day.

This immediacy of recording change runs counter to those comments referred to earlier on the re-use of printings in later editions and seems to refute Cowan's findings. As a check on this, it was considered relevant and worthwhile to note how long it took to add certain features to the plans (Fig. 14).

The more general indication of the growth of the national rail network is particularly relevant in the assessment of the maps of the whole of Scotland included in the national directories produced by Slater, Kelly and MacDonald, where re-use of earlier versions is certainly apparent.

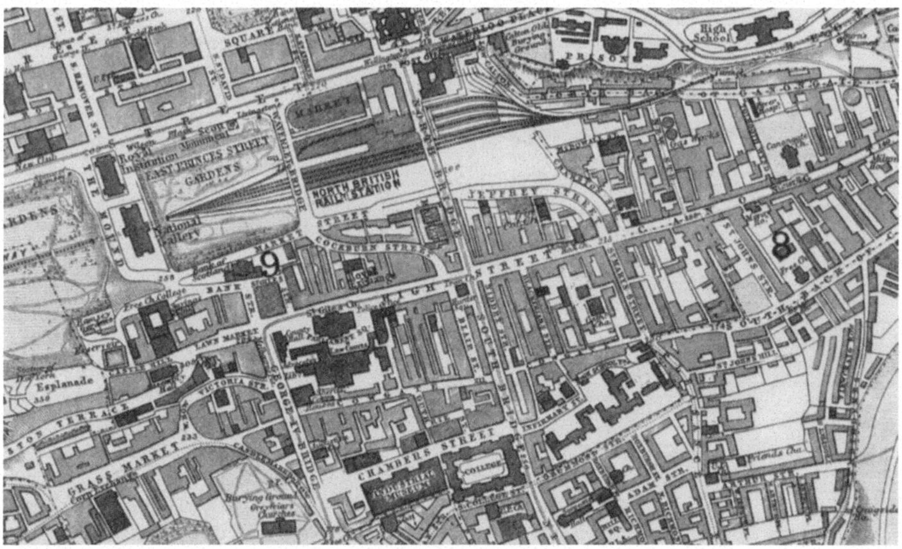


**Fig. 13** Tramlines indicated on the Bartholomew *Post-Office Edinburgh & Leith Directory* plan (1882) (Courtesy of the National Library of Scotland)

Many plans also reflect a growing involvement by local and central government in the care and welfare of citizens by the depiction of hospitals (such the new Edinburgh City Hospital for Infectious Diseases) and schools, particularly after the passing of the Education (Scotland) Act of 1872. An example of this is Kelvinside Academy in Glasgow, first shown in 1877, although founded a year later. Another theme is the identification of military establishments, most notably barracks, such as those at Maryhill in Glasgow. In the run-up to the outbreak of the Great War, other signs of increased military activity can be seen in the identification of a torpedo factory in Greenock and the marking of a seaplane station in west Dundee in 1914 (Fig. 15).

Perhaps the most interesting social phenomenon recorded is in the depiction of recreational and sporting facilities and the changes over time. Bowling greens have long been a feature of Scottish burghs and, as a matter of commercial promotion,

| | |
|---|---|
| Burns Monument, Edinburgh - completed 1831 | appears on 1833 plan |
| National Gallery, Edinburgh – foundation stone laid 1850 | appears on 1852 plan |
| Sir John Moore statute, Glasgow – unveiled 1819 | appears on 1825 plan (first after) |
| Knox Monument, Glasgow – erected 1825 | appears on 1826 plan |
| Free Church College, Glasgow – built 1856-7 | appears on 1857 plan |
| Albert Monument, Perth – inaugurated 1864 | appears on 1865 plan |

**Fig. 14** Immediacy of recording new features on directory plans

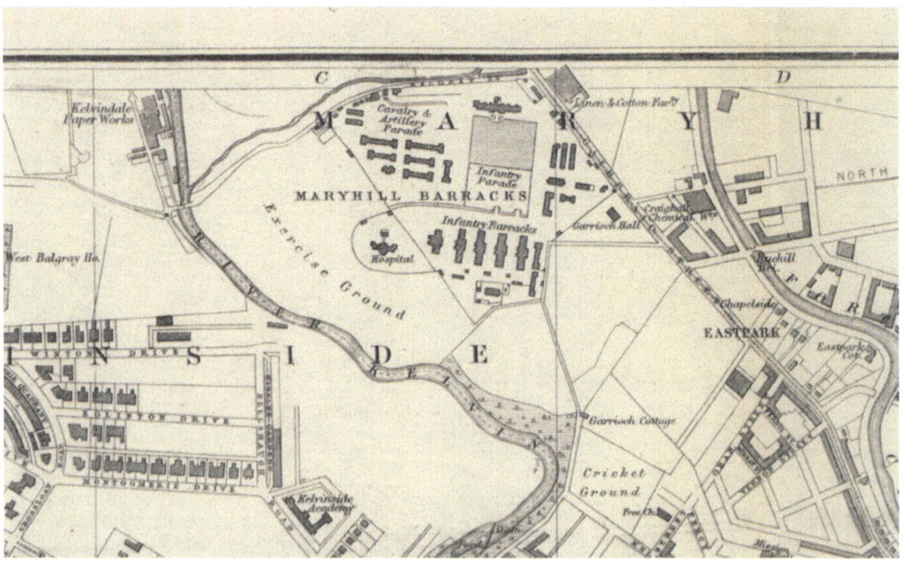


**Fig. 15** Detail of Maryhill Barracks and the newly opened Kelvinside Academy on the Bartholomew *Post Office Glasgow Directory* plan (1888) (Courtesy of the National Library of Scotland)

golf courses are detailed in the plans covering such seaside resorts as Carnoustie but they appear in many other burghs as well. The growth in popularity of cricket in the later nineteenth century can be discerned in the marking of pitches in both Edinburgh and Glasgow from 1865 but it is in the recording of the various parks where the 'Beautiful Game' of football was played that the plans come into their own for currency and relevance, especially in a country such as Scotland. Having said this, one would need to know a fair bit about football to work out the teams playing at the various grounds. For example, in Dundee, both Dens Park and Tannadice are located almost immediately after the founding of the Dundee and Dundee United football clubs but the Bartholomew plans fail to identify the home side at either ground. While Tynecastle is marked initially on the Edinburgh plan for 1886, the first year that Heart of Midlothian played there, it was only when Johnston resumed publishing the plans in 1896 that the ground was identified as theirs on the map. This is in stark contrast to Bartholomew's naming of the cricket teams on the various Edinburgh grounds. Directory plans, like so many other cartographic items, are prone to the interests and selectivity of their producers (Figs. 16 and 17).

**Postscript** While this has given an indication of the major features of directory map production in Scotland, this is only a first step in the overall assessment of provision. Already the research has led to an increased availability of this specific type of map on the NLS website, thereby hopefully guaranteeing their survival for future generations. The initial findings for Wales suggest marked contrasts with Scotland in production and availability. Beyond this, the opportunities for a wider more international approach are there for development. There is much more to discover but any judgement of validity is left to the reader.

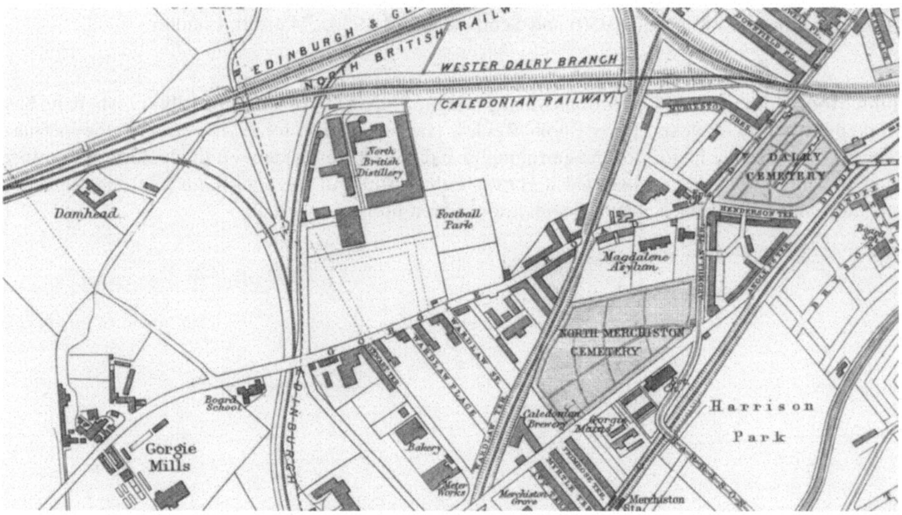


**Fig. 16** Indication of the Heart of Midlothian football ground on the Bartholomew *Post-Office Edinburgh & Leith Directory* plan (1891) (Courtesy of the National Library of Scotland)

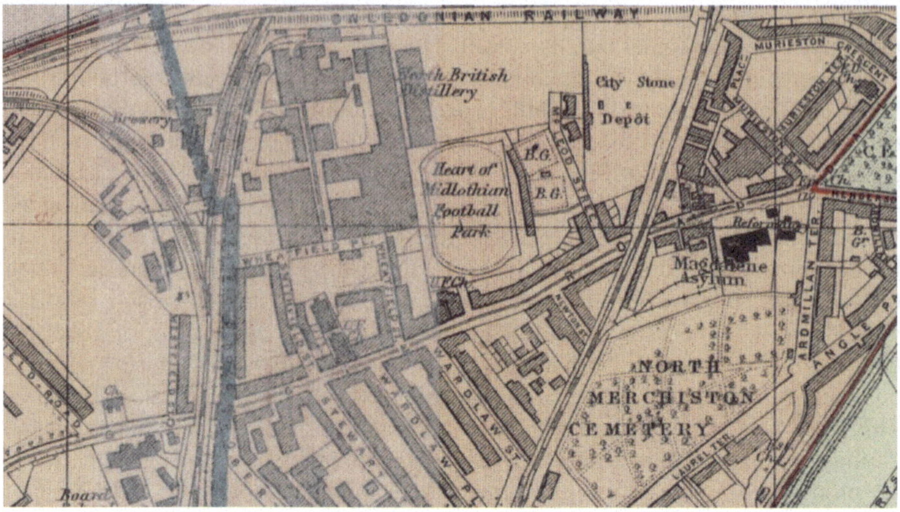


**Fig. 17** Indication of the same ground by W. & A.K. Johnston in the Post Office Edinburgh & Leith Directory (1905) (Courtesy of the National Library of Scotland)

## References

Cowan W (1932) The maps of Edinburgh, 1544-1929. Edinburgh Public Libraries, Edinburgh

Devine TM (1988) Urbanisation. In: Devine TM, Mitchison R (eds) People and society in Scotland. 1760-1830, vol 1. John Donald, Edinburgh

Harley JB (1972) Maps for the local historian. Standing conference for local history, London

Shaw G, Tipper A (1996) British directories: a bibliography and guide to directories published in England and Wales (1850-1950) and Scotland (1773-1950). Mansell, London

**John Moore** is currently Collections Manager at the University of Glasgow Library where he has worked for nearly 40 years. He is Books Review (Europe) Editor for *Imago Mundi*, the international journal for the history of the cartography and has written widely on the history of Scottish mapping. In 2016, he was awarded a Harley Fellowship and his book *Glasgow: mapping the city* was short-listed for a Saltire Society non-fiction literary award.

# Revising, Rectifying and Regulating the Danube: Cartographic Reconstructions of the River and the History of Maps of South-Eastern Europe in the Sixteenth to Eighteenth Century

**Zsolt Győző Török**

**Abstract** The representation of the Danube was always important for makers of maps representing Central and Southern Europe. This paper suggests that, instead of a steady progress, the improvement of cartography characterized by increasing accuracy was actually a slower and complex process. Cartographically, the depiction of the Danube Bend, a spot where the direction of the river takes a sharp turn, was highly important. From the mid-sixteenth century the region was occupied by the Ottoman Empire and could not be surveyed. After the more realistic representation on the Lazarus map (1528) the alternative images compiled by Lazius (1566) and Sambucus (1571), as well as by the Angelini family of military engineers distorted the spatial structure. Western European map makers in the sixteenth to seventeenth centuries misrepresented the river's course and distorted the spatial structure (e.g. Mercator, Blaeu, Coronelli). As exemplified by the map of the Habsburg military engineer Stier, field work only could not improve the entire structure. The river running across the former battlefield was first surveyed by Marsigli and his assistant, Müller. They calculated the coordinates of very few places from data collected by astronomical observation. Based on the new information alternative representations were published by Delisle (Paris, 1703) and Müller (Vienna, 1709). The revised and rectified river course in Marsigli's *Danubius* (1726) eliminated earlier problems—but brought new local distortions. European cartographers in the eighteenth century reconciled different representations for their compilations. The situation changed after Lipszky's map (Pest, 1806), which was based on a network of reference points, and the systematic hydrographic survey of the Danube (1823).

Z.G. Török (✉)
Eötvös Loránd University, Budapest, Hungary
e-mail: zoltorok@map.elte.hu



M. Altić et al. (eds.), *Dissemination of Cartographic Knowledge*, Lecture Notes in Geoinformation and Cartography, DOI 10.1007/978-3-319-61515-8_11

## 1 Introduction

The '*Fontana dei Quattro Fiumi*', designed by the famous Baroque sculptor, Lorenzo Bernini in 1651, is a fountain in the Piazza Navona in Rome, Italy. Of the four famous rivers, representing the four continents, the Danube is the symbol for Europe. Indeed, from the ancient times it has been considered a major river, which was described and mapped by early modern European map makers starting with de Jode, who published a two-sheet work under the title '*Tractus Danubii tabula nova exactissima descriptio*' in his atlas (1578). The river remained a most important communication line of primary strategic importance during the centuries of the Turkish Wars.

In 1664 Jacob Sandrart published Sigismund Birken's large map in Nuremberg, which depicted the entire river from its source to the Black Sea. The map appeared in the year of the Habsburg army's victory at Szentgotthárd, Hungary, and this event revived the hope in Europe that the Ottomans can be defeated. The large map, as it is clear for its long dedicatory cartouche calling the Christian rulers to defend the faith, served primarily Christian propaganda. Birken considered the Danube as the principal river of the continent: '*Danubius—Fluviorum Europea Princeps*'. A second edition of the same map was published almost two decades later in a different situation. After 1529, when they besieged the imperial city unsuccessfully, in 1683 the large Ottoman army marched again along the river to reach the walls of Vienna.

In this paper we give a historical overview of the mapping of the Danube from the sixteenth to the early nineteenth century. In this period the river's geographical representation was obviously considered a significant issue in different contexts of European culture, and we are to demonstrate that the cartography of the river was not a separate, scientific-technical problem related to map making techniques only. It is proposed that this historical subject should be interpreted in wider, social-historical contexts and would require the study of cartographic processes regarding the production and use of maps. These were the interrelated, intertwined and networked actions of various makers and user that resulted not always, and not only in graphic representations of increasing accuracy and detail. Although looking back to their history we can see development, e.g. in terms of increasing geometric accuracy of maps, we should keep in mind that historical processes always produced alternative mappings and maps for different readership and serving different functions.

## 2 'Fluviorum Europea Princeps'

Stretching over 2800 km from its source in the Black Forest, Germany to the mouth at the Black Sea, and connecting countries, people and cultures in Central and South-Eastern Europe the Danube was always a communication highway of primary importance (Wawrik 1993: 139–146). Geographically, the historical material supporting our argumentation covers the middle section of the Danube, the part of the rivers' course between two cities: Vienna, Austria and Budapest, Hungary.

The reason of this focus is obvious on a modern map: there is a sudden change of the direction of the course of the Danube here, the famous Danube Bend, today a major tourist attraction. This is the important spot where the steady west-east course of the river takes a radical turn, and from here downstream the water flows towards the south. Because of a mountain range the Danube Bend is a not one simple turn, but a series of curves. In a few kilometers, the meandering river flows in opposite directions. For any surveyor or map maker this is a confusing section and, we argue, this natural feature could cause misrepresentation and serious cartographic errors on early maps.

Before the eighteenth century map construction, especially in the plains, was mainly based on the survey of the rivers. This is what Luigi Ferdinando Marsigli and Johann Christoph Müller did for their "*anatomy*" of the river, the careful scientific description of the nature of the Danube (1726) (Deák 2004). In 1732 Sámuel Mikoviny, a pioneer of Enlightenment cartography in the Kingdom of Hungary, still listed the 'hydrographic' method, the survey of the direction of waters with compass, among the fundamentals of modern map making. In pre-Enlightenment cartography, for its accessibility during the survey and also in lack of a uniform, abstract geometrical basis, the hydrographic network served as the backbone for smaller scale, chorographic or geographic maps. For the vast extension of the Danube's hydrographic system, the geometric structure of early maps of the region depended ultimately on this backbone. For this reason, the mapping of the Danube is actually a key issue for the better understanding of the history of the early modern cartographic representations of Central and South-Eastern Europe.

## 3 The Danube on Sixteenth Century Maps

Once the *limes* of the Roman Empire, marking the limits of civilized world, from the fifteenth century the course of the Danube directed the expanding Ottoman Empire towards its final goal: Vienna, the capital of the Habsburg Empire. From the early sixteenth century the representation of the Danube became important for the military defense of the region.

The first printed map of Hungary, the 1528 map by Lazarus was based on an authentic collection of geographic information. The original structure was constructed mainly from itineraries, route surveys, and this fact explain the exceptional geometric accuracy of smaller districts on the restructured, printed edition (1528), of which a single copy survived. The manuscript was found in the ruins of the Hungarian royal castle in Buda after its Turkish occupation, was edited in Vienna and printed from four wood blocks with stereotyped names in the workshop of Apianus in Ingolstadt, Bavaria. This was a remarkable early regional map in

**Fig. 1** North oriented detail of the "Lazarus map", printed in 1528, depicting the region of the Danube Bend, Hungary. Note the significant change in the direction of the river's course at the town Szentendre (*Andre*). Courtesy of the Department of Cartography and Geoinformatics, Eötvös Loránd University

Europe, especially if its content and accuracy is considered (Török 2007: 1821–1827). Unfortunately, for the consequences of the editorial work, the transformation of the structure resulted in an upright format with incorrect and misleading orientation marks. The printed version of the maps is not easy to read because modern the modern reader is used to the now traditional north orientation (Fig. 1).

However, if the general hydrography is considered it shows high similarity to the modern map. The rectangular turn of river is significant, but the depiction of the Danube Bend is rather problematic, unless one considers the possibility that the original manuscript was somehow misinterpreted by the editors. Namely, if we suppose that what was indicated on the original manuscript was not the river but the survey route, the misinterpretation is explicable. One should keep in mind that this map was constructed from itineraries, although these are not represented on the printed version. The editors in Vienna could believe that the road along the Danube was taken by the surveyor and could draw the water stream accordingly. This may explain that, while they missed the Danube Bend, they placed a rectangular turn of the river at the town Szentendre (*Andre*). Anyhow, we can claim that the general hydrography is much reliable and accurate on the Lazarus map than any later map up to the late seventeenth century.

Wolgang Lazius, a physician and historian in the Habsburg court, reported that he had revised the earlier map and published his own work in Vienna in 1552–1556.

Lazius "corrected" the wrong orientation of the printed Lazarus map of 1528 and placed the Kingdom of Hungary into a horizontally expanded rectangle, with margins corresponding to the cardinal directions. The new format, which actually resulted in the serious deformation of the spatial structure, greatly contributed to the popularity of the new map: it was simply easier to read for Lazius' sixteenth century European contemporaries.

It is important to note that the Danube Bend was actually not represented on the large wall map by Lazius (Fig. 2). From Vienna the river runs to the east to reach Esztergom, Hungary, the city of the Archbishop, and the gate to the Danube Bend. However, from this city downstream slight curves and some islands appear but the river gradually takes a new direction to the south. The crucial section, the significant series of rectangular turns is completely missing, instead of geographic reality we have a graphic solution to the problem of true direction. Apart from this important omission, the other cartographic 'novelty' Lazius introduced was the distorted course in the central part of the former kingdom, then a Sandjak under Turkish rule.

To direct the river towards the mouth at the Black Sea south of Buda, the former capital city, the river flow was drawn diagonally, in north-west to south-east direction, across the Great Hungarian Plain on his deformed map. Why Lazius did not follow the earlier hydrographic model preserved on the 1528 Lazarus map is as unclear as some other curious features of his map, e.g. the Tihany peninsula connected to the southern coastline of Lake Balaton instead of the northern one.

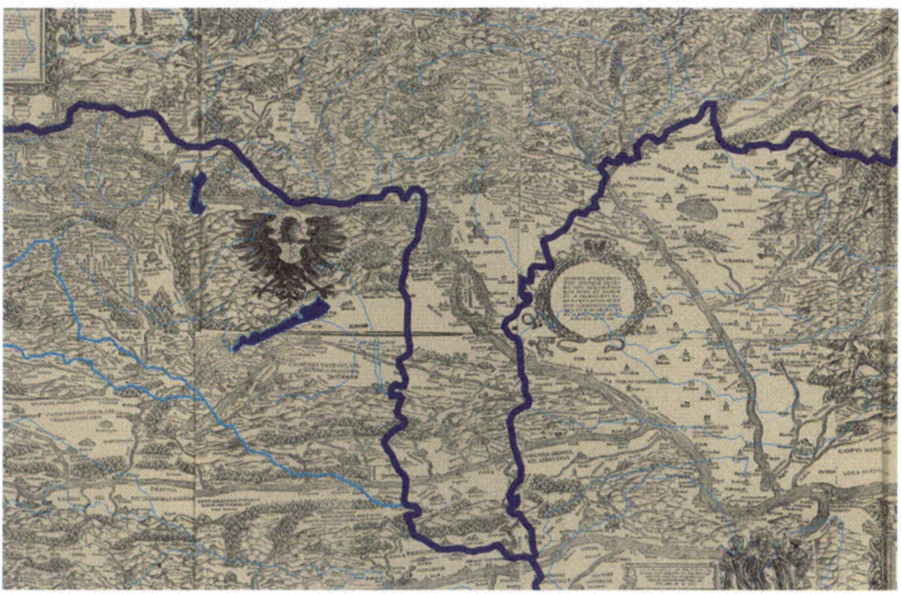

**Fig. 2** The misrepresentation of the Danube on Wolfgang Lazius' 1556 map of Hungary is demonstrated by a visual comparison where the right course of the major rivers, Danube and Tisza, are emphasized in blue to demonstrate the structural problems

Obviously, the new format of his map required the transformation of the total geometric structure of the earlier work and this concluded in serious distortions and errors. Despite all representational problems, some of them was clearly realized already the sixteenth century, the reduced version of Lazius map, which was included in Ortelius' atlas *Theatrum Orbis Terrarum* from 1570, became a cartographic archetype for later European cartography.

A much lesser known, different from Lazius and a few years later published map was Johannes Sambucus' work, also printed in Vienna (1571). Although earlier he published a new edition of Lazarus' map (1566), this new image of the country was a different, alternative representation. As he was not a map maker, his map was a compilation, based on cartographic sources different from those Lazius had access to. Today we can argue that the humanist Sambucus, also in Habsburg imperial service, could well use some military map, unknown for map historians. His model could be similar to the general map in the manuscript fortification atlases produced by Nicolo Angelini, an Italian master builder, and his family members in the 1570s in Vienna (Török 2015: 63–83). One of their country maps has survived in a copy of a fortification atlas, preserved in the archives in Dresden. This manuscript map has very similar geometric structure and almost identical title to the work published by Sambucus in 1571.

The cartometric analysis and comparison of the content of these maps demonstrated their common origin, but also revealed characteristic differences. For example, if the Danube is taken into account one must observe that on the Angelini manuscript after a short diagonal section south of Buda the river takes significant, north-south direction again. On the other hand, on Sambucus' printed map there is no proper Danube Bend. Instead, we have a series of smaller turnings and some islands, represented in rather arbitrary way. This map maker directed the river at approximately the right spot towards the south. However, apparently influenced by the Lazius' model, south of Buda the river took again the wrong, diagonal course on his map.

Although distorted, Lazius' 1556 map of Hungary became a model for most western European commercial map publishers in the sixteenth to seventeenth century, starting with Ortelius atlas (Antwerp, 1570). In contrast, Sambucus' map was used rarely, but it was republished in smaller format by Matthias Quad (Nuremberg, 1592) and also remained in circulation. Similarly, the later Italian editions of Lazarus' map were also reissued or still available in the map market (e.g. a version issued by Orlandi (Rome, 1602). Even the best map compiler of the sixteenth century, e.g. Gerard Mercator could not reconcile the rather different, contradicting sources, and the critical compiler could not construct a better map representing the actual situation of the Danube and East-Central Europe. His posthumously published atlas map, *Hungaria (1595)*, was a compromise where Mercator tried to reconcile Lazarus and Lazius, the main sources, and put the structure into his own, general European framework.

## 4 Visions and Revisions in Seventeenth Century

Before the hydrographic surveys, in lack of reliable information collected in the field or survey, the course of the river south of Vienna, the imperial capital, remained very different from reality. The wrong course of the middle section of the Danube, how we can see it on Blaeu's decorative atlas sheet, became a strong cartographic tradition in the rather uncritical compilations produced by commercial map makers. As it is apparent when looking at the similar contemporary printed maps of Hungary, the problematic hydrography distorted the spatial structure of the whole region.

For its importance, a more reliable and accurate representation of the Danube became a pressing task for Habsburg imperial, military and state administration. More realistic representations of shorter sections of the river were mainly produced by military engineers in connections with military campaigns. These larger scale representations, which were based on field observation and occasional surveys, usually remained in manuscript. In some exceptional cases printed works were based on these military sketches or plans, usually considered as confidential documents with strictly limited access. Vincenzo Coronelli published a view of the 1685 siege of Esztergom, "*Territorio di Naiasel, e Gran Sin'a Comorra descritto, e dedicato dal P. maestro Coronelli, Lettore, e Cosmografo publico. etc.*" which appeared in his island book, *Isolario* (Venice, 1689). Beyond abundant geographic detail and a description of the siege the unusual, south-west orientation is a most telling evidence suggesting the source sketch was made on the spot. Unfortunately, the section of the river beyond Esztergom is not depicted and we can not see its portrayal.

After the successful recapture of Buda in the year following the siege of Esztergom, in 1686 Vincenzo Coronelli, with Jean-Baptist Nolin in Paris, published a new map of Hungary (1687). Although it was to serve as a reliable reference map, the representation of the crucial section of the Danube brought little improvements (Fig. 3). Unfortunately, the false, diagonal course of the river was maintained by the Venetian cosmographer. The reason could have been again the scarcity of local information—although for a time Coronelli was the general of the Franciscan order in Hungary and, as we could see above, he could access to information collected in the field.

As a basis of the general structure, the hydrography was copied by Coronelli from an earlier map (Fig. 4). This source, considered a reliable basis, was constructed by a Styrian military engineer, Martin Stier. As a return for his services, including some military intelligence, the engineer was granted imperial privilege to publish a general map of Hungary in 1664. Earlier Stier had inspected the border districts in Croatia and Sclavonia and produced military atlases for the Emperor and the War Council in Vienna. He certainly knew well the south-western military districts and could use other Habsburg sources for the northern part of the country. But Central Hungary under Turkish rule remained inaccessible, and he could use

**Fig. 3** Detail of the map *Royaume de Hongrie*... by Coronelli (Paris, 1687) with the middle section of the Danube, running diagonally across Central Hungary (Courtesy of the Department of Cartography, Eötvös Loránd University)

fragments of information and outdated works for his compilation work. As a result, he depicted mountain ranges instead of humble sand hills in the great plain.

A new map was so badly needed that Stier's map was reprinted 20 years later in Nuremberg (1684). In turn, Coronelli's version was copied and published in the form of sections, sheets of a pocket atlas, *Parvus Atlas Hungariae* (Vienna, 1689). The product celebrated the Habsburg victories three years after the recapture of Buda (1686), at the same time, it opened the region for Western travelers, diplomats and merchants. When looking at Stier's map, however, the popular map is disappointing: the earlier problems regarding the Danube Bend remained. The meandering is missing, and—except for a short north-east section at Buda—the river still runs diagonally towards the south-east, distorting the geometric structure of the entire map.

Fig. 4 Coronelli's map referred to a version of Stier's map (Vischer, 1685) to reveal highly similar geometric structures as demonstrated here by the distortion grid

## 5 Seventeenth Century Traveler's Opinion: Maps Are Imperfect

During wartime conditions military intelligence was always an important source of information, although observation or survey was a dangerous activity as the map maker could face execution. Travelers or diplomats, sometimes taking the rather risky continental route from Vienna to Belgrade or even further to Istanbul, were not trained map makers, but their accounts were important sources for anyone studying geography.

The English physician, George Brown visited Hungary in 1669–1670, and three years after his return to London he published his travelogue. For the vivid descriptions of the unknown European regions south-east of Vienna soon it became a popular piece of travel literature not only in England but, as a French edition appeared in 1674, in continental Western Europe as well. Brown travelled extensively and described his field observations and experience. This is demonstrated by a note related to the river: "*altho I have seen the Danubius, for about seven hundred*

*Miles; yet I cannot tell whether I may certainly say, I have not seen any part of Ister*" (Brown 1687: 2).

Brown reedited his book a few years later and published an enlarged, second edition in 1687 that included new chapters (e.g. *Some occurrences and observations of this journey*) where the author commented on contemporary maps and, in particular, the representation of the Danube. Brown reported: "*When I found occasion, I used to look upon some Maps, which I carried with me: whereas the Osman Chiaus smiled, saying, there is no depending upon Maps; they set down only great Towns, and often falsely. Ciauses are able to make the best Maps, who pass their Days in travelling Countries; and take notice of Places, and their Turkish names; and in many particulars I found some truth in his word for the maps of Hungary are not exact, those of Servia, Bulgaria, Macedonia and Thessaly very imperfect... The Danube seems to fetch about too much to the South-west, before the Tibiscus enters into it.*" (Brown 1687: 51).

This historically important remark makes it clear that not only was the imperfect nature of the contemporary European maps clear for the English traveler but it had been realized by Ottoman officials as well. The reference to this intercultural dialogue, between the East and the West, suggests that by the late seventeenth century European maps were absorbed by Ottoman culture—although with reservations as to their reliability.

## 6 A Rectification Attempt: Marsigli and Müller

The former battlefield, the central part of the Kingdom of Hungary could be surveyed only in the late seventeenth century after the 1699 Treaty of Karlowitz, when the territory was regained by the Habsburg Empire. The cartographic work of the military architect and early Enlightenment naturalist, Luigi Ferdinando Marsigli of Bologna, and his assistant, Johann Christoph Müller of Nuremberg, resulted in a revised course for a long section of the Danube. Marsigli's cartography was modern but it is the location of the production of new cartographic knowledge, in other words the "where" of their map making that should be also considered a significant issue. Without field work and actual surveys those modern methods and tools only could not result in any better maps of the region.

Following the terms of the Treaty of Karlowitz, from 1699 to 1701 Marsigli oversaw the mapping of Habsburg-Ottoman border. Hundreds of manuscript maps and site plans, resulted from his efforts and his reports included remarkable thematic maps (Török 2006: 403–412). The border map, *Mappa Geographico-Limitanea in qua Imperiorum Caesarei et Ottomanici Confinia...* (1702–1703) was the graphic summary of the work, and the international border was depicted in detail on 39 large-scale, sectional maps (Deák 2006: 353–354). Marsigli, who was well aware of the importance of cartography for military purposes, also realized the importance of maps for state administration.

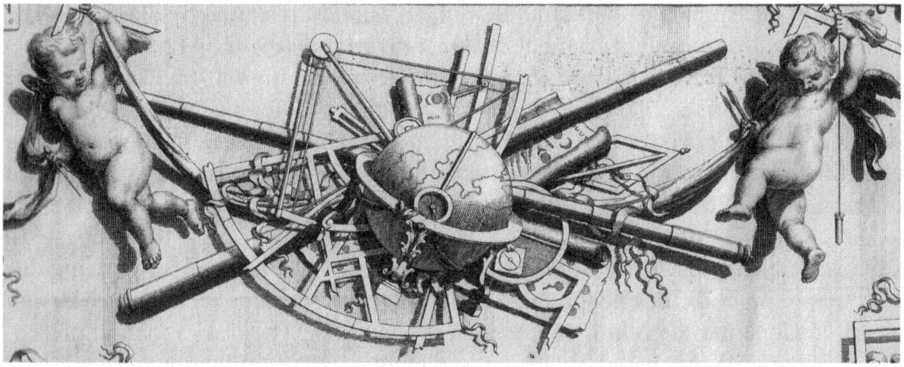

**Fig. 5** The scientific instruments Marsigli and Müller could use in the field were prominently represented in the *Danubius Pannico-Mysicus* (Amsterdam, 1726)

He was aware of the limited accuracy of contemporary surveying methods and contacted the famous astronomer, Giovanni Domenico Cassini, the director of the observatory in Paris. Cassini advised Marsigli how to start his mapping project based on astronomical measurements. For more precise measurements Marsigli needed modern surveying instruments (Fig. 5). He contacted a Nuremberg astronomer, Georg Christoph Eimmart, and asked for his assistance. Eimmart sent him the instruments and he recommended an excellent assistant, Johann Christoph Müller. In 1696 Müller went to Vienna, and from this time on he and Marsigli worked together on the scientific exploration of the region (Deák 2004: 104).

Müller kept a detailed diary during the 1699 border demarcation, including field notes and sketches, which were bound into one volume, and has been preserved at the University Library in Bologna (BUB Mss di Marsigli, Vol. 66). From this collection ("*NOTITIAE GEOGRAPHICAE ORIGINALES, circa Lineam Limitaneam Cisdanubialem, sive DIARIA, Itinere Limitaneo Cisdanubiali, collecta, Primis ac Originalibus suis schedis expressa, ac in III. Libros, digesta. LIBER, I. A SALANKEMEN, USQUE AD CONFLUENTES Savi et Unnae, fll.*") we know they used the method of route survey. In the plains they travelled by boat and recorded compass directions, and from the measured time and the speed of the water they calculated the distances. Based on the traverses rivers were drawn on maps from topographic field sketches.

The diary of Müller's 1696 astronomical expedition is another important documentary source for the study of the astronomical-geographical basis of the Marsigli-Müller maps. The handwritten title in Latin ("*OBSERVATIONES ASTRONOMICAE HUNGARICAE habitae a Comite Ludovico Ferdinando Marsili in Cactris Hungaricis, Anno 1696 Circa Solem, Lunam, Planetas, ac Stellas Fixas cum Figuris*") attributed the document to Marsigli for the reason that it belonged to his collection, but the author of the manuscript was certainly Müller. The same title explains the content, that is Müller original field notes. He observed the celestial position of the Sun, the Moon, the planets and some of the fixed stars and his systematic measurements are structured in tables and illustrated by diagrams. The analysis of this important historical document is beyond the scope of this paper.

What should be noted in connection with the rectification of the Danube is the very limited number of the astronomical positions. After the calculations only seven–eight points could be located in the geographical space. They marked the expedition route and were Müller's observation points: Buda, Baja, the confluence of Danube and Drava rivers, Senta (Serbia), a military camp at Titel (Serbia) close to the confluence of the Danube and Tisza), Szeged, Eger and Buda (repeated). Obviously, these locations were selected mainly along the Danube. Perhaps this was simple practical consideration, but the selection was not accidental: Marsigli clearly understood the utmost importance of the river. The determination of geographical latitude was based on the measurements of the culmination of fixed stars, and, more rarely, based on the altitude of the Sun (which was observed in Vienna). The accuracy of the calculated latitudes was very good. For longitudes Cassini's method, the observation of the Jovian satellites was used (at Senta only these were observed). Although they experimented with the Galilei's method but Marsigli was not satisfied with its accuracy and it was eventually abandoned.

In 1700 Marsigli published a book in Nuremberg, the *Prodromus,* which was a specimen of the great work. The new geographical coordinates regarding Hungary became public. This early publication may explain the curious fact, that the new, rectified representation of the Danube appeared first on a map published by a *Géographe du cabinet*, the excellent French map compiler, Guillaume Delisle (Paris, 1703). In the cartouche of the map there is a reference that his rectified image was based on Marsigli's observation ("*Rectifiez pa les Observations du C(om)te Marsilii.*") (Fig. 6). Johann Christoph Müller, who was the surveyor and draughtsman for Marsigli's project and who later became a cartographer of the Habsburg court could also use the new coordinates when he his own map of Hungary, published in 1709. This was the first official country map of the Kingdom of Hungary.

The revised and rectified image of the river was published by Marsigli with almost 20 years of delay. This was the backbone of a major work and was included in the first volume of the *Danubius Pannico-Mysicus* (The Hague – Amsterdam, 1726). In this work a new, more realistic representation of the Danube Bend appeared, and, after two centuries, the general course of the river on the Carpathian Basin is very close to the actual situation.

The Danube Bend is represented, for the first time more or less realistically on Marsigli's sectional map (Fig. 7). As a result, the Danube took a new, north-south direction in Central-Hungary. After the centuries of the diagonal course this change was an important one and resulted in more accurate geography of a vast, continental region.

On the other hand, for the limited number of control points and due to the limited accuracy of the measurements Marsigli's general map of the Danube was not as revolutionary as it has been considered by historians of cartography. Although in

Fig. 6 The cartouche of Guillaume Delisle's "rectified" map of Hungary (1703) with a reference to Marsigli's observations Image courtesy of the Department of Cartography and Geoinformatics, Eötvös Loránd University

general it was a 'better' depiction of the middle section of the Danube, it was still a cartographic compromise—not a new survey but a compilation work using some new coordinates and field sketches. The result was acceptable but not radically new—and not without errors.

To demonstrate these remarkable, local errors let us look at the representation of the river south of Buda, where the great Danube island (today Csepel island) is depicted. It is not far from the capital city, whose environment was certainly known by Marsigli and Müller, who took celestial measurements here. Although we would expect reliability and higher accuracy here, as one can see from the distortion grid, the representation of the large island is highly distorted. We also compared Marsigli's representation of the island with the actual geography and at this point the situation became really confusing. Marsigli's (1726) outlines are actually almost exactly coincide with the mirror image of the outlines on a modern map. This match is improbably to be accidental, but what happened here is another cartographic mystery we should solve in the future.

**Fig. 7** Detail of Marsigli's general map *Mappa Generalis in qua Danubii Fl.*... published in the first volume of the monograph on the Danube (Courtesy Department of Cartography and Geoinformatics, Eötvös Loránd University)

## 7 Outlook and Conclusions

European cartographers in the late eighteenth century could still produce different images of the river, and only the systematic surveys based on triangulation could eliminate most of the problems of compilation cartography. But this was a rather slow process before systematic surveys, and the generations of map makers could gradually improve the maps in the century starting with Samuel Mikoviny's pioneer work (Török 2003). In 1769 the military engineer Ignác Müller constructed a new general map from the best available sources (*Mappa Geographica novissima Regni Hungariae: divisi In suos Comitatus cum Districtibus...*, Vienna, 1769). His huge and detailed wall map was still a compilation with a large, figurative representation of the river god following the iconography appearing in Marsigli's *Danubius* (Fig. 8).

The representation of the Danube and the entire region dramatically changed only in the nineteenth century, after the publication of a new map of Hungary by János Lipszky (Pest, 1806), an imperial military officer. This work was based on an extensive network of 400 reference points, based on the astronomical measurements of Imre Bogdanich (Fig. 9). His surveying expedition started in Rijeka, Croatia in 1798, and the first results were reported by Franz Xaver von Zach's journal in the following year indicating European interest.

By the early nineteenth century the further economic development of the Kingdom of Hungary made the systemic survey of the Danube necessary, and the

**Fig. 8** Traditional iconographic representation of the Danube on Ignác Müller's wall map of Hungary (Vienna, 1769) Image courtesy of the Department of Cartography and Geoinformatics, Eötvös Loránd University

**Fig. 9** The Danube Bend on Lipszky's map (Pest, 1806) Image courtesy of the Department of Cartography and Geoinformatics, Eötvös Loránd University

hydrographic survey and mapping of the river started after preliminary surveys with the leadership of Mátyás Huszár in 1823. In the following decades the whole section of the river in Hungary was surveyed, both horizontally and vertically at large scale (1: 3600) and with unprecedented accuracy. As exemplified by László Vörös' map of Pest and Buda (1833) a new cartographic era began (Török and Hillier 2014) when the geometrical accuracy of the maps, both military and civilian, dramatically increased. The river was not only revised and rectified on the map: the regulation of the Danube changed nature and also heralded the new age of industry in early nineteenth century East-Central Europe.

## References

Brown E (1687) A brief account of some travels in divers parts of Europe, viz. Hungaria, Servia, Bulgaria, Macedonia, Thessaly, Austria, Styria, Carinthia, Carniola, and Frivli, through a great part of Germany, and the Low-Countries, through Marca Trevisana, and Lombardy on both sides the Po. With some observations on the gold, silver, copper, quick-silver mines, and the baths and mineral waters in those parts. As also, the description of many antiquities, habits, fortifications and remarkable places. Benjamin Tooke, London

Deák AA (2004) A Duna fölfedezése. Dicovery of the Danube. Luigi Ferdinando Marsigli: Danubius Pannonico-Mysicus Tomus I. Vízügyi Múzeum, Levéltár és Könyvgyűjtemény, Esztergom

Deák AA (2006) Maps from under the crescent Moon. The Manuscript maps of L.F.Marsigli and J.C. Müller. Duna Múzeum, Esztergom

Marsigli LF (1726) Danubius Pannonico-mysicus. Observationibus geographicis, astronomicis, hydrographicis, historicis, physicis perlustrarus et in sex tomos digestus. P. Gosse, R. Chr Alberts, P. de Hondt, Apud Herm. Uytwerf & Franç. Changuion, The Hague – Amsterdam

Török Z (2003) Bél Mátyás, Mikoviny Sámuel és a honismereti iskola (Mátyás Bél, Sámuel Mikoviny and the study of homeland). Országos Pedagógiai Múzeum, Budapest

Török Z (2006) Luigi Ferdinando Marsigli (1658-1730) and Early Thematic Mapping in the History of Cartography In: Zentai L, Győrffy J, Török Zs (eds) Térkép – tudomány: tanulmányok Klinghammer István professzor 65. születésnapja tiszteletére. Studia cartologica, 13

Török Z (2007) Renaissance Cartography in East-Central Europe c. 1450-1650 In: Woodward D (ed) Cartography in the European Renaissance: the history of cartography, vol 3, Part 2. University of Chicago Press, Chicago

Török ZG (2015) 16th century fortification atlases of the Habsburg-Ottoman border zone. In: Holzer G, Newby N, Svatek P, Zotti G (eds) A world of innovation: cartography in the time of Gerhard Mercator. Cambridge Scholars Publishing, Newcastle upon Tyne

Török ZG, Hillier D (2014) Exploring and mapping the Danube: reading a hydrographical map of Buda and Pest (1833). In: Liebenberg E, Collier P, Török ZG (eds) History of cartography, Lecture notes in geoinformation and cartography. Springer, Berlin

Wawrik F (1993) Die Donau im Kartenbild. In: Zögner L (ed) Flüsse im Herzen Europas. Rhein – Elbe – Donau. Kartenabteilung der Staatsbibliothek zu Berlin, Berlin

**Zsolt Győző Török** is Associate Professor of Cartography at the Department of Cartography and Geoinformatics, Eötvös Loránd University, Budapest, Hungary. He served as Chair of the *International Society for the History of the Map*(ISHMap) since 2011, and also in the board of directors of *Imago Mundi Ltd.* He studied cartography and philosophy and his research interests lie in relations and processes of Renaissance and Enlightenment cosmography, map-making and military architecture; Enlightenment cartography in Central Europe and the colonial mapping of the Libyan Desert. He studies modern cartographic visualization as a cultural and cognitive tool. In his private workshop he has revived traditional practices with his experimental map and globe making projects. He is national representative and member of the ICA Commission on the History of Cartography and organized the 2012 Budapest symposium.

# The Fortune of the Cartographer Ptolemy Preserved in the Libraries of Rome

**Cosimo Palagiano and Lamberto Lesti**

**Abstract** The geographical work by Ptolemy (Γεωγραφική Ύφήγησις) had a considerable fortune until 1730, even when modern atlases such as Ortelius and Mercator were published. Our contribution intends to examine the spread and the interest that the Atlases of Ptolemy had between Italians scholars and the reading public. Therefore, continuing a study already faced by Italian geographers, we propose to go further, to checking not only the presence of the various editions of Ptolemy, the state of preservation, the presence or absence of old and new tables, and especially—when we could—checking the number of users who have taken in reading these atlases. We found it useful to point out the places of issue of Ptolemaic Atlases, to highlight the area of distribution of this type of cartographic product in the libraries of Rome and consequently in the cities of publication of editions of atlases.

The examined Libraries are: Biblioteca Accademia Lincei e Corsiniana (BALC), Biblioteca Alessandrina (BAL), Biblioteca Angelica (BA), Biblioteca Apostolica Vaticana (BAV), Biblioteca Casanatense (BC), Biblioteca Fondazione Marco Besso (BFMB), Biblioteca Nazionale Centrale Roma (BNCR), Biblioteca Osservatorio Astronomico Romano (BOAR), Biblioteca Società Geografica Italiana (BSGI), Biblioteca Vallicelliana (BV).

Our intention is also to edit a tentative list of comments on geographical work by Ptolemy. In conclusion we wish to consider the reasons for the geographical diffusion of the work of Ptolemy, even after the knowledge of the world had widened after the discovery of the Americas.

C. Palagiano (✉)
Sapienza University of Rome, Rome, Italy
e-mail: cosimo.palagiano@uniroma1.it

L. Lesti
University of Cassino, Cassino, Italy
e-mail: lambertolesti@alice.it

 
M. Altić et al. (eds.), *Dissemination of Cartographic Knowledge*, Lecture Notes in Geoinformation and Cartography, DOI 10.1007/978-3-319-61515-8_12

# 1 Context and Origin

## *1.1 Introduction*

The fortune of Ptolemy goes on from the first known translations and editions with amendments till the recent times. It is interesting to note that the geographical work by Ptolemy (the γεωγαφικὴ ὑφήγεσις) was held in great consideration also when the geography of the world had extended its knowledge all over the continents, including the Americas and Australia. This is because the work by Ptolemy has its validity in the cartographic methodology. For this reason he can be considered a master.

## *1.2 The Aim of the Research*

First of all, the aim of this research is to analyse the most significant codices, manuscripts, and comments on the Ptolemy's work. Some of these codices and manuscripts had been lost. But we have information on a tradition from the first Syrian codices and the Arabic manuscripts which can be referred to. We cannot come back before the eighth century. The history of these translations is told by many books and papers. Many authors, like Fischer (1932), Lelewel (1957), Nallino (1896), Mžik (1926), Ruska (1918) etc. are involved in this ancient history.

## *1.3 The Kitab by Abū Ǧaʿ far Muḥ ammad ibn Mūsā Al-Ḫuwârizmî and by Other Arab Geographers*

The main Arabic author of the translations and emendations of the geographical work by Ptolemy was Abū Ǧaʿ far Muḥ ammad ibn Mūsā Al-Ḫuwârizmî, a Persian mathematician, astronomer and geographer (Baghdad ? c. 780-c. 850, during the Abbasid Caliphate). His geographical work is entitled:

كتاب صورة الأرض (*Book of the shape of the earth*).

Hans von Mžik (Mžik 1926: IX–X) published this book (the *Kitāb*) according to the manuscript Cod 4247 of the Bibliothèque de l'Université et Régionale in Strassburg. Hans von Mžik distinguishes two groups of names in the *Kitāb*: those of Greek origin come mostly from Ptolemy's Geography and in small part by the Romance of Alexander. Generally they refer to a literary tradition, mediated by Syrian geographers (Mžik 1926: IX). Because the Arabs, according to Mžik the first group had no direct knowledge of the Greek texts, a second group comes from the Islamic tradition. In general the two groups can be identified clearly. Mžik reproduces five tables. One of them shows the Nile (Fig. 1).

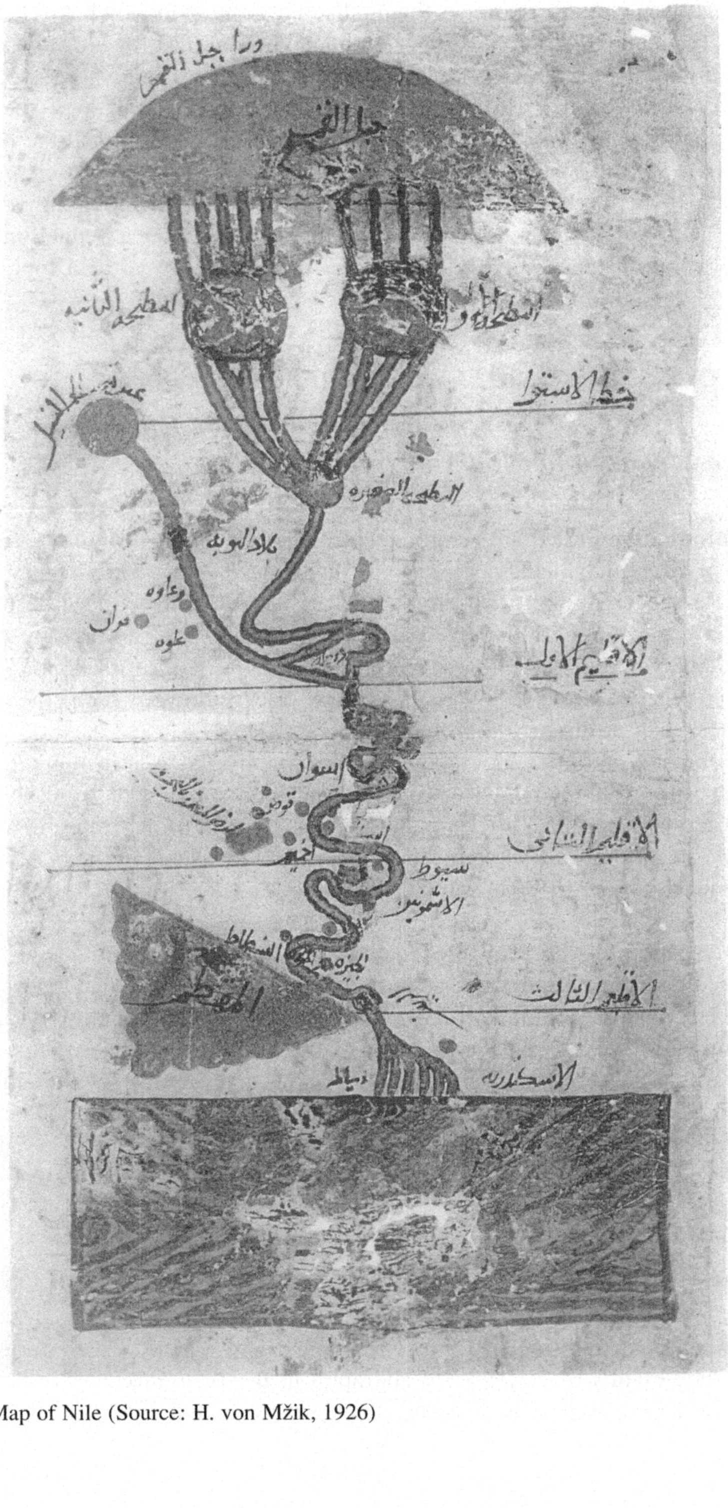


**Fig. 1** Map of Nile (Source: H. von Mžik, 1926)

Carlo Alfonso Nallino examines the intellectual context in which Al-Ḫuwârizmî was formed. A scientific school had been founded by Harunar-Rashid and expanded by al-Ma' mūn. The school was equipped with a rich library, where special and clever scholars were destined to continually increase the budget of the books. Scholars of great fame, as Salmâ, Abū Ḥayyān, Sahl ben Hârûn had the direction of that vast scientific establishment. Nallino claims that al *Kitab* is a restoration of the Ptolemaic text, with differences in the coordinates. Among other things, between Tunis and Alexandria al-Ḫuwârizmî had operated a reduction of too great Ptolemaic distance. In Italy he moved towards the west at a longitude of about 2°.

## 1.4 The History of the Culture in Its Passage from the Medieval to Modern Times

The tradition of the Ptolemaic geographical work tells the history of the culture in its passage from the medieval times to the modern ones. This history can be followed in the etymologic, cosmographic and geographic works like those by Saint Isidore of Seville (560–636). We are able to refer to the works by Albateni (eighth century), Abulfeda (1273–1331). The last wrote *The astronomic determination of places* (Arabic: تقويم البلدان). He followed both the Islamic geographers, like al-Idrisi, and Ptolemy.

An Arabic patron of the geography and cartography was Al-Ma'mun (786–833), who is credited with having started the translation movement from the Greek scientific and philosophical works, with the establishment of the House of Wisdom, a library, a cultural center and meeting place for scholars and translators. Al-Muqaddasi (945/946–991) sought to provide a systematic basis for Arab Geography.

The translation movement was used, among other things, to provide al-Ma'mun of the ideological tools to fight against the Byzantines, as judged not only as misbelievers, but also culturally inferiour, darkened by the irrationality of Christianity, too unworthy to be considered heirs of the Greeks. For the Arabic scholars Ptolemy was a cornerstone. They generally showed reverence for Ptolemy as an authority, even as their own theories and travels caused his information to be revised and expanded.

## 1.5 The Classical Tradition in Latin

Also the classical tradition in Latin was represented e.g. by Cassiodorus (sixth century) who claims to Ptolemy's Geography in the *Institutiones*. We have read many manuscripts, codices and books, which are preserved in the Roman libraries, particularly at the Biblioteca Apostolica Vaticana. This Library is rich in Greek and Latin manuscripts. Most editions refer to modern times, after the translation in Latin by *Jacobus Angeli* (born about 1360). We present some selected pictures, in particular those which reproduce the Mediterranean sea, that is the focal point of the Ptolemaic's work. The breath of the Mediterranean sea, in Ptolemy's work, is wide, about 20° wider than in reality, while in many late editions it is shortened, as it is more realistic. But the major problem of the Ptolemaic work is the presence or absence of maps. Some manuscripts and editions are without maps, others (the most recent) added some new maps (*Tabulae novae*) to the ancient ones.

### *1.6 The Methodology*

There are many ways of examining the geographical work by Ptolemy. It can be considered from the points of view of linguistics, philology, toponymy, history, cartography etc. We have considered it from a geographical point of view. Our issue took in consideration over all the Mediterranean sea, as it is drawn on the maps.

## 2 The Editions Preserved in the Libraries of Rome

The codes of the Ptolemy's *Cosmographia* are preserved exclusively in the Biblioteca Apostolica Vaticana (BAV). We have not analyzed all the archive resources that form the heritage of the library, but only those in which we had the expectation of finding translations of the work of the Greek geographer. The assets of the "Manuscript Section" analyzed are: Archivio del Capitolo di S. Pietro; Barberiniani greci; Barberiniani latini; Barberiniani orientali; Borghesiani; Borgiani arabi; Borgiani greci; Borgiani latini; Capponiani; Chigiani; Ottoboniani greci; Ottoboniani latini; Palatini greci; Palatini latini; Reginensi greci; Reginensi greci di Pio II; Reginensi latini; Rossiani; Urbinati greci; Urbinati latini: Vaticani arabi; Vaticani greci; Vaticani latini. We have found the following manuscripts.

### *2.1 The Greek Codes*

**Barberiniano Greco 163**
Parchment codex of the end of the fifteenth century (1476–1500), cm 24.7 × 15.6, ff. 1r-231v. There are no maps. Among its owners appears the Cistercian monk Ilarione Rancati, who is connected with the collection of books and manuscripts which is the original library, in Rome, of the Cistercian Congregation of "S. Bernardo in Italia."

**Palatino Greco 261**
Miscellaneous collection of several authors, among whom Dioghénēs Laértios and Strábōn, compiled in the fifteenth century. In the text only "Excerpta ex initio Ptolemaei geographie," at f. 305, are included. The code belonged to the Cretan humanist *Marcus Musurus* and contains a dedicatory poetry of his own.

**Palatino Greco 314**
Paper manuscript, copied in Crete, between 1465 and 1470, by the Byzantine humanist and theologian *Michael Apostolius*, ff. 1–224. It is only the text of Geography, without maps. Inside the code there are several attempts of drawing and developing the different types of Ptolemaic projections, even one quite strange almost cordiform. Some drawings are made by hand without the use of a compass. The text is preceded by a note that shows the origin of the manuscript: *Sum de Bibliotheca quam Heidelberga captam, spolium fecit et P. M. Gregorio XV trophaeum misit. Maximilianus utriusque Bavariae dux et S. R. I. Archidapiser et princeps elector. Anno Christi 1623.*

**Palatino Greco 388**
Paper codex, end of the fourteenth century, ff. 1–280. The exposition is preceded by the "*Maximi Planudis in tabulas geographicas versibus eroicis*". Without maps. The first printed Greek edition of Ptolemy's *Geography* (Basel 1533) is based on this code.

**Reginense Greco 82**
Paper codex, leather covered, app. cm 40 × 29, on the back of the manuscript the Papal Coat of Arms. Without maps. At the beginning of f. 1r a decoration depicting some white swirls on a red background. In the part regarding Northern Europe, beside the Greek place names the German ones have been added, while at f. 147r the third Ptolemaic projection has been drawn, the *circularis sphaera* (armillary sphere). It is a typical code for studying, as attested by the several mathematical and geometric observations.

**Urbinate Greco 80**
Paper codex of 279 folios, cm 28.7 × 20.2. This is a miscellaneous collection of the fourteenth and fifteenth century, including Ptolemy's *Geography* and the Plátōn's *Dialogues*. The work of the Greek geographer is followed by the hexameters on *Agathodaimon* and contains only three maps in cylindrical projection, including the first of Europe: the British Islands. It presents another code for studying, as the several drawings and testings of cartographic grids demonstrate.

**Urbinate Greco 82**
Parchment codex of cm 57.5 × 41.8, eleventh/twelfth centuries. It contains the *Agathodaimon* subscription and 27 maps: 1 planisphere in conical projection (Fig. 2) and 26 regional maps in cylindrical projection (Fig. 3). Together with Seragliensis 57 and Fabricianus graecus 23 it is the oldest manuscript with Ptolemaic maps (Diller 1983: 99–100). It was brought from Costantinople to Florence, in 1397, by Chrysoloras at the expense of the Florentine humanist Palla di Nofri

**Fig. 2** Map world, Urb. gr. 82, ff 60v-61r (Courtesy 2016 of BAV)

**Fig. 3** Map of Italy, Urg. gr. 82, ff. 71v-72r (Courtesy 2016 of BAV)

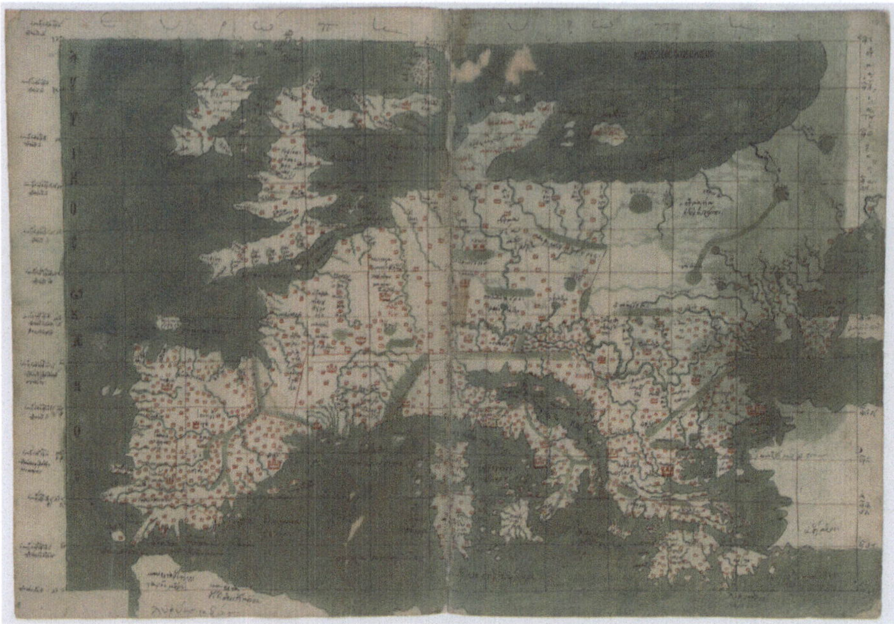

**Fig. 4** Map of Europe, Urb. gr. 83, ff 112v-113r (Courtesy 2016 of BAV)

Strozzi, collector of rare books. It is the archetype from which the following manuscripts with 27 maps (redaction A) derive. It even belonged to the family of Montefeltro Dukes of Urbino.

**Urbinate Greco 83**

Paper manuscript of cm 41 × 29, end of fifteenth century. This belongs to the redaction B of the Ptolemaic cartographic apparatus, with 64 maps. Inside the code 60 regional maps have been drawn, more detailed than those of the Latin redactions. Four more large-format maps are enclosed: (1) Western Europe countries up to Costantinople (Fig. 4); (2) Africa; (3) Eastern Europe and part of Asia; (4) Eastern Africa and Middle and Far East. All the maps are in cylindrical projection.

**Vaticano Greco 176**

This is a miscellaneous code, that contains only the first two chapters of *Geographia* at the beginning. On the margins of the text "*extant scholia Nicephori Gregorae, ut docet notula f. 1v, imo margine adposita.*" At f. 28r of the code, the Ptolemy's *Harminicorum libri* start.

**Vaticano Greco 177**

Paper codex of cm 24 × 16, thirteenth century. At f. 1v the owner of the code is annotated into elegant Latin letters of the fifteenth century: *Claudii Ptolemaei liber geographie et est. proprius domini Maximi philosophi greci ac monaci in monasterio Chore in Constantinopoli emptus a quadam AndronicoYneote.* The code belonged to Maximus Planudes (approx. 1260–1310), the leader of the literary renaissance under the Paleologi. Without maps.

**Vaticano Greco 178**
Miscellaneous paper code of the fourteenth century, cm 18.3 × 12.5, ff. 216. In the manuscript are included books I–VII only, missing the eigth although there are geographical indices derived from it, without maps. The work of the Greek scholar is followed by the *Excerpta librorum medicinalium* by *Aetius Amidenus*.

**Vaticano Greco 191**
Paper code of the thirteenth century, cm 36.5 × 23.7. It is a collection of texts on mathematics and astronomy, including Ptolemy's *Geography* at ff. 130r-169v. The manuscript belonged to Chrysoloras as we infer from some notes of his own. It was brought by the same Chrysoloras to Florence together with Urbinate Greco 82. Without maps, although a subscription of the manuscript mentions a set of 26 or 27 maps in some archetype (Diller 1983: 105). The manuscript was recorded together with two other codices, the Laurenziano XXVIII, 9 and the Laurenziano XXVIII, 38, as the source on which the first Latin version of *Cosmographia* was realized (1406 or 1409), the finished one by the closest student of the Bizantine master, the Florentine humanist and Greek scholar *Iacobus Angeli*.

**Vaticano Greco 193**
Paper code of the fifteenth century, cm 29.2 × 20. This is also a collection of scientific works, among which some chapters of *Geography*. Without maps.

## 2.2 *The Latin Codes*

**Archivio Capitolo San Pietro H. 32**
Parchment codex of 191 ff., cm 29 × 21, fifteenth century. The code is finely decorated *cum prima initiali littera miniata, ac exornata cum stemmmate card. Ursini, et cum titulis rubeis maioribusque litteris coloratis*. Without maps. The manuscript is the work of the copyist Pirrus de Noha and it was commissioned by Cardinal Giordano Orsini.

**Ottoboniano Latino 1771**
Paper code of cm 30 × 22, fifteenth century, without maps. The volume contains till f. 58r the *Plinii Secundi Cosmographia* and from f. 61r Ptolemy's work starts. It is a code with a humble guise, without decorative or aesthetic elements, the only one is the red ink used for the titles. We read in the colophon: *finitus est. liber ipse per me fratrem Symonem Degrymmis ordinis heremitarum Sti. Augustini de provincia Saxonie et Turingie sub anno domini MCCCCXI, in octava corporis chri<sti>*. At f. I: *Ex Codicibus Joannis Angeli ducis ab Altaemps* is specified. Together with Vat. lat. 2974 it is the most antique code and the date of its transcription demonstrates the deep interest that was very soon born on the work of Ptolemy.

**Ottoboniano Latino 1837**

Parchment codex of cm 27 × 20, sixteenth century. Without maps. In the lower part of f. 1r an image of a scantily dressed native of the newly discovered lands, seemingly agitating a bunch of flowers, was drawn. Numerous glosses or comments to the text are contained in the theoretical chapters. This code too belonged to *Joannis Angeli ducis ab Altaemps*. In the colophon: *Si copiò el dicto libro Iulij<quattro>. Adì 22 de Genaro 1545*.

**Reginense Latino 1555**

The code contains only a *Ptolemei divisio orbis* at f. 222.

**Urbinate Latino 273**

Parchment codex of cm 45.8 × 31.2, ff. 204, 1481/1482. It is the remake in verses of Ptolemy's *Geography*, by the Florentine humanist and cosmographer Francesco Berlinghieri. The code contains books I–VII, 27 Ptolemaic maps and 4 *tabulae novae*: Spain, Gaul, Italy and Palestine. All the maps, with the exception of the planisphere in homeotere projection, are in cylindrical projection. The cartographic apparatus is attributable to *Nicolaus Germanus*, although the type of projection of the maps, cylindrical, is typical of Piero del Massaio. Probably, Berlinghieri used models which mixed technique and style of the two cartographers.

**Urbinate Latino 274**

Parchment codex of cm 44.4 × 29.9, ff. 135, between 1468 and 1471. The maps are 31: 27 Ptolemaic, 3 new, 1 planisphere in oval projection, approx. 1528/1530, added to the codex after the discovery of the Strait of Magellan and other American territories. The map of the world of the Ptolemaic apparatus, in homeotere projection, has been modified in the upper edge to insert Northern European countries and Greenland, reported in the map of the Danish cartographer *Claudius Clavus* (Claus Claussøn Swart). All the other maps, even the new ones of Spain, Northern Europe and Italy, are in trapezoidal projection, some of them modified, with the exception of the map of *Taprobana Insula*, in cylindrical projection. The text is preceded by the *Nicolai Germani in tabulas geographicas a se pictas epistola nuncupatoria ad Paulus II pontificem*. In the dedicatory letter, the same contained in the code for Borso d'Este (Lat. 463, of the Bib. Est. Un. in Modena, approx. 1466), *Germanus* exposes the first and the second Ptolemaic projection to which he contrasts his own, in which the more the meridians are inclined, the further they lean from the centre of the figure, this way acquiring the reticulate the characteristic trapezoidal shape.

**Urbinate Latino 275**

Parchment codex of 129 ff., cm 42.1 × 28.5, between 1468 and 1471. The code is identical to the previous for content and cartographic apparatus and it was made in the same Cosimo Rosselli's Florentine workshop (Gabrielli 2005: 54–55); only some elements of the decorations differ. Cartographer *Nicolaus Germanus* (f. 69r). The code contains 30 maps: 1 planisphere in modified conical projection, 26 regional maps and 3 *cartae novae*.

**Urbinate Latino 277**

Parchment codex of 135 ff., cm 59.5 × 43, dated 1472 as we read in the colophon at f. 70r: *Claudii Ptolomei Cosmographie textus explicit per me Ugonem Comminellis de Maceriis supra Mosam in Francia. Florentie die quinta Ianuarii 1472. LAUS DEO*. The type of script is ancient cursive also called *umanistica*. Although not indicated, the author of the maps is the Florentine painter and cartographer Piero del Massaio, because the type of projection and the cartographic realization are typical of his production. In those years, besides, he collaborated with Comminelli in the realization of other codices. The manuscript contains 1 planisphere in conical projection (Fig. 5), 26 regional maps in cylindrical projection (Fig. 6), 7 modern maps (*descriptiones novae*) of Spain, Gaul, Italy, Etruria, Peloponnese, Crete, Egypt, and 10 views of the most important cities of the past: Milan, Venice, Florence, Rome, Costantinople, Damascus, Jerusalem, Alexandria, Cairo and Volterra. The modern maps of Spain, Gaul and Italy are very interesting, because they were especially drawn of shorelines on the ground of the coeval nautical cartography, as well as on updated topographic surveys. The city views are not original but they had just been collected and copied by del Massaio; that of Rome, for instance, belongs to a series of views realized as a bird's eyes view referable to the fourteenth century, while the realization of the plant shows some similarities to an afresco by Taddeo da Bartolo dated 1414. In the view of Venice the Rialto Bridge and the Basilica of Saint Mark stand with evidence, while in that of Rome on the right bank of the Tiber the church of Saint Peter appears, still on its medieval structures, together with the papal residence at *Castrum Sancti Angeli*. The code is similar to Vat. lat. 5699.

**Fig. 5** Map world, Facsimile Urb. Lat. 277 f. 72v-73r

**Fig. 6** Map of Italy, Facsimile Urb. Lat. 277 f. 124v-125r

### Vaticano Latino 2052

Parchment codex of cm 23.1 × 15.4, ff. 208, fifteenth century. Without maps because "*Quamvis in libro VIII tabulae precipue describantur, nulla earum in codice delineatur vel folii inseritur.*" Code of limited size, with drop caps in gold and the initial letters of the paragraphs in red and light blue, surrounded by racemes and vine leaves. It belonged to Niccolò V's library. "*In tegumenti dorso stemmata Pii IX et A. Mai card. bibliotecarii.*"

### Vaticano Latino 2053

Parchment codex of cm 27.3 × 18.2, ff. 177, end of fifteenth century; without maps. Richly illuminated manuscript belonging to Pope Giulio II. In the lower margin of f. 1r the coat of arms of the family Della Rovere appears, surrounded by oak twigs with acorns and floral motifs, topped by Papal Miter. In the lower part of f. 3r two gold colored Putti and with light blue wings hold up a picture, in which a medallion is inserted with the image of Julio II with the Pontifical Miter and the inscription: "*Iulius II Pontifex max., Sixti IIII nepos, MDVII.*" The volume was restored on 20 nov. 1998.

### Vaticano Latino 2974

Paper codex of 156 ff., without maps. It belonged to Niccolò V's library. We read in the colophon at f. 156r: "*Liber Cambii Nicholai de Salviatis quem dei gratia scripsi anno domini MCCCC nono Inditione tertia die XXII mensis decembris.*" It is the oldest Latin existing manuscript, probably an apograph, and the Florentine Cambius Nicolai de Salviatis, the owner, was a notary. In 1414, he redacted *quoddam publicum instrumentum* concerning the legal protection and the inheritance of Neri and Angelo Acciaiuoli. He is found cited, besides, at the end of the XCVI section of the V book of the Charter of Florence of 1415, since de Salviatis was among those who had collaborated to putting in writing the Florence Charter, a reform of the previous 1409 Charter. To better identify the informations that interested him more, he drew small stylized hands with the index finger pointing towards them.

**Vaticano Latino 3810**
Parchment codex of 68 ff., fifteenth century. The text ends at the beginning of the eighth book with this inscription: "*His igitur prefinitis que restant incipienda sunt. Hinc secuntur tabule.*" At f. 1r the usual image of Ptolemy is painted with a yellow and green hat and with a patchwork yellow, blue and green dress; in the lower part two Putti with an ambiguous expression between enigmatic and irritated hold a frame of branches and leaves in which is outlined the shape of a coat of arms without insignia. At f. 2v once more stands the bust of Ptolemy with an astronomer's hat, beard and long hair. He is holding a compass and a book in his hand, on the cover of which two triangles are drawn.

**Vaticano Latino 3811**
Parchment codex of cm 42.2 × 28.6, 64 ff., fifteenth century. It is a collection of 32 *tabulae* that, in the past, was considered the cartographic apparatus of the previous code (Vat. lat. 3810), representing originally a single volume, written on Sisto V′ request. From some receipts of payment (Ruysschaert 1986: 95–97) and an examination of the two codices it seems, however, that the only collection of maps, made by Nicolaus Germanus in 1477, was requested by the Pope and it was added later to a code that was lacking. The maps are 32 and constitute the so called third redaction of *Cosmographia* by German cartographer, on which is based the printed edition of Ulm in 1482. Among the 5 *tabulae novae*, Spain, Gaul, Northern Europe, Italy and Palestine, that of Italy is preceded by a history of the country including the description of the populations, while that of Palestine shows the territorial divisions among the various ethnic groups: *tribus Iuda, tribus Simeon, tribus Benjamin* etc. At f. 64v the name of the cartographer is written: *opus donni nicolai germanij.*

**Vaticano Latino 5698**
Parchment codex of app. cm 57 × 40, of 50 ff., beginning of the fifteenth century. It is the collection of the 27 maps only, that have been modelled with extreme precision directly on Urb. gr. 82. The place names and the other written parts is in a *littera textualis* countextualized with difficulty, but quite close to Florentine models of the early 1400s (Gentile 1992: 295). On the planisphere in conical projection the winds are represented with a particular iconography: men's bearded faces blowing in a kind of horn rather than blowing putti. To the right of the map the zodiac signs with the relative months have been drawn. In the map the scientific interest prevails rather than the aesthetic one and the ecumene is covered with a dense geographic grid. The other 26 regional maps are drawn in cylindrical projection. At f. 50r we read: *Die XXVIII martij 15 < 5 > 8 paulus pontifex IIII misit hunc librum nobis custodibus cum aliis codicibus.*

**Vaticano Latino 5699**

Parchment codex of 130 ff., cm 59.5 × 44, dated 1469. It is one of the most sumptuous and awe-inspiring codes, the realization of which is work of the painter and cartographer Piero del Massaio and of the French copyist Hugues de Comminellis de Mazières, as we can read in the colophon at f. 64r: *E numero scriptorum minimus: Hugo Nicolai de Comminellis, natione Francus me una cum tabulis sequentibus ad instantiam Petri del Massaio Florentini qui me picturis decoravit. Anno domini Millesimo Quadrigentesimo Sexagesimo Nono, Florentie scripsit finitique die Vigesima octava mensis Novembris*. Except for the view of Volterra, the cartographic apparatus is the same of Urb. lat. 277 and it is composed from 27 Ptolemaic maps, 7 *tabulae novae* and 9 views of important cities of the past. The codes are virtually identical, except in the map of world of Vat. lat. 5699 a different iconography to represent the winds was used and the zodiac signs were drawn; in some of the maps, in addition, orography is made with a more pronounced dotted line. Regarding the city views there are some differences in that of Milan, because some of the buildings have not been reported in the same position.

**Vaticano Latino 7289**

Parchment codex of 77 ff., finely illuminated, end of fifteenth century (1480/1485). In the code the 27 Ptolemaic maps are included: 1 planisphere in modified conical projection with 12 blowing putti who symbolize the winds, 25 regional maps in trapezoidal projection and 1 (*Taprobane insula*) in cylindrical projection. It should be noted that the map of the world and the eighth map of Europe (*Sarmatia et Taurica Chersonesus*) have been revised and supplemented on the ground of the indications resulting from the map of Northern Europe by the cartographer Claussøn Swart. As often happens, at the beginning of the code an image of Ptolemy was painted with beard, long hair and a colorful cloak. The scholar holds in his hand a sphere on which the parallels are depicted. The code does not report the name of the cartographer nor of the copyist, but the realization of the maps has been attributed by Fisher to the German cartographer Erik Hammer, although the projection type of the maps may even refer to the other German cartographer, *Nicolaus Germanus*. Only another code of the *Cosmography*, the Magliabechiano XIII, 16 of the BNC of Florence, is attributed to Hammer. This code, dated 1496, beyond the 27 Ptolemaic maps, contains 12 *tabulae novae*, constituting the sum of geographic and cartographic knowledge of the fifteenth century.

**Biblioteca Vallicelliana, Ms. R 116**

At the Biblioteca Vallicelliana is preserved a paper code, which is a collection of the 27 Ptolemaic maps only. The specimen, dated mid sixteenth century measuring cm 42 × 56, is devoid of any indication that attest the author, but since the regional maps are developed with a trapezoidal projection, it could be assimilated to the first drafting of *Cosmographia* by *Nicolaus Germanus*. Due to the lack of paper support, for the absence of miniatures and decorative elements and the reason that the map plains were left uncoloured, it could be assumed that we are dealing with a study specimen.

## 2.3 The Incunabula

**VICENZA 1475: *Cosmographia*. (by Hermanus Levilapide Coloniensi)**
It is the *editio princeps* of the Latin version of the work, finished around 1409 by one of the most linked to the Byzantine master Manuel Crysoloras students: *Jacobus Angeli* from Scarperia. Without maps. In the dedicatory letter to Alessandro V, Angeli points out that following the tradition of the Latin authors he decided to translate the title of the work with the word *Cosmography*, rather than with the more philologically correct *Geography*.

BAV: Inc.III.198; Stamp.Ross.1669. BALC: 50.C.26.

**BOLOGNA 1477: *Cosmographia*. (by Domenicus de Lapis Bononiensis)**
This edition is the reprint of that of 1475, although it is the first that contains the maps, 26 totally: 1 planisphere, 10 maps of Europe, 4 of Africa, only 11 of Asia being united the second and third map of this continent. The low quality maps are, with the exception of the planisphere, in trapezoidal projection and they take us back to the distinctive technique of *Germanus*. The cartographer knew the miniaturist Taddeo Crivelli, one of the main creators of this edition.

BNCR: 70.4.G.7.1. BALC: 49.G.1.

**ROME 1478: *Cosmographia*. (by Conradus Sweynheym and Arnoldus Buckink)**
The Rome edition of 1478 is one of the best. It is the result of the work of Conrad Sweynheym and after his death it was finished by one of his assistant: Arnold Buckinck. The edition includes 27 maps that are a masterpiece of copper-plate printing: 1 planishere in conical projection, 25 in trapezoidal projection and 1 (*Taprobana Insula*) in cylindrical. The cartographic apparatus should be brought back to the first drafting of *Cosmography* of *Germanus*, in particular, on Fisher's suggestion, to the codex *Ebnerianus* MS MA 97 preserved at the New York Public Library.

BAV: Inc.Chig.S.171. BA: INC.746. BALC: 50.G.15.

**FLORENCE 1482: *Geographia di Francesco Berlinghieri fiorentino*. (by Niccolò di Lorenzo)**
The text is a singular remake in Italian language triplets (aba, bcb, cdc, etc) of Ptolemy's work. Beyond the 27 Ptolemaic maps it contains 4 modern ones: Spain, Gaul, Italy, Palestine. The cartographic apparatus, 1 map world in homeotere projection and the other maps in cylindrical, the same of Urb. lat. 273, could be attributed to *Germanus*, although the considerations reported for the Urbinas code are the same. It seems possible that Berlinghieri, who was reputated a cosmographer by his contemporaries, had resorted to mixed technical models produced in the Florentine workshops.

BNCR: 70.1.G.5; 70.1.G.7; 70.1.G.10. BSGI: E.7 base I.20. BAV: Inc.S.120; Stamp.Barb.AAA.IV.15; Stamp.Ferr.S.37; Stamp.Ross.301. BALC: 51.G.1.

**ULM 1482: *Cosmographia*. (by Nicolaus Germanus and Leonardus Hol)**
It is the first edition of the *Cosmographia* published out of Italy and it is based on the third redaction of Ptolemy's work by *Germanus* (see Vat. lat. 3811). The woodcut maps are 32, because beyond the 27 Ptolemaic ones 5 *tabulae novae* have been added. They are those of Spain, Gaul, Palestine, Italy and Northern Europe: the first four are the same of the Florence edition but in trapezoidal projection, while the last derives from the map of the Danish cartographer Claus Claussøn Swart, which is inserted in the manuscript 441 (*Europe tabula XI*) of the Bibliothèque Municipal in Nancy (Dalchè 1992: 375).

BALC: 50.G.11. BAV: Inc.S.154. BC: VOL.INC.732.

**ULM 1486: *Cosmographia*. (by Iohannes Reger et Iustus de Albano)**
It is the reprint of the previous work. For the first time, some additions, like the small treatise *De Locis ac Mirabilibus Mundi* which follows the maps, have been included; they will become more consistent in the following editions.

BAV: Inc.Chig.S.60; Stamp.Barb.AAA.IV.9. BNCR: 70.1.G.6; 70.4.G.14; 70.4.G.7.2. BC: VOL.INC.733.

**ROME 1490 (by Petrus de Turre)**
This is a reprint of the Rome edition of 1478, but the text was amended. The 27 maps included were engraved from the same copper plates (Figs. 7 and 8). The treatise *De Locis ac Mirabilibus Mundi* of the 1482 Ulm edition is present too.

BNCR: 70.4.G.8; 70.44.G.7.3. BC: VOL.INC.737; VOL.INC.925.

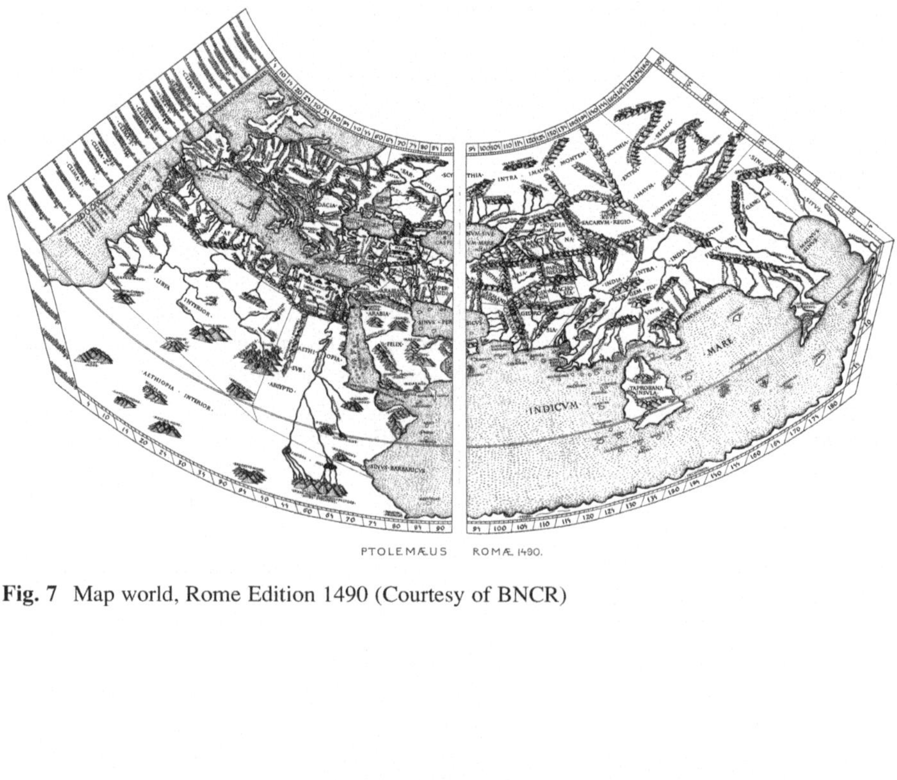


**Fig. 7** Map world, Rome Edition 1490 (Courtesy of BNCR)

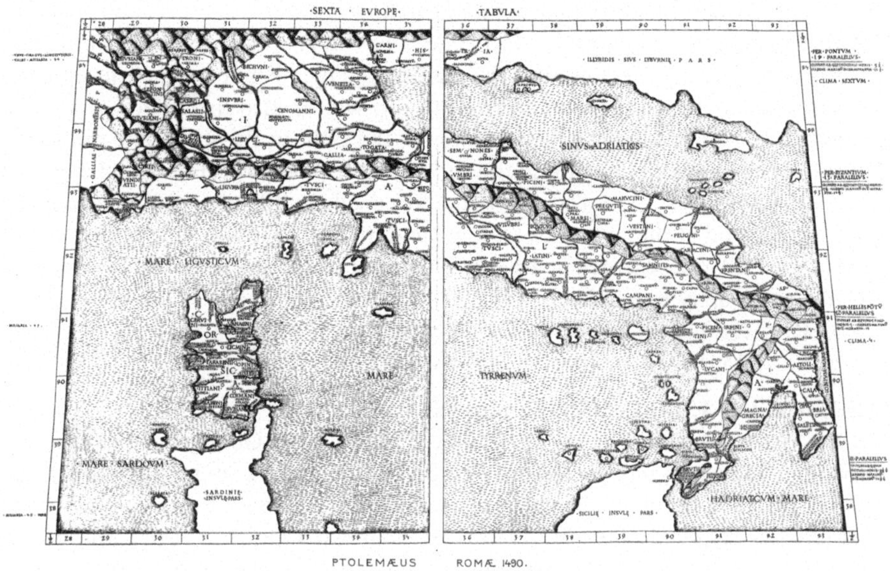


**Fig. 8** Map of Italy, Rome Edition 1490 (Courtesy BNCR)

## 2.4 *The Later Editions*

**ROME 1507 (by Evangelista Tosino Brixiano)**
Edition with 33 maps: 27 Ptolemaic and 6 new. Among the new maps, selected by the philosopher, mathematician and cosmographer Marco Beneventano, has a sizeable interest only that of *Poloniae, Ungariae, Boemiae, Germaniae, Russiae, Lithuaniae*, because it is original and it shows the first cartographic representation of Poland. The map is the work of the Polish humanist and cartographer Bernard Wapowski, who was a friend of Copernico and resided in Rome at that time.

BV: S.BORR. P VI 146. BSGI: E.5 base I.4.

**ROME 1508 (by Evangelista Tosino Brixiano)**
It is the first Ptolemy on which a planisphere with the Columbus' and Vespucci's discoveries appear. It is the *Universalior cogniti orbis tabula* by the German cartographer Ioannes Ruysch, reputed by Beneventano the most skilled cartographer of his time. In this map of the world the old limits of the Ptolemaic grid, 63° north and 16° and 25 min south and from the meridian of the *Fortunatae Insulae* to the 180° east of the Chinese territory, go literally shattered. To represent the ecumene Ruysch draws a map, it is a map drawing which is an extension of the first Ptolemaic projection, which stretches from the North Pole to 38° south and includes the 360° of the circumference of the earth. The world was enlarging at a dizzying speed and hitherto unthinkable.

BNCR: 69.2.G.1. BC: BB.II.42. BA: F.ANT.BB.22.2.

**VENICE 1511 (by Iacobus Pentius de Leucho)**
This edition contains 28 maps: the Ptolemaic ones and a new map of the word in an imperfect cordiform projection. It is the planisphere by the humanist, cosmographer and cartographer Bernardo Silvano from Eboli, who is also the author of a small manuscript Ptolemy, dated 1490, preserved at the BNF in Paris. Silvano draws the planisphere adapting and enlarging the second Ptolemaic projection and the grid of the coordinates goes from 80° north to 40° south, while the part between the meridians located at the longitude of 250° and of 290° is missing. It must be even stressed that Silvano corrected the old Ptolemaic maps on the basis of Nautical Coeval Cartography indications.

BNCR: 6.25.O.23. BAV: Racc.Gen.Geografia.S9. BALC: 63.L.1; 63.L.2. BV: S.Borr.P.VI.143. BSGI: E.5 base I.2; SGA Z 4 IV 1.

**NUREMBERG 1514 (by Johann Stuchs)**
It is the *Nova translatio primi libri Geographie*, by the German mathematician, astronomer and cartographer Johann Werner. Although the volume is without maps it is of capital importance, because it contains the treatise *De quatuor terrarum orbis in plano figurationibus*. In the treatise is drawn a globe in cordiform projection, made developing and expanding the second Ptolemaic projection so as to cover with the projective grid the entire terrestrial surface, while Silvano, 3 years prior, just reached only 320° of extension.

BNCR: 8.50.E.3. BA: F.ANT. BB.17 13.

**STRASBOURG 1520 (by Ioannes Scotus)**
This is a reprint with few changes of the Strasbourg edition of 1513, in which the placenames of the text in Greek characters have been inserted by Giovanni Francesco II Pico della Mirandola. It contains the same 47 tables of the previous ed.: 27 Ptolemaic maps and 20 modern ones. In the second group 1 *Hydrographia sive charta marina* (a planisphere), 10 maps of Europe, 2 of Africa derived from Portuguese maps, 3 of Asia and 4 regional chorographies are included; they can be considered the first embryo of a modern atlas.

BNCR: 9.BANC.3.19. BA: F.ANT. BB.22.3.

**STRASBOURG 1522 (by Iohannes Grieninger)**
In this edition are contained 50 maps, three more of the previous including a new map of the world by the French mathematician and cartographer *Laurentius Frisius* (Laurent Fries), on which the New World is designed with the name of America. The name America appeared for the first time on the 1507 Waldseemüller's planisphere.

BNCR: 69.9.I.6.

**STRASBOURG 1525 (by Iohannes Grieninger e Iohannes Koberger)**
This edition contains the translation of the humanist Willibald Pirkheimer and the *Annotationes Ioannis de Regio Monte in errores commissos a Iacobo Angelo in translatione sua,* of the previous century. In the VIII book the hexameters on Agathodaimon have been added. The maps are 50 and are the same of the edition of 1522.

BNCR: 6.25.O.33. BAV: PALATINA.679.S.4. BC: O.I.38. BA: F.ANT. BB.22.4. BSGI: E.5 base I.14.

**INGOLSTADT 1533 (by Petrus Apianus)**
It is Werner's incomplete edition of Ptolemy's work, printed in Nuremberg in 1514, with an *Introductio geographica* by the German cartographer and astronomer *Petrus Apianus* (Peter Bennewitz) and some changes. Without maps.

BNCR: 69.3.D.22. BV: S.Borr. H IV 77.

**BASEL 1533 (by Hieronymus Froben e Nikolaus Episcopius)**
It is the *editio princeps* in Greek, due to the humanist Erasmus of Rotterdam. The text of this edition is based on the code Pal. gr. 388 preserved at the BAV (Diller 1983: 130). Without maps.

BNCR: 69.6.F.1. BAV: BARBERINI.J.VIII.21. BC: BB.VIII.8. BA: F.ANT. BB.12 20. BAL: B f.44 f2.

**LYON 1535 (by Melchiori et Gasparis Trechsel Fratres)**
It is the Pirkheimer's translation revisited by the Spanish theologian and physician *Michael Servetus* (Miguel de Villanueva). The Servetus' notes on the margins of the text and the insertion of the modern names of towns and countries reported in Ptolemy's text are worthy of observations. The 50 maps which constitute the cartographic apparatus are engraved from the same blocks of the editions of 1522 and 1525.

BNCR: 6.6.O.2. BAV: RACC.GEN.CLASSICI.S.12 (int.2). BA: F.ANT. BB.22 5.

**COLOGNE 1540 (by Ioannes Ruremundanus)**
A new Latin translation by the mathematician and philosopher *Ioannes Noviomagus* (Johann Bronkhorst), with Greek and Latin place names. Without maps.

BA: F.ANT. BB.9 19.

**BASEL 1540 (by Henricus Petrus)**
It is the first edition of the *Geography* edited by the humanist and cosmographer Sebastian Munster. The maps are 48, 21 new ones, and generally inferior to those of previous editions. The Munster's main merit is to have updated the *Tabulae novae* with the latest geographical discoveries and to have indicated the sources of which he had served.

BNCR: 6.21.Q.23.

**LIONE-VIENNE 1541 (by Hugonis a Porta and Gaspar Trechsel)**
It is a new edition of the 1535 Ptolemy of *Servetus*, but of a lower quality; text and maps are unaltered.

BC: O.III.20. BA: F.ANT.BB.22.6. BAL: N h 100. BSGI: E.5 base I.17; SGA Z 9 III 4. BNCR: 6.25.O.28.2; 6.5.O.2.

**BASEL 1545 (by Henricus Petrus)**
This edition is a reprint of the Basel former ones of 1540 and 1542. Of the 54 maps included 42 are the same of the 1540 ed., while 12 are amended maps of Munster or new ones. Among the latter, must be noted that of Bohemia, because conventional signs to point out the different religious and political conditions of the country have been inserted.

BNCR: 6.32.M.5; 6.37.N.27. BAV: R.I. II.64. BC: BB.IX.68. BALC: 63.K.7. BAL: RARI430.

**PARIS 1546 (by Christianus Wechel)**
A reprint of the Greek edition Basel 1533. Without maps.

BAV: RACC.GEN.CLASSICI.IV.684. BA: F.ANT. BB.21.4. BALC: 22.L.12. BV: S.BORR. Q.II.96(1).

**VENICE 1548 (by Gioan Baptista Pedrezano)**
A fine Italian edition with 60 maps: 26 Ptolemaic (missing the planishere) and 34 new ones, all copper engraved by the Piedmontese cartographer Jacopo Gastaldi. To replace the Ptolemaic map of the world the *Universale novo* and the *Carta marina nova tabula* have been inserted. The translation "in volgare italiano" is work of the Sienese physician and humanist Pietro Andrea Mattioli.

BNCR: 6.11.I.35. BAV: BARBER.J.III.15; ROSSIANA.5596; R.G.CLASSIC. V.93. BA: F.ANT.BB.10 23. BV: S.BORR. H.I.107. BALC: 348.A.45.

**BASEL 1552 (by Henricus Petrus)**
A new edition of Munster's Ptolemy with 54 maps. They are the same of the 1545 edition, except the insertion of the map of Pomerania in place of that of the Lake of Constance.

BNCR: 6.15.N.24; RC434; 6.21.Q.6. BALC: 64.M.16. BAL: I p 43. BSGI: E.7 base I.16.

**VENICE 1561 (by Vincenzo Valgrisi)**
A new translation into Italian, from Greek, by the Italian scholar academic and polygrapher Girolamo Ruscelli, with some *addenda* of his own. It contains 27 Ptolemaic maps and 37 modern ones. They are, with few exceptions, enlarged copies of Gastaldi's maps of the 1548 edition. Among the new, the one of Arctic Regions, copied from the map edited in the account of the journey of Zeno brothers, must be remembered, in which nonexistent lands are depicted, as the islands of *Frislandia* and *Icaria*, that, however, have remained for more than a century on the Atlantic maps.

BNCR: 6.11.L.13. BAV: R.G.GEOG.IV.614; RACC.I. IV.2099. BA: RARI I.5 2; F.ANT. BB.215. BC: Y.XI.35. BSGI: SGA Z 4 I 31; E.5 CASS.4.

**VENICE 1562 (by Vincenzo Valgrisi)**
It is the Latin edition of Pirckheimer with the revision of the Sicilian mathematician and astronomer Giuseppe Moleti. The maps are the same of the Italian edition of the previous year.

BNCR: RB655; 6.10.L.26; RB697. BAV: RACC.GEN.CLASSICI.IV.261. BA: F.ANT. BB.21 8. BV: S.BORR. H.III.95. BAL: B g.35 f2. BOAR: BPS X D 15.

**VENICE 1564 (by Giordano Ziletti)**
It is a new edition of the 1561 Ptolemy by Ruscelli, with just few changes in the structure of the text. It contains 64 maps, printed from the same blocks of the previous edition.

BC: G. V60.

**VENICE 1574 (by Giordano Ziletti)**
Another new edition of the 1561 edition, with the revision of the geographer and humanist Giovanni Malombra, born in Cremona. The maps are the same of the 1561 and 1562 edd., except for the Ptolemaic planisphere, drawn by Malombra in conical projection. One map concerning the *Territorio of Roma* has been added.

BNCR: RB 656; RB 657; 6.12.L.11; 6.15.K.11; 8.43.D.6. BAV: RACC.I. IV.2238. R.G.SCIENZE. IV.1635; BARBER. J.III.114. BC: Y.XI.27. BV: S.BORR. H.III.82. BAL: A e 34. BSGI: SGA Z 4 II 13; SGA Z 4 II 14; E.7 base I.17.

**COLOGNE 1578 (by Gottfried Kempen)**
It is a collection of the 27 Ptolemaic maps and of a smaller one of the Delta of Nile reworked by the Flemish mathematician, astronomer and cartographer *Gerardus Mercator* (Gerhard Kremer). The edition is without text, but some notes and a list of nearly 7000 place names have been included.

BNCR: 6.25.O.27. BAV: PALATINA. 2498. S.3; PALATINA. 2499. S.10. BA: F.ANT. BB.22 7.

**COLOGNE 1584 (by Gottfried Kempen)**
It is the 1578 edition, to which the Ptolemy's text has been added.

BNCR: 6.43.O.11. BAV: RAFF.GEN. GEOGRAFIA S.67. BAL: N k 101. BA: F.ANT. BB.22 8/3. BALC: 64.L.6.

**VENICE 1596 (by Haeredes Simonis Galignani)**
This edition is due to the Paduan astronomer, mathematician and cartographer Giovanni Antonio Magini. The tables included are 64, 27 Ptolemaic ones and 37 *tabulae recentiores*, and are all engraved in copper by Gerolamo Porro. The work is divided in two parts: the first volume contains the Ptolemy's text and the second the cartographic apparatus, in which is included a map of the two hemispheres of the world, *Orbis terrae descriptio compendiosa*, in a stereographic projection. The modern tables are accompanied by very detailed commentaries of geographical, historical and social nature of the mapped countries. This edition is very important, as it marks the definitive moment of divergence between modern maps and the Ptolemaic maps, the second ones being destined from that period to perform a mere historical function, as it emerges from the words of Magini: "*...unde non minus necessaria Geographica descriptio terrae hodierna videri debet pro praesentibus temporibus, quam sint Geographicae Ptolemaei tabulae pro praeteritis*."

BNCR: 6.19.K.1.1; RB.654. BAV: RACC.GEN.GEOG. IV.138.

**COLOGNE 1597 (by Petrus Keschedt)**
It is a reprint of the Venetian previous year edition, the maps too are reproduced without changes.

BAV: RACC.GEN. CLASSICI IV.639. BA: F.ANT. BB.21 9.

**VENICE 1598 (by Gio. Battista e Giorgio Galignani Fratelli)**
It is the translation into Italian of the edition of *Geography* by Antonio Magini (1596).

BNCR: 6.8.N.3. BC: BB.IX.15. BA: F.ANT. BB.17 9. BV: S.BORR. H.IV.65. BSGI: SGA Z 4 II 5: E.5 CASS.2.

**VENICE 1598 (by Eredi di Melchior Sessa)**
It is a new Italian edition by the Friulian physician, cosmographer, cartographer and traveller Giuseppe Rosaccio. The text is the same of the edition Venice 1561, as the maps are practically a copy of those of the same edition. The only original part is the *Breve descrittione di tutta la terra* by Rosaccio.

BNCR: 6.19.L.1; 6.33.M.2; 6.43.E.11. BAV: RACC.GEN.GEOG. IV.860. BA: F.ANT. BB.21 7. BFMB: G.03.I 003. BC: BB.IX.15.

**FRANKFURT 1605 (by Iodocus Hondius and Cornelius Nicolai)**
This edition is in Greek and Latin, with 28 maps from 1578 Mercator's plates. The revision of the text, without consulting the manuscript codes, is work of the cartographer, engraver and publisher *Petrus Montanus* (Pieter van den Keere), while the Flemish theologian, mathematician, cartographer and cosmographer *Petrus Bertius* (Pieter de Bert) has collaborated in the publication of this edition.

BNCR: 6.25.O.19. BC: BB.II.46. BALC: 18.K.4.

**AMSTERDAM 1605 (by Iodocus Hondius and Cornelius Nicolai)**
It is the same edition of Frankfurt.

BA: F.ANT. BB.22 9.

**ARNHEIM 1617 (by Ioannes Ianssonius)**
It is a new edition of the *Geography* by Magini and most of the maps are the same of the 1597 Cologne edition.

BNCR: 7.3.I.39. BAV: RACC.GEN.GEOG. IV.223. BV: S.BORR. H.III.74.

**AMSTERDAM 1619 (by Iodocus Hondius)**
Edition in Greek and Latin, by *Petrus Bertius*. The text, both Latin and Greek, has been revised and corrected on codes Pal. gr. 314 and Pal. gr. 388, by the German classical scholar of Greek Friedrich Sylburg (Diller 1983: 131); the maps are still those elaborated by Mercator.

BA: F.ANT. BB.22 8/1–2. BC: BB.II.44. BV: S.BORR. H.VIII.44. BNCR: 68.3. H.3.

**PADUA 1621 (by Fratelli Paolo e Francesco Galignani)**
It is a very faithful reprint of the Magini's Italian edition published in Venice, in 1598, by the brothers Gio. Battista and Giorgio Galignani.

BNCR: 6.21.Q.8.2; 6.21.Q.8.1; 6.16.N.22.2; 6.16.N.22.1. BAV: RACC.GEN. GEOG. II.13; MAI XI.D.IX.44. BA: F.ANT. BB.17 10. BALC: 64.L.15; 344.G.8.

**UTRECHT 1698 (by Franciscus Halmam and Guiljelmus van de Water)**
It is another reprint of the maps of Mercator without text.
BA: F.ANT. BB.22 10.

**AMSTERDAM 1730 (by E. et J. Wetstenios et Guil. Smith)**
It is the last reprint of the Ptolemaic maps of Mercator without the text, but with an extensive index. This edition is expecially for *usum geographie veteris studiosorum*.
BC: M.I.8. BALC: 85.G.15.

## References

Dalchè PG (1992) L'oeuvre géographique du Cardinal Fillastre: representation du monde et perception de la carte a l'aube des découvertes. Archives d'Histoire Doctrinale et Littéraire du Moyen Age 59(1992):319–383

Diller A (1983) Studies in Greek manuscript tradition. A. M. Hakkert, Amsterdam

Fischer J (1932) Claudii Ptolemaei Geographiae Codex Urbinas Graecus 82. Tomus Prodromus, De Cl. Ptolemaei vita operibus Geographia praesertim eiusque fatis. Pars Prior. Commentatio. Apud Brill EJ, Harrassowitz O, Lugduni Batavorum, Lipsiae

Gentile S (1992) Emanuele Crisolora e la Geografia di Tolomeo. In: Cortesi M, Maltese EV (eds) Dotti bizantini e libri greci nell'Italia del sec XV. M. D'Auria, Napoli

Lelewel J (1957) Épilogue de la Géographie du moyen Age, étudiée par Joachim Lelewel, accompagné de huit plances. Chez V[e] et J. Pilliet, Libraires, Succ[rs] de P.-J. Voglet, Bruxelles

Nallino CA (1896) Al-Ḫuwârizmî e il suo rifacimento della Geografia di Tolomeo. Atti della R. Accademia dei Lincei CCXCI (1894): 4–53

Ruska J (1918) Neue Bausteine sur Geschicte der arabischen Geographie. Geogr Zeit 24:77–81

Ruysschaert J (1986) Du globe terrestre attribué à Giulio Romano aux globes et au planisphère oubliés de Nicolaus Germanus. Monumenti, musei e gallerie pontificie Bollettino 6:93–104

von Mžik H (1926) Das Kitāb ṣurat al-'arḍ des Abū Ǧa'far Muḥammad ibn Mūsā al-Ḫuwârizmî. Bd. 3. Harrassowitz, Leipzig

**Cosimo Palagiano** since 2011 is emeritus professor of La Sapienza University of Rome. He is corresponding member of the Accademy of Arcadia, the Accademia Nazionale dei Lincei, and of the Istituto di Studi Romani. In 2011 he was appointed Co-Chair of the IGU-ICA Joint Commission on Toponymy. In 1980 he founded the Medical Geography in Italy and organized 1982–2010 about every 3 years ten International Seminars on Medical Geography. Professor Palagiano actively collaborates with public and private institutions in the field of cartography and GIS. He has been member of the Editorial Board of the journal *Geography*, Environment and Sustainability, published by the Russian Geographical Society and of the magazine Espacio y Tiempo of the University of Seville. The fields of researches by Professor Palagiano are: (1) Medical Geography; (2) The History of Geography, and (3) Toponymy.

**Lamberto Lesti** is PhD in Historical Geography. He has discussed a thesis on Development of the Battlements and Fortifications in the Territory which sourrounds the Trasimeno Lake. His main fields of research are Medical Geography and History of Nutrition. In 2015 he wrote a book on the Nutrition from the beginning of Modern Age to rhe end of the eighteenth century in the Perugia Province.

# Colonialism in the Bosatlas

**Ferjan Ormeling**

**Abstract** For the purpose of making available scans of all maps in all 55 editions of the Bosatlas (1877–), the Dutch school atlas, on the Utrecht University Library website, all these maps are compared to each other, in order to reconstruct the editorial process. The purpose of this web service will be to allow Dutch school children to follow recent history in their own school atlas and to study, diachronically, the recent changes in particular areas.

The first part of the project is to scan all 36 editions published up to World War II, and this period, 1877–1945, can be regarded as the heyday of colonialism. For this paper, the atlas is analysed from the point of view of depicting colonialism, trying to find changes in attitude in the language and the symbology used in the portrayal of colonial areas. For this analysis, the following aspects of the atlas have been researched: the sequence of the maps and the emphasis on specific areas, together constituting the atlas structure; the colouring of the atlas; naming (both the imposition of European names in the nineteenth century and the references to indigenous groups, their realms and settlements; symbolization (portrayal of settlements, boundaries) and finally the subjects mapped on thematic maps of colonial areas.

The results of this analysis of the colonial portrayal will be part of the accompanying explanatory texts offered to the users of the website, together with texts on the reasons for the changes on the same map from one edition to the next.

## 1 Introduction

Colonialism is the extension of a state's sovereignty over areas outside its national boundaries, (and thus a system in which one nation dominates others) and it implies its political dominance and economic exploitation of those areas. Despite all possible ideological foundations (like humanitarian or civilizing missions) it is a system based on inequality, and this has been recognised during the post WWII

All the maps from the Bosatlas shown here have been reproduced with the consent of the copyright holder, Noordhoff Publishing Company, Groningen, Netherlands.

F. Ormeling (✉)
University of Amsterdam, Amsterdam, Netherlands
e-mail: F.J.Ormeling@uva.nl



M. Altić et al. (eds.), *Dissemination of Cartographic Knowledge*, Lecture Notes in Geoinformation and Cartography, DOI 10.1007/978-3-319-61515-8_13

decolonisation. But before that happened, in many colonizing countries it was very much regarded and accepted as a normal relationship between more and less advanced nations.

These views also extended to science, to the arts and to education, and the educational publications in which these views would be most conspicuous, as they would portray both the colonizing and the colonised nations, would be school atlases. For this paper I have analysed the main Dutch school atlas, the Bosatlas, so called after its first editor, Pieter Roelf Bos, first published in 1877 and still going strong (Bos 1877). Currently, I am also contributing to a project aimed at making available all 55 editions of the Bosatlas on the Utrecht University Library website (Ormeling 2016). There, it will be possible to compare all maps in all editions to each other, in order to find the differences between editions, helped by commentaries to highlight these changes. The purpose of this service will be to allow Dutch school children to follow recent history in their own school atlas and to study, diachronically, the recent changes in particular areas. The first part of the project is to scan all 36 editions published up to 1945, and this period, from the congress of Berlin till the end of the Second World War, can be regarded as the heyday of colonialism.

For this paper, these 36 editions of the atlas were analysed from the point of view of depicting colonialism and its implied inequality between colonizer and colonized, trying to find differences in attitude in the language used and in the names, as well as differences in type-fonts; then there are different colours for the portrayal of different types of colonies (spheres of influence, protectorates, colonies), different symbols for towns in colonized areas, differences in the sequence or emphasis in the depicted colonies. As the same maps figured in the atlas in nearly all 36 editions, consecutive versions of these maps have been compared, and changes in the differences in portrayal have been noted and will be explained.

## 2 Analysis of the Bosatlas Maps

So, for this analysis, different aspects of the atlas have been researched, such as:

1. the sequence of the maps and the emphasis on specific areas, together forming the atlas structure;
2. the themes portrayed for colonial areas, as deviating from the themes for the metropolitan areas;
3. the contents of the maps, and the difference between similar kinds of maps for Europe and for colonized areas;
4. discerning between different types of habitat;
5. the colouring of the atlas (such as the colours reserved for European powers, and the way native states were portrayed);
6. other aspects of symbolization (such as the portrayal of settlements and the rendering of boundaries);

7. the naming (the rendering of native names and the imposition of European names in the nineteenth century).

After analysing the differences in these aspects that could be perceived, it will be tried to discern eventual trends in the depiction/rendering of the colonised areas.

## 2.1 Atlas Structure

In the first edition of the Bosatlas, some 60% of the maps are on European countries, the remaining maps are on the world and the other continents. This strong Eurocentrism is long maintained, and is only broken by a more extended coverage of the colonies, to the extent that more information on the colonised areas became available over time. In the first edition there is one map spread devoted to the colonies (4% of the total map space), while in the 36th edition nine spreads (15–18%) cover the colonies. In Fig. 1 one may see the demise of the Dutch colonial empire as expressed in the percentage of map space on Indonesia in the Bosatlas, rising to 15% in 1939, then decreasing to 14 after the transfer of sovereignty to newly independent Indonesia in 1949, to 13% when the Dutch were thrown out in 1956 due to Sukarno's confrontation policy because of the Dutch refusal to hand over Dutch New Guinea as well, to 6% due to the New Guinea armed conflict (1959), and the disappearance of the Netherlands from the

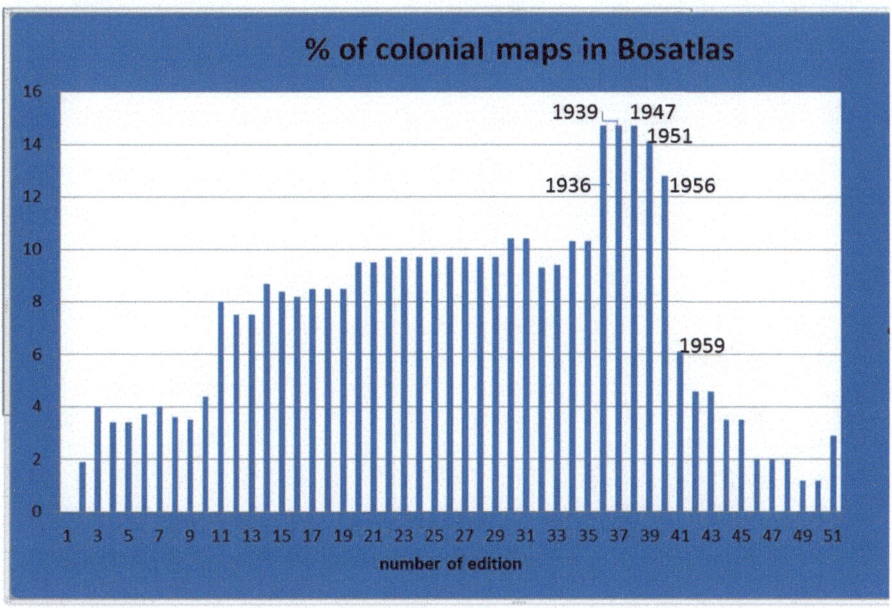


**Fig. 1** Percentage of map space allotted to the colonies in the Bosatlas, per edition

archipelago in 1962, after its transfer of sovereignty over New Guinea to the United Nations in 1961.

## 2.2 *Map Themes*

Apart from physical and administrative overview maps and geographical overview maps, school atlases contain town plans, physical and socio-economic thematic maps and typological maps, for instance portraying different landforms. Amongst the thematic maps in the Bosatlas, economic maps or maps of industrial landscapes (like the Midlands or the Ruhr area) are only incorporated for European countries and North America. For Asian and African countries there are agricultural maps, showing dominant subsistence crops, or dominant cash crops, like the rubber, tea and coffee estates in Sumatra (Fig. 2). For the colonies, geological maps would show the mineral resources as well as the threats to agriculture such as volcanic eruptions. The inset map of Africa shows religion, dominant ways of life such as hunting and gathering, pastoralism or sedentary agriculture, topics not covered for other continents. One may object that there was no manufacturing industry in colonized areas, so it could not be rendered on maps either, but there also were

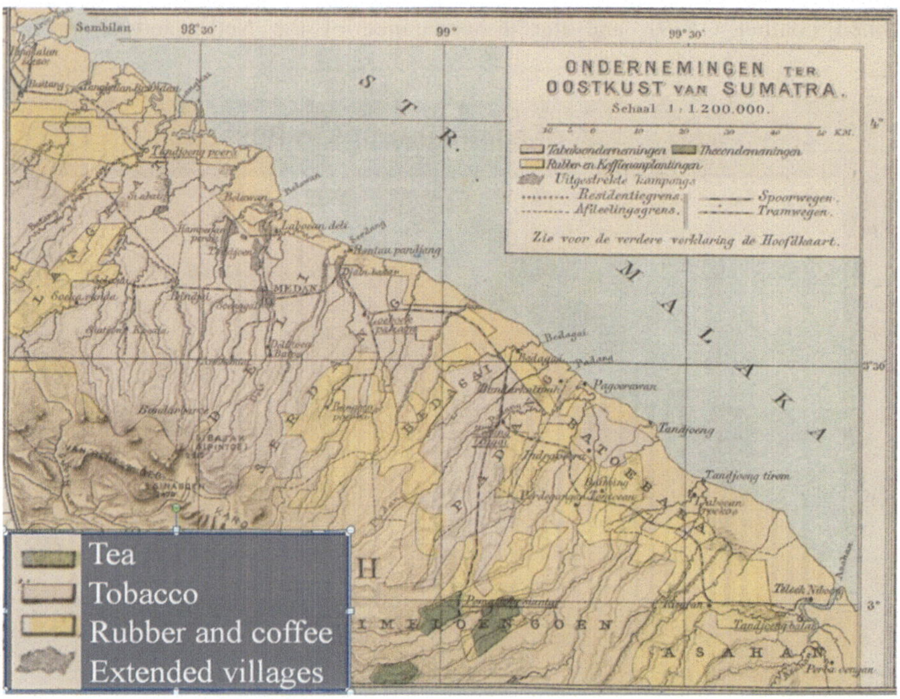


**Fig. 2** Map of the plantation area on Sumatra's east coast (1936) in the Dutch East Indies

no attempts to portray the incipient manufacturing industry in Japan, China or India in the atlas. In the 12th edition the shipping network that held the Indonesian archipelago together was included, showing the frequency of the shipping services. In Africa, after the map showing the ineffectual Congo free-trading zone was abolished, a map on alternative currencies and transportation was incorporated, showing the area where goods could be bartered for cowri shells or Maria Theresia-Thaler, linked to caravan routes.

As important as the themes depicted are of course the themes not depicted. With the Dutch East Indies, the Kingdom of the Netherlands was the state with the largest number of Muslims in the world—but there is no single map or symbol that refers to that aspect amongst the maps depicting this colony—although at the same time, maps are included of the protestant and catholic missionary activities in the Dutch East Indies. The world map of religion in so far reflects the colonial attitudes, in that it refers to heathens as well—we would term them nature worshippers nowadays. On this world map, amongst the Muslims, apart from Shiites and Sunni's, the Senussi are also discerned, as a sort of precursor of the Muslim brotherhood.

## *2.3 Map Contents*

Similar kinds of maps, like town plans, would have different contents for colonial and non-colonial areas. When comparing town plans, the colonial infrastructure (consisting of buildings like the town hall, banks, jail, military hospital, arsenal, shooting range, residencies of governors, theatre, even a state opium factory) would stand out, although theatres, educational establishments (also for indigenous medical personnel) and museums also are indicated. There is hardly anything in these colonial town plans that reminds of the native population. In town plans of the Dutch East Indies the only native building rendered is the kraton, the old Sultan's court, in Palembang. This is due to the fact that the colonial cities, at least in Indonesia and Suriname, were new cities, laid out by the colonizers (Fig. 3).

On the plan of Paramaribo, the capital of the Dutch colony of Suriname, although there would have been Hindu temples and mosques for those that came over as contract labour, only churches are rendered. In 1932 a Mosque was built in the town centre, but it is not rendered on this plan from 1939 (Fig. 4). The Paramaribo plan includes the jetties of the various shipping companies, the one railway line to the interior, the industrial infrastructure (saw mill and gas plant), 11 churches, the sanitary infrastructure (with the military and the mental hospital) as well as the administrative infrastructure (governor's residence, barracks, telegraph and post office).

The mapping of Suriname in the Bosatlas was characterised by the various booms that occurred in that colony (Ormeling 2006). It started with a gold rush in 1876, which even caused a railway line to be constructed from Paramaribo to the

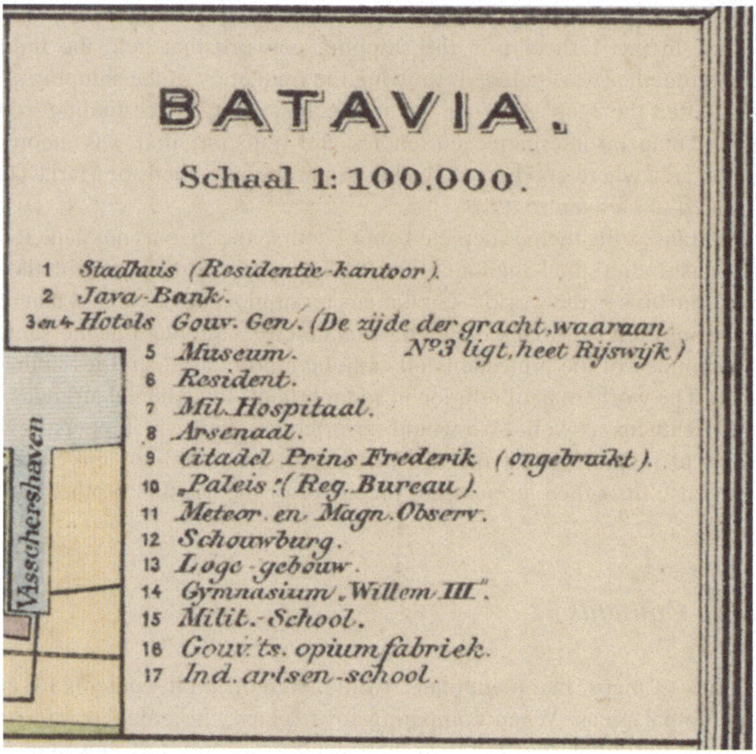


**Fig. 3** Legend of Batavia's town plan (1925)

goldfields in the Lawa river area, 150 km southwards. The concessions for gold mining would be represented, totalling some 500,000 hectares. After 1900 gold production gradually decreases, and instead the population turns en masse to the collection of balata, related to rubber. In the 1920s an agricultural boom takes its place: in the 28th edition (1923) a large area on the north coast is coloured red, in order to show the new agricultural concessions. Later, in 1967 and thus beyond the scope of this project, the bauxite open pit mines would be shown, the next resource on which Suriname depended.

## 2.4 Different Types of Habitat

This aspect illustrates that the concepts adhered to in mapping in the metropolis, did not always fit the colonies. The population in Indonesia was and is often concentrated along rivers or on coastal ridges in extended elongated settlements, which may stretch for tens of kilometres. Nucleated settlement symbols as used in school atlases do not do justice to this kind of habitat. In the Bosatlas it was only taken

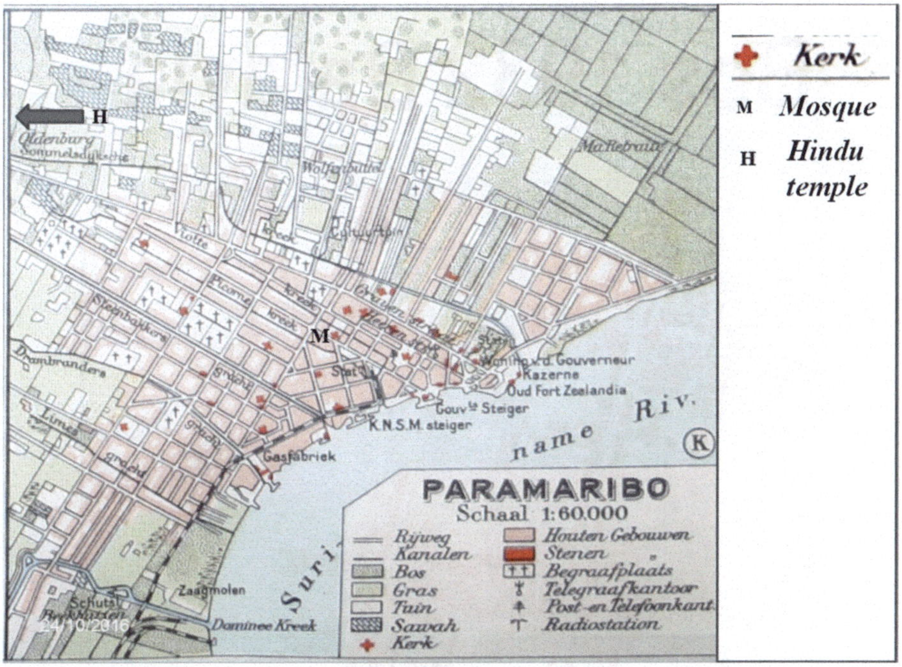


**Fig. 4** Paramaribo town plan (1935); the letter symbols for mosque and Hindu temple have been added by the author

account of on the agricultural map of the Deli plantation area on Sumatra (Fig. 2): ‘uitgestrekte kampongs’ means ‘elongated villages’.

Another aspect of habitat mapping was the customary differentiation between stone or brick buildings and wooden buildings. This was a usual aspect of European military topographic mapping, relevant for the artillery. Within the shooting range of defence positions, it was not allowed for instance to construct stone or brick buildings, only wooden ones would be permitted. This mapping practice was also adhered to in topographic maps in Indonesia, but there, over time, it got another connotation: it came to signify the difference between European and native habitation. The more affluent or more influential European population, consisting of civil servants, plantation owners, traders and shopkeepers, would be able to afford the cooler stone houses, while the native population was used to wooden buildings. After the transfer of sovereignty in 1949, when the Dutch publishers tried to hang on to their traditional market outlets, they changed this kind of presentation for a unified built-up area symbol, which was much less informative from the standpoint of showing urban morphology, however (Figs. 5 and 6).

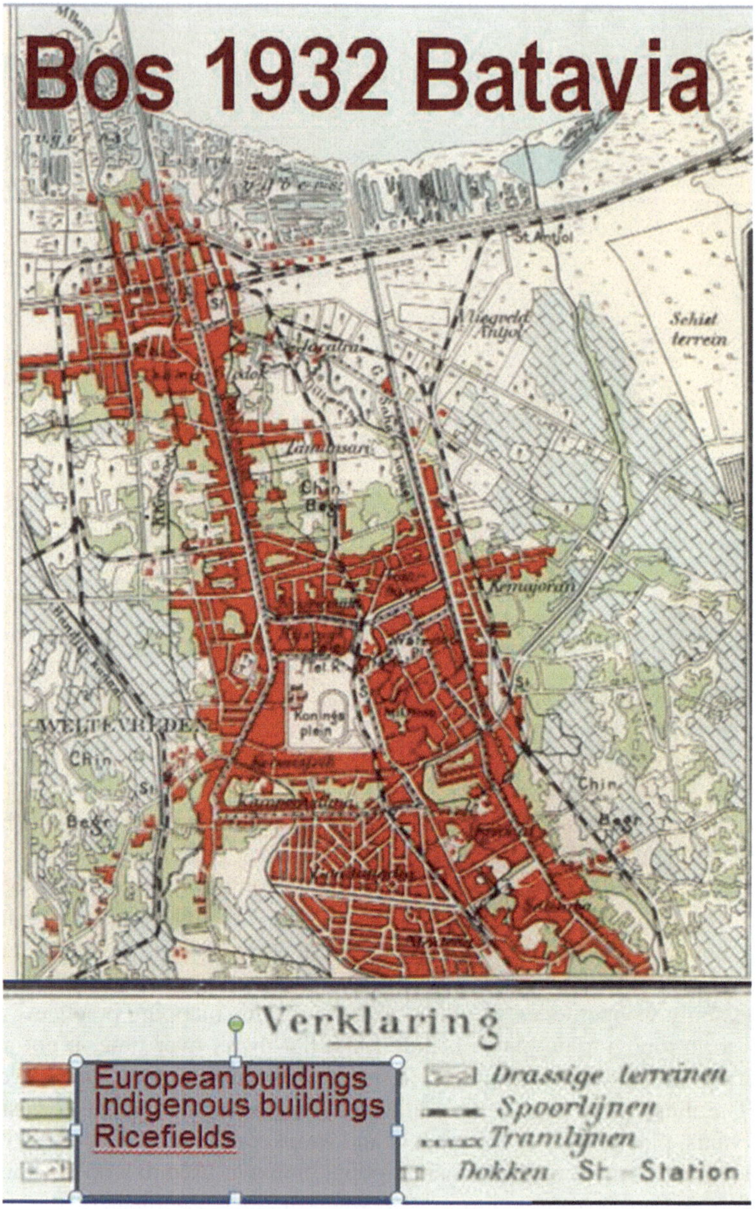


**Fig. 5** Town plan of Batavia from the 1932 edition of the Bosatlas

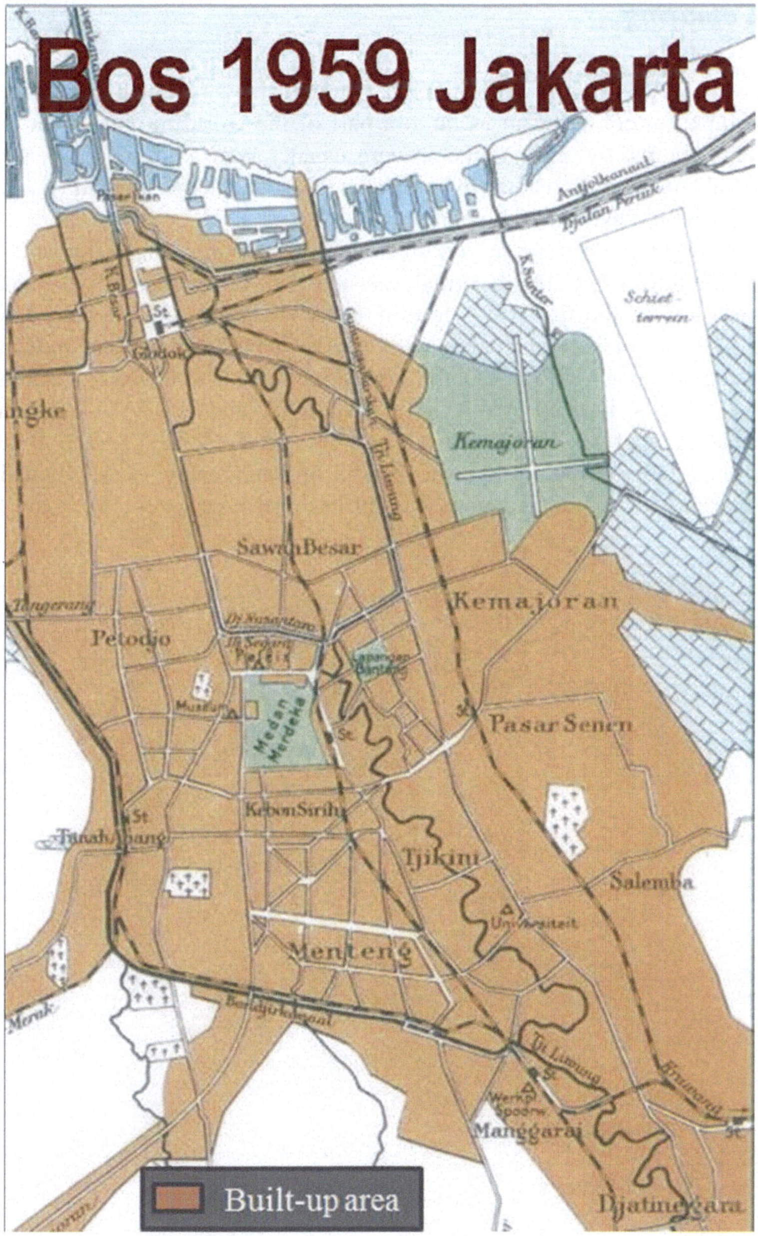


Fig. 6 Town plan of Jakarta from the 1959 edition of the Bosatlas

## 2.5 Colouring

From the point of view of cartography, colonialism very much seems to be an activity aimed at colouring in as large a part of the world in one's own colours. Indigenous realms, by cartographers, were usually kept white (Fig. 7), probably with the hidden denotation: 'still to be colonized' or 'still to be pacified', as was also the case in the Indonesian archipelago.

On the larger scale maps a certain hierarchy in portraying these uncoloured indigenous realms might be discerned, consisting of different types of boundaries and type-fonts. In the first 1877 edition of the Bosatlas, on the map of Africa (Fig. 8), there are indigenous realms with clearly demarcated borders and names in capitals, like those of MOROCCO, the Abyssinian realms of GONDAR and SHOA, in central Africa BORNU, WADAI, BAGIRMI and ADAMAWA, and the South African republics TRANSVAAL and ORANGE FREE STATE. Then there were demarcated sub-realms like the Fellata or Peul realms of Massina, Hausa, Gando and Nijffe, with their names written in lower case letters; there are realms

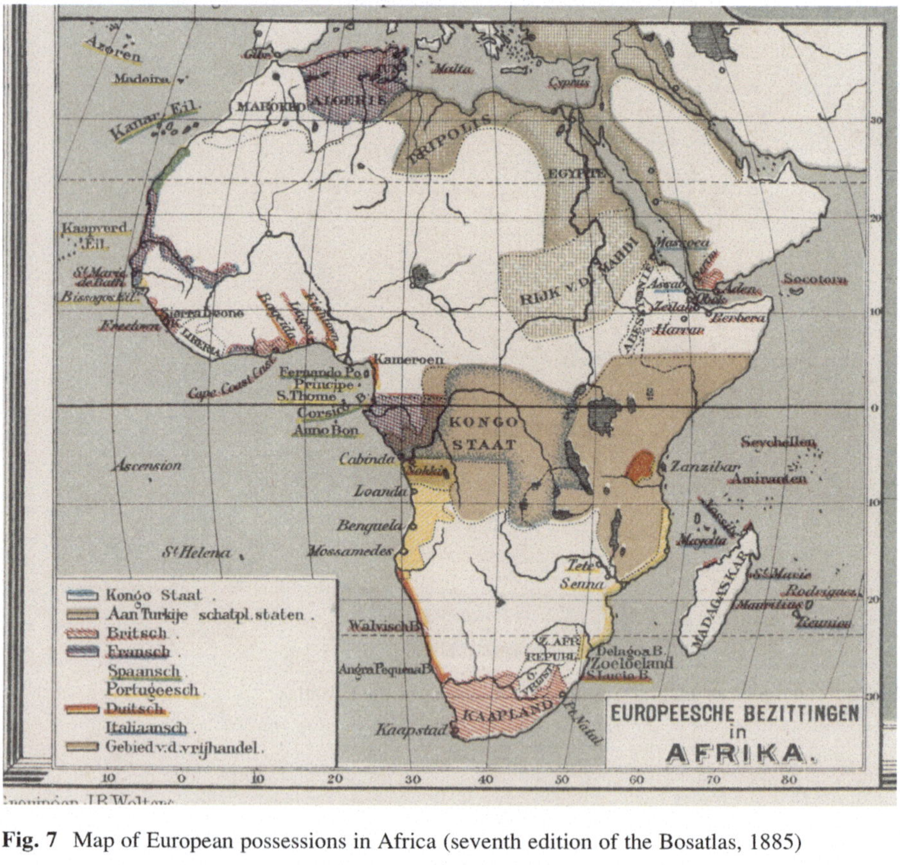


**Fig. 7** Map of European possessions in Africa (seventh edition of the Bosatlas, 1885)

**Fig. 8** Detail of the 1877 map of Africa

with names written in upper case letters, but without demarcated boundaries like those of the YORUBA or ASHANTI, as well as there are realms indicated by lower case letter names, without boundaries, like the Bambarra or Witoe. Lowest in the hierarchy are the lower case italic names, with the connotation of tribal names, like those of the Moors, Tuaregs, Tibbus, Shillooks, Mandingo's, Fan, Kaffirs, Bushmen, Hottentots, Bechuana's or Tswana, Makololo's, Matabele, Gallas, Masai and Somalis, Dinka, Djoer, Bongos and Niam Niam. This indication of the original supposedly tribal population in colonised areas is not restricted to Africa: even on the map of Russia in Europe, the names of the Samoyeds and Kirgiz peoples are indicated in italics, more so on the administrative maps of Asia and North America, where Mongols, Khalkas, Yukagir, Bashkir and Ostyak, Koryak, Yukagir and Chukchen, Eskimo, Sioux, Apaches may roam.

Then **colouring** is used to discern between the colonial powers: The colonial colours are conventionalized to some degree: British areas in pink, French in purple, Italian in dark green, Portuguese in light green, German in (Prussian) blue, the Dutch possessions in orange, and Russian areas in green (after 1917 in red). But within these colour-differentiated territorial claims of the colonial powers, different types of dominance are distinguished: there may be outright colonies or possessions, and there may be protectorates, indicated by lighter shades of the colonial colours. Such distinctions at first were not even explained in the legends, and apparently were regarded as self-explanatory. Finally, in the third edition these lighter shades were explained as *Vassal states*. This phenomenon of protectorates or vassal states discerned by lighter shades of colonial colours can also be seen on the map of Africa, before and after Tunisia was made a French protectorate (1881): On the 1880 map of Africa Tunisia was blank, while on the 1882 map of Africa it has a lighter French shade as compared to Algeria which was regarded a full part of France, and therefore had a full purple colour.

Possessions too small to be given distinctive colouring are indicated as colonial possessions by having their names underlined in the respective colonial colours.

At some point, in 1887, there was a reaction in the Bosatlas against the colouring in of European nations, by giving indigenous realms their own colours, according to their language groups. Now, all of a sudden the Sudan and Sahel are full with these indigenous realms, with their own boundaries. The same goes for southern Africa, with the realms of the Tswana, Matabele, Mambunde, Umzila, Lunda, Kazembe and Urua in Central Africa. Over this, incongruously, the blue rectangle of the Congo Free Trade area is laid. But this colour scheme favouring the native states (Fig. 9) did not last long and was suppressed after three editions, when the atlas returned to showing only the possessions of colonial states in Africa in colours.

In the 14th edition (1899), in a world map on colonies and transportation, a special legend box (Fig. 10) was introduced for 'Staten zonder koloniën' (states without colonies). Apparently, it was clear that colonies were regarded as the norm. The following states were included as colonial states in 1899: Belgium, Britain, Denmark, France, Germany, Italy, Netherlands, Portugal, Russia, Spain, Turkey and the United States. The latter was regarded as a colonial state by the atlas editor

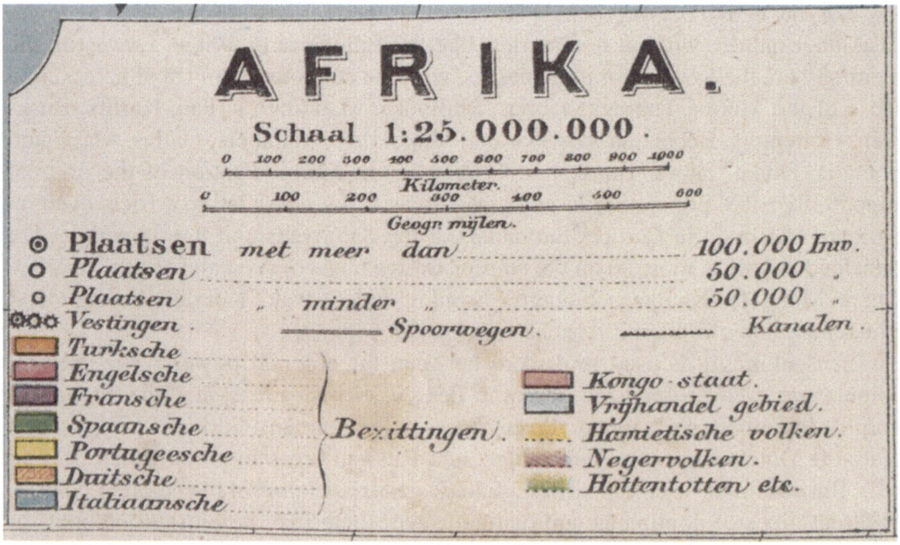


**Fig. 9** Indigenous realms inhabited by Hamitic, Negro and Hottentot (Khoisan) language groups are indicated with distinctive, non-colonial colour bands (see lower right in this legend for the 1887 administrative map of Africa in the Bosatlas).

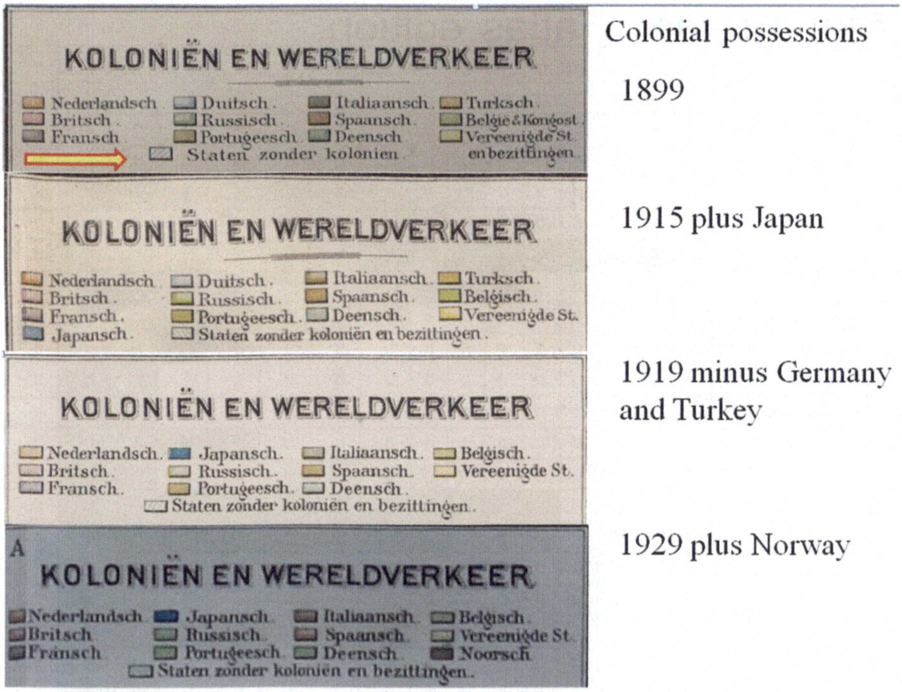


**Fig. 10** Consecutive legends of the same world map in various editions of the Bosatlas

after the war with Spain in 1898. It was interesting that Russia was also seen as a colonial state. Although Japan's expansion in Asia dated from the first Sino-Japanese war 1894–1895 when it acquired Formosa, and although Korea was conquered already in 1910, it was only in 1915 that Japan was first acknowledged as a colonial power on this world map. Apparently, from the point of view of the atlas editor, colonialism was something regarded at first as reserved for European or Caucasian powers. After the first World War, Germany and Turkey were dropped from the legend as colonial powers; to make up for the original number, Norway was included in 1932, although we would nowadays hardly regard Spitsbergen and Bouvet island as colonies.

## 2.6 *Symbology*

It was in the third edition of the atlas that places were first differentiated on the basis of their population numbers—this started (Fig. 11) for the maps of the Dutch provinces in the third edition, was continued in the sixth edition for geographical overview maps of Britain and Spain, and by the 15th edition in 1902 all

**Map theme - atlas edition**

| ¤ | 1¤ | 2¤ | 3¤ | 4¤ | 5¤ | 6¤ | 7¤ | 8¤ | 9¤ | 10¤ | 11¤ | 12¤ | 13¤ | 14¤ | 15¤ | 16¤ | ¤ |
|---|---|---|---|---|---|---|---|---|---|---|---|---|---|---|---|---|---|
| halfronden¤ | ¤ | ¤ | ¤ | ¤ | ¤ | ¤ | ¤ | ¤ | ¤ | ¤ | ¤ | ¤ | ¤ | ¤ | ¤ | ¤ | ¤ |
| zeestromen¤ | ¤ | ¤ | ¤ | ¤ | ¤ | ¤ | ¤ | ¤ | ¤ | ¤ | ¤ | ¤ | ¤ | ¤ | ¤ | ¤ | ¤ |
| isothermen¤ | ¤ | ¤ | ¤ | ¤ | ¤ | ¤ | ¤ | ¤ | ¤ | ¤ | ¤ | ¤ | ¤ | ¤ | ¤ | ¤ | ¤ |
| Europa¤ | ¤ | ¤ | ¤ | ¤ | ¤ | ¤ | ¤ | ¤ | ¤ | ¤ | ¤ | ¤ | ¤ | ¤ | ¤ | ¤ | ¤ |
| Nl-overz¤ | ¤ | ¤ | ¤ | ¤ | ¤ | ¤ | ¤ | ¤ | ¤ | ¤ | ¤ | ¤ | ¤ | ¤ | ¤ | ¤ | ¤ |
| Nl-prov¤ | ¤ | ¤ | ¤ | ¤ | ¤ | ¤ | ¤ | ¤ | ¤ | ¤ | ¤ | ¤ | ¤ | ¤ | ¤ | ¤ | ¤ |
| Alpen¤ | ¤ | ¤ | ¤ | ¤ | ¤ | ¤ | ¤ | ¤ | ¤ | ¤ | ¤ | ¤ | ¤ | ¤ | ¤ | ¤ | ¤ |
| Duitsland¤ | ¤ | ¤ | ¤ | ¤ | ¤ | ¤ | ¤ | ¤ | ¤ | ¤ | ¤ | ¤ | ¤ | ¤ | ¤ | ¤ | ¤ |
| Belg+Zwits¤ | ¤ | ¤ | ¤ | ¤ | ¤ | ¤ | ¤ | ¤ | ¤ | ¤ | ¤ | ¤ | ¤ | ¤ | ¤ | ¤ | ¤ |
| Frankrijk¤ | ¤ | ¤ | ¤ | ¤ | ¤ | ¤ | ¤ | ¤ | ¤ | ¤ | ¤ | ¤ | ¤ | ¤ | ¤ | ¤ | ¤ |
| Oost-Hong¤ | ¤ | ¤ | ¤ | ¤ | ¤ | ¤ | ¤ | ¤ | ¤ | ¤ | ¤ | ¤ | ¤ | ¤ | ¤ | ¤ | ¤ |
| Rusland¤ | ¤ | ¤ | ¤ | ¤ | ¤ | ¤ | ¤ | ¤ | ¤ | ¤ | ¤ | ¤ | ¤ | ¤ | ¤ | ¤ | ¤ |
| Scand¤ | ¤ | ¤ | ¤ | ¤ | ¤ | ¤ | ¤ | ¤ | ¤ | ¤ | ¤ | ¤ | ¤ | ¤ | ¤ | ¤ | ¤ |
| Groot-Brit.¤ | ¤ | ¤ | ¤ | ¤ | ¤ | ¤ | ¤ | ¤ | ¤ | ¤ | ¤ | ¤ | ¤ | ¤ | ¤ | ¤ | ¤ |
| Spanje¤ | ¤ | ¤ | ¤ | ¤ | ¤ | ¤ | ¤ | ¤ | ¤ | ¤ | ¤ | ¤ | ¤ | ¤ | ¤ | ¤ | ¤ |
| Italie¤ | ¤ | ¤ | ¤ | ¤ | ¤ | ¤ | ¤ | ¤ | ¤ | ¤ | ¤ | ¤ | ¤ | ¤ | ¤ | ¤ | ¤ |
| Balkan¤ | ¤ | ¤ | ¤ | ¤ | ¤ | ¤ | ¤ | ¤ | ¤ | ¤ | ¤ | ¤ | ¤ | ¤ | ¤ | ¤ | ¤ |
| Azië¤ | ¤ | ¤ | ¤ | ¤ | ¤ | ¤ | ¤ | ¤ | ¤ | ¤ | ¤ | ¤ | ¤ | ¤ | ¤ | ¤ | ¤ |
| ZO-Azië¤ | ¤ | ¤ | ¤ | ¤ | ¤ | ¤ | ¤ | ¤ | ¤ | ¤ | ¤ | ¤ | ¤ | ¤ | ¤ | ¤ | ¤ |
| Insulinde¤ | ¤ | ¤ | ¤ | ¤ | ¤ | ¤ | ¤ | ¤ | ¤ | ¤ | ¤ | ¤ | ¤ | ¤ | ¤ | ¤ | ¤ |
| Java¤ | ¤ | ¤ | ¤ | ¤ | ¤ | ¤ | ¤ | ¤ | ¤ | ¤ | ¤ | ¤ | ¤ | ¤ | ¤ | ¤ | 29¤ |
| Sumatra¤ | ¤ | ¤ | ¤ | ¤ | ¤ | ¤ | ¤ | ¤ | ¤ | ¤ | ¤ | ¤ | ¤ | ¤ | ¤ | ¤ | 36¤ |
| Afrika¤ | ¤ | ¤ | ¤ | ¤ | ¤ | ¤ | ¤ | ¤ | ¤ | ¤ | ¤ | ¤ | ¤ | ¤ | ¤ | ¤ | ¤ |
| N-Amerika¤ | ¤ | ¤ | ¤ | ¤ | ¤ | ¤ | ¤ | ¤ | ¤ | ¤ | ¤ | ¤ | ¤ | ¤ | ¤ | ¤ | ¤ |
| Ver.St.¤ | ¤ | ¤ | ¤ | ¤ | ¤ | ¤ | ¤ | ¤ | ¤ | ¤ | ¤ | ¤ | ¤ | ¤ | ¤ | ¤ | ¤ |
| Z-Amerika¤ | ¤ | ¤ | ¤ | ¤ | ¤ | ¤ | ¤ | ¤ | ¤ | ¤ | ¤ | ¤ | ¤ | ¤ | ¤ | ¤ | ¤ |
| Australië¤ | ¤ | ¤ | ¤ | ¤ | ¤ | ¤ | ¤ | ¤ | ¤ | ¤ | ¤ | ¤ | ¤ | ¤ | ¤ | ¤ | ¤ |

**Fig. 11** Table showing in which edition a legend for the settlement based on population numbers was introduced for the maps of various countries or continents. For all of them by the 15th edition (1902) such legends were incorporated except for the maps of the Dutch East Indies where the practice of basing the differentiation of settlements on their administrative status persevered until WWII

geographical overview maps and all administrative maps of continents included such legends. In 1902 even the geographical overview map of Indonesia differentiated places on the basis of their population numbers in combination with their administrative status, which was shown by an underlining of the names. But on the more detailed maps of the Dutch East Indies, place symbols continued to be differentiated between only on the basis of their administrative hierarchy, such as places where colonial governors, residents or commissioners were stationed. For Java this practice ended in the 29th edition of 1924, for the other areas only in the 36th edition of 1939.

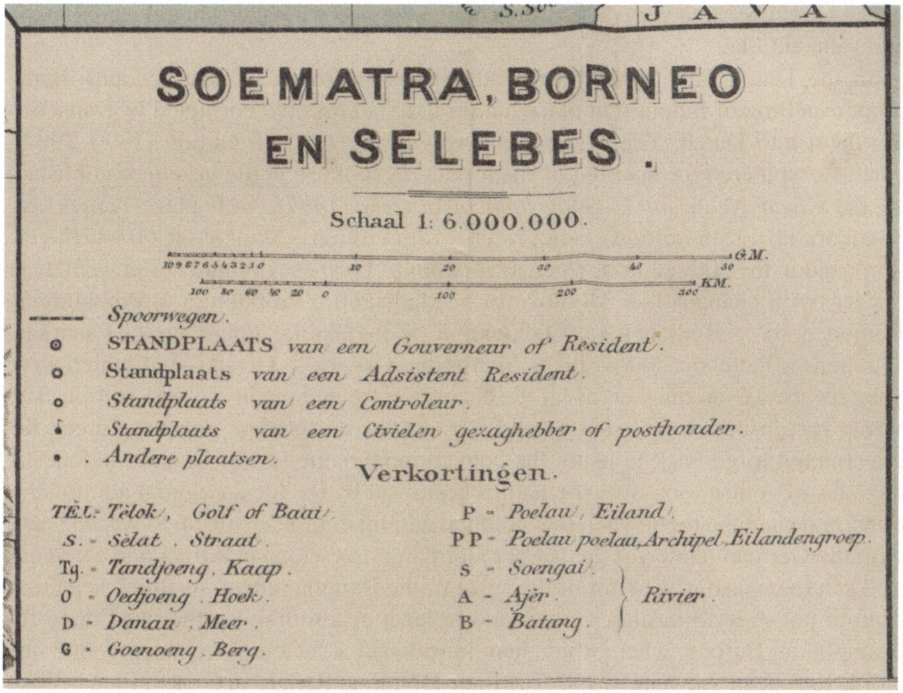


**Fig. 12** Legend for the map of the Indonesian isles from the 11th edition (1893) with administrative differentiation of the settlement symbols

In Fig. 12, showing the legend of the map of the Indonesian outer islands, places are differentiated as places where a governor or resident was stationed, stations of assistant-residents, stations of controllers, stations of minor civil administrators and 'other places'.

## 2.7 Naming

Here of course the fact that European or Japanese names were imposed, stands out, not only for colonial settlements, like Leopoldville, Brazzaville, Victoria, Santa Isabel, Nouvel Anvers, Coquilhatville, Ponthierville, Albertville, Elisabethville, Port Francqui, Abercorn, Broken Hill, Johnston, Batavia, but even for countries, like Rhodesia, Kamerun, Sierra Leone, Liberia, the Philippines, etc. and of course the honours bestowed on explorers like Livingstone and Stanley are highlighted on the map (Livingstone Falls, Livingstone Mountains, Stanley Pool, Stanley Falls, etc.). And even if the colonizers were willing to accept indiginous names on their maps, these more often than not were maltreated, and misspelled, and once codified on the map it became very difficult to change them to what they should have looked and sounded like.

In the Dutch East Indies, twice attempts were made to take account of the proper spelling of Indonesian place names that had become corrupted by transcribing them into Dutch. The first attempt was in the 10–15th edition (1890–1902), when Bos adhered the spelling of the toponyms to those in the gazetteer published by the *Royal Netherlands Geographical Society* (1897), with place names like Soematra (10–15th edition), Selebes (10–12th edition), Mangkasar (10–12th edition), and in the 30th edition, when adaptation to Javanese pronunciation went even further, with changes like Modjokerto > Madjakerta; Probolingo > Prabalingga, Bodjonegoro > Bodjanegara, Djokjakarta > Jogjakarta; Ponorogo > Panaraga, Tjirebon > Tjerebon, Sidoardjo > Sida'ardja, Grissee > Gresik—but which were to be reversed even quicker, as all the changes were undone in the next edition. The White backlash, colonial society and most of the colonial newspapers forced the government to go back to using the conventional name forms. Outside Indonesia, the atlas of course took over the names bestowed by the local colonial administrations, and changed them when that colonial administration changed, as was the case with the German names in the former German possessions in Africa and Oceania.

Then there also is the fruit of the spirit of the Enlightenment, in which scientists tried to put everything in a system and order it accordingly as they had done for example in Europe where they had introduced educational concepts like the North-European Lowlands, the Castilian Dividing Range, and the Finnish Lake Plateau, for the benefit of learning the geographical structure of the continent. In the same vein European Geographers divided African mountain chains into North and South Equatorial ranges, and divided the Indonesian archipelago in the Greater and Lesser Sunda isles, the Sunda and the Sahul shelfs, and in the Caribbean discerned between the Greater and Lesser Antilles. For the sake of geographical order, these encyclopedists or educationalists also brought order to the seas, and each stretch of the seas got its distinct name.

In this spirit, the waters between the Indonesian isles got baptised, a process in which first names like Java Sea had the connotation of the waters in which Java was situated, or the sea to be traversed in order to get to Java, but later on these named seas became strictly demarcated. In the course of the nineteenth century the name Java Sea came to be reserved for the waters in-between Java and Borneo. In the same way (Ormeling 2010) the Celebes Sea was established as the sea in-between Celebes and the Philippines, after this name had first been consigned to the waters in-between Celebes and Flores. Those became the Flores Sea however, and Celebes Sea got its present place, forcing the name Sulu Sea to the waters northwards of the Sulu archipelago. The name *Sunda Sea* was used in geological texts for the combination of Java and South China Sea (covering the Sunda shelf); it has also been used as an alternative for the combination of Java Sea and Flores Sea or for the Java Sea only (see Van der Aa, Nederlandsch Indië 1849), and also as an alternative name for the sea area now called Flores Sea. Its disturbing presence disappeared in the twentieth century.

## 3 Missing Maps or Symbols

What was not shown? Which aspects of colonial society were not expressed on maps? We have until now seen how colonialism expressed itself in the atlas as a benign process, this colouring in of the tropical parts of the earth that before were white, unexplored, and after exploration white came to mean not colonized, with the connotation uncultured. Afterwards there came towns, roads and railways, dams and canals for irrigation. What we have not seen is the process of subjugation, however much it may have been based on humanism, idealism, the desire to improve the living conditions of the natives or their souls. We have not seen the theatres of war, with the exception of one map, of Aceh, in the 19–30th edition, but even that map bore no signs of the continuous military actions (Fig. 13). This is opposed to Europe, or other areas outside the Dutch colonies where, even if the Dutch themselves were able to keep their neutrality, the Bosatlas maps bore many reminiscences of military actions, to Napoleonic wars (like Austerlitz, Waterloo, Quatre Bras, La Belle Alliance (a farm!), the Berezina, Wagram and Aspern), French colonial wars (with Bac-ninh in Tonkin where the French army had a

**Fig. 13** Map of the Aceh theatre of war in the 1924 edition

decisive victory over the Empire of Annam in 1884), British colonial wars in Sudan, in South Africa (with Rorkes's drift in Zululand, and Majuba), the First World War (with the Doberdo plateau in Istria and Doiran north of Thessaloniki, and Gallipoli, besides the trenches and fortifications in France (the 1916 map of France contained 84 fortification symbols in Northern France). Military actions or pacification actions continued in the Dutch East Indies, the last one in Bali in 1908, but there were no traces of them on the map.

## 4 Trends

The following trends can be discerned when looking at all the consecutive maps in the atlas on which colonies were displayed:

1. A gradual partition or filling of all tropical areas between or by European colonial powers. This starts as commercial trading stations, taken over by state authorities, exchanged for other areas, more conveniently located to control resources. These colonies, begun as single ports or railway heads, gradually extend themselves inland until their advance is blocked by other colonies.
2. A general disregard of native states.
3. Areas can be designated on the map as being part of a colonial state for decades, without any actual control by that state whatsoever.

### *4.1 Gradual Extension of the Areas Controlled by the Colonial Powers*

The places Assab and Massua on the Red Sea are indicated as Italian in the seventh edition of 1885—They were acquired in 1880 by Italian shipping firms, although located on the Red Sea coast nominally possessed by Turkey. In 1882 these commercial rights were converted into territorial rights when the Italian government took them over, thus starting the colony Italian East Africa. Elsewhere commercial interests did likewise: The Germans had commercial bases or lodges in Nokki on the Congo mouth, in Lüderitz on the Angra Pequena bay in Namibia, in East South Africa's St Lucia bay and in the Usagara area in present Tanzania. There was much juggling with such claims that easily could be exchanged for better territorial prospects elsewhere: St Lucia Bay for instance was exchanged for a British trading post on the coast of Cameroon which the Germans fancied as well.

When France went after Morocco, it appeased Spain and Germany, the first by allotting them the Rif area in the north and the Sahara coast in the South of Morocco, the Germans, by dismembering their own colony of Equatorial Africa and so doubling German Cameroon in size, that so gained access to the Congo river. By projecting and comparing all the different editions of this map this colonizing process would be visualized.

## 4.2 Disregard of Native States

In this whole colonizing process native states or realms were disregarded on the map, even if they completely blocked the advance of the colonial powers. The realm of the Mahdi in the Sudan is an interesting case. This charismatic preacher had revolted against Egypt in 1880 out of discontent over the growing British influence in Egypt; he had defeated an Egyptian army at El Obeid in 1881, and by 1885 had become master of the whole Sudan. After his death that year his successor was able to hold the country together until 1899 when defeated by Kitchener. But even in the 13th edition, of 1897, the area was still labelled as Egyptian Sudan, with no trace of the Mahdi.

In the 17th edition of 1906, a mysterious yellow band is rendered along the coast of the Indian Ocean in present Somalia. It is explained on the map by the text '*Gebied van den Moellah*', the realm of the Somali freedom fighter Mohammed Abdullah Hasan, the 'Mad Mullah' according to British newspapers, who was able to keep the independence of his Darwish state between 1900 and 1920, when due to

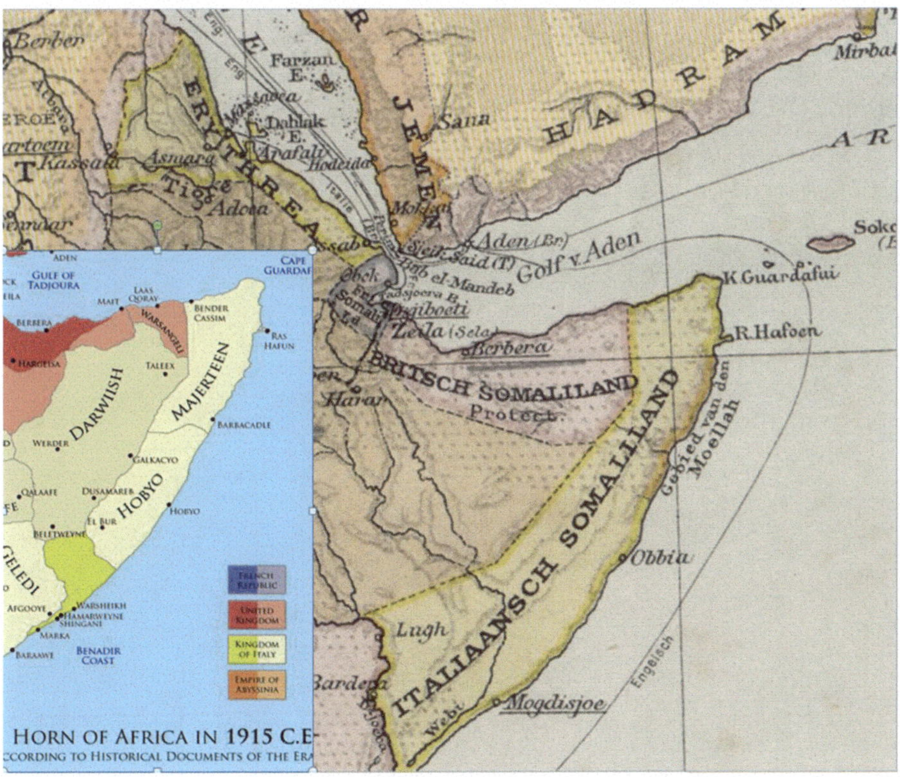


**Fig. 14** The Horn of Africa in the 1914 edition of the Bosatlas. The inset map showing the extent of the Darwish state in 1915, has been taken from the Wikipedia website 'Dervish State', accessed 27-10-2016

the launching of aeroplanes he no longer was a match for British and Italian forces. Before his demise the area he controlled equalled that of British Somaliland in size, but cartographically he is only accounted for by this insignificant yellow band along the coast in Fig. 14.

## 4.3 *Paper Colonies*

The 19th edition of the Bosatlas published in 1910 (Fig. 15) shows the Kufra oases in the Libyan Desert as being under British control, because of the pink band along its boundaries. This was the area were the Senussi's were in charge, and there was no, not even nominal British authority over it, it sided with Turkey in the First World War. It would only be subjugated by the Italians in the 1930s, also by the use of aeroplanes and bombs, without any objections by the British, who certainly would have protested had they been in control.

**Fig. 15** Kufra oasis (Koefra Oasen) in the 1910 edition of the Bosatlas

## 5 Colonialism and Bias

So these are the traces of colonialism in this school atlas that were discovered—there may be even more, to which those conditioned by the same colonialism have been blind. As indicated above, scans of all the editions of the atlas will be made available/accessible through the internet, in a website aimed at high school pupils, and this analysis or deconstruction of the maps was undertaken in order to make them aware of these colonial attitudes, as codified in the maps.

But these attitudes should not be confused with the almost normal (because it occurs in practically all school atlases) bias inherent in school atlases: these atlases tend to be nationalistic, self-centred, emphasizing their national relationship with the world, highlighting the topics they stand out in, worldwide, disregarding achievements by other countries and often also reflecting the hobby horses of the consecutive editors (Ormeling and Van der Vaart 2005). Countries farther away from us tend to get less emphasis than countries closer by, unless for some reason (language, religion, way of life, etc.) we feel more related to them. As thus the emotional distance to them would be smaller, they would get more emphasis than other countries at the same distance from us. These countries, perceived as more related, may even be former colonies!

## References

Bos PR (1877) School atlas der geheele aarde. J.B.Wolters, Groningen

Ormeling F (2006) De weergave van Suriname in de Bosatlas 1877–1940. Caert Thresoor 25 (2):25–32

Ormeling, F (2010) Changing Indonesian Sea names. In: Proceedings of the 16th international seminar on Sea names, Den Haag 2010. The Society for East Sea and the Northeast Asian History Foundation, Seoul, pp 133–147

Ormeling, F (2016) A total history Atlas: the Bosatlas editions online 1877–1939. In: 2016 conference on the making of historical atlas. Northeast Asian History Foundation, Seoul, pp 285–303

Ormeling F, van der Vaart R (2005) Biografie van de Bosatlas. Wolters-Noordhoff, Groningen

van der Aa AJ (1849) Nederlandsch Indië of Beschrijving der Nederlandsche bezittingen in Oost-Indië. J. F. Schleijer, Amsterdam

**Ferjan Ormeling**, from 1985–2009 held the cartography chair at Utrecht University, specializing in thematic cartography, atlas production and toponymy. As a member of the Explokart research group he later transferred from Utrecht to Amsterdam University. He has written extensively about the history of cartography of the Netherlands East Indies (Indonesia) and on the history of school atlases in the Netherlands. His latest project (2015–2017) for Utrecht University Library was the preparation of a website in which all maps of all 55 editions of the Bos-atlas can be compared one to another.

# The Dutch Commemorative Toponyms in the Seventeenth Century East Asia, Based on the Cartographic Works Left by the Dutch East India Company (VOC)

**Peter Kang**

**Abstract** The paper mainly aims at discussing the commemorative toponyms in East Asia used by the Dutch East India Company (*Verenigde Oostindische Compagnie* or VOC) in the seventeenth-century mainly by examining contemporary Dutch cartographic works. The VOC cartographic works include those for maritime exploration of the coastal areas and off-shore islands of China and Japan, as well as the only territorial colony of the VOC, the island of Taiwan, then being called Formosa. The paper classifies the VOC commemorative place names into those of ship names, personal names, geographical names, and the highlighted nature of certain incidents. In addition to exploring the nature of the VOC commemorative place names, the paper also concludes with the way that toponyms were displayed on VOC maps which reveals the language boundary in the southeastern coast of China.

## 1 Introduction

The Dutch East India Company (*Verenigde Oostindische Compagnie*) or VOC, the European newcomers to the lucrative Asian maritime trade at the turn of the seventeenth century, vainly attempted to establish direct trading linkages with Ming China in the 1600s. After the establishment of its headquarters in Batavia, Java, the Dutch launched an abortive attack on the Portuguese at Macau and later moved to occupy the Pescadores, the archipelago between China and the island of Taiwan, then called Formosa, in 1622. From there the VOC started to survey the southeastern coast of Ming China. The result of their hydrographic survey was later turned into reports and cartographical works, as well as a large number of toponyms being assigned to various visited or detected places.

The map entitled "Chart of a part of the China Sea and Coast from Punta Lampacao to the Overveens Islands, the island of Formosa" (*Kaart van een*

P. Kang (✉)
National Donghwa University, Hualien, Taiwan
e-mail: kang@gms.ndhu.edu.tw

 
M. Altić et al. (eds.), *Dissemination of Cartographic Knowledge*, Lecture Notes in Geoinformation and Cartography, DOI 10.1007/978-3-319-61515-8_14

*gedeelte der Chinesche Zee en Kust, van Punta Lampa cao tot de Overeens Eilandens, het Eiland Formosa ens. Hierop komt mede voor: de rivier van Canton, van Linkian de mond tot Fangsucis*) witnesses the toponyms being allocated by the VOC along the southeastern coast of China.[1] It summarizes the toponyms baptized by the VOC personnel during their hydrographic mission.

The VOC place naming system could be generally categorized into four types: Dutch commemorative, Dutch landscape-descriptive, Dutch spatial oriented and spellings based on the local languages. If we use the VOC naming practices in Taiwan as examples, place names such as *Gouden Leeuwseylandt* or Golden Lion Island, *Amsterdams polder*, *Verraders of Moordenaar Rivier* or Traitors or Murderers River, and *Van der Burch polder* all belong to the Dutch commemorative place names. The aforementioned four cases are the name of a VOC sailing vessel, a place name from the Netherlands, the highlighted nature of an incident, and a contemporary prominent figure, respectively.[2]

The Dutch landscape-descriptive place names include *Appenbergh* or Ape Mountain, *Schilpads Baai* or Turtle Bay, *Duyvels hoek* or Devil's Corner, and *Dorre Eiland* or Arid Island. *Suydooster Eylandt* or Southeastern Island is classified as a Dutch spatial oriented place name. Place names such as *Sinckan* (Sin-káng) or New Port and *Tampsuy* (Tām-chúi) or Fresh Water are derived from Hokkien, while *Pimaba* and *Dorkop* come from Formosan Austronesian and are spellings based on local languages.

There may happen to be two of the aforementioned four types of place names for a designated geographical feature on a single cartographical work. It mostly happens with one example of Dutch commemorative or landscape-descriptive nomenclature, followed by one based on the spelling of the local language. For instance, the modern Chinese toponym Jinghaiwan (靖海湾) or Bay of Pacified Sea was recorded as *Wieringens bai of Zinhaitou*. *Wieringens bai* stands for the Bay of *Wieringen*, after the name of a VOC yacht (*jachten*) which found shelter there in a typhoon. *Zinhaitou* is Jinghaitou (靖海头) or the Head of Pacified Sea, the local name for Jinghaiwan. *Tangsoa of Victorie bay* is today's Dongshan (铜山) or Copper Mountain. *Tangsoa* is the local Hokkien pronunciation *Tâng-soa*$^n$ for Dongshan whereas *Victorie bay* is the name of a VOC yacht that participated in the hydrographic survey of 1622 and which rescued six VOC sailors who were once trapped on land (Cheng 2014). It is *Wieringen* and *Victorie* that constitute the Dutch commemorative nature of toponyms upon which the current paper is based.

[1]4 VEL 272, National Archief, the Netherlands. See: http://www.gahetna.nl/collectie/afbeeldingen/kaartencollectie/zoeken/weergave/detail/q/id/af99516c-d0b4-102d-bcf8-003048976d84. The map is originally produced by Isaak de Graaf. A list of geographical names on that map and their modern Chinese equivalents is also available in *Grote Atlas van de Verenigde Oost-Indische Compagnie* (Schilder et al. 2006: 353).

[2]Some would classify the place names on the VOC cartographic works into descriptive names, commemorative names (i.e., place names at home), the names of contemporary prominent figures, explorers, ships, and the conceptual names (Ormeling 2006: 33–37). The current paper treats the names of contemporary prominent figures, explorers, ships as commemorative ones.

## 2 Ship Names of the VOC

The names of VOC sailing vessels were mostly utilized to commemorate the events associated with the ships themselves. They were mostly named after islands and coastal water bodies such as bays. For instance, in the case of Taiwan, the VOC cartographical works mostly marked Lamey Island off the southwestern coast of Taiwan as *Gouden Leeuwseylandt* after the murder of the crew members of the *Gouden Leeuw* or Golden Lion in 1622 by the inhabitants of Lamey Island when the former went ashore to fetch fresh water and search for provisions.[3]

The VOC hydrographic survey on the coast of Southeast China conducted in 1623 and in the subsequent years produced many Dutch commemorative topomyms named after ships (Cheng 2014). From the north we had *Domburch* for modern Huangwanyu (黄湾屿) or the Islet of Yellow Bay, *Zirik Zee* (*Zierikzee*) for modern Beigan (北竿) or North Bamboo Pole, *Texel* for modern Jingwei (井尾) or Rear of the Well, *Victorie* for modern Dongshan (东山), *Wieringen* for modern Jinghaiwan (靖海湾), *Groeningen* for modern Tianwei (田尾) or Rear of the Farmland, and *Haarlem* for modern Tai Pan Wan (大鹏湾) or the Bay of the Roc.

In addition to the ship names *Victorie* and *Wieringen* that are mentioned above, *Domburch* was one of the yachts being assigned to inspect the coastal China north of Min Jiang (闵江) or Min River in 1629. It reached Huangwanyu as the northernmost point and thus left its name there. *Zirik Zee* was the flagship of Cornelis Reijsersen, who led the VOC fleet to East Asia in 1622 and later constructed a stronghold in the Pescadores.[4] *Groeningen* was one of the yachts led by Upper-merchant Cornelis Neyenrode in 1622 to coastal China. It failed to anchor at the Pescadores during its return trip and drifted during a storm to Tianwei, where the crews regained the control of the ship. Thus Tianwei was named after *Groeningen* to commemorate the rescue. *Haarlem* was the name of another ship that drifted during a storm to coastal China in 1622. It ran into the aforementioned *Groeningen* and thus obtained supplies and reinforcements from the latter. The place where Haarlem was rescued later acquired the name *Haarlem* (Cheng 2014).

[3]The *Gouden Leeuw* was later wrecked on March 3, 1634 on a reef in Borneo. The reef was also named *Gouden Leeuwseiland* afterwards. Source is from De VOC Site: Gouden Leeuw (1615). Available at: http://www.vocsite.nl/schepen/detail.html?id=11714.

[4]In addition to *Zierikzee*, According to Dapper (1670: 73), *Overveen*, or modern Gaodeng Dao (高登岛) and *Naerden*, were ship names employed for the three islands close to the city of modern Fuzhou (福州). However, the reasons for naming the islands after *Overveen* and *Naerden* are unknown to us.

## 3 Place Names from the Netherlands

Place names from the Netherlands were principally used for the naming of islets and reclaimed polders. The Dutch geographical names used for place-naming normally have no direct relationship with the baptized places. The naming after place names from the Netherlands are mostly for the creation of symbolic landscapes of nostalgia, and is normally associated with the naming of a group of small islets (Ormeling 2007: 106–108). Nevertheless, there are few cases for the glorification of VOC achievement. The place names from the Netherlands for this purpose are where the six VOC Chambers (*Kamers*), namely *Amsterdam*, *Delft*, *Hoorn*, *Enkhuizen*, *Rotterdam*, and *Middelburg*, are located, and the six aforementioned place names would appear altogether exclusively.

The place names from the Netherlands used for symbolic landscapes of nostalgia on the islets along coastal China included *Amsterdam*, *Rotterdam*, *Vissingen* (*Vlissingen*), *Haarlem*, *Hoorn*, *Delft*, *Enchuissen* (*Enkhuizen*), *Middelburg,* and *Durkerdam* (*Durgerdam*). They are concentrated on the area north of *Stad kita* or today's Luo Yuan (罗源) and south of the Banchao Archipelago (半招列岛), which was named as *Walvis* or Whale.

The place names derived from the Netherlands for the glorification of VOC achievement mostly appeared on the territories where the VOC had relatively stronger control such as the colonies. Normally this type of naming is different from the naming of a construction such as *Fort Zeelandia* in South Taiwan after the westernmost province of the Netherlands, Zeeland, or *de punt Noortholland* for a fort in North Taiwan bearing the same name. A place solely named after geographical names in the Netherlands could not be easily detected and portrayed as a visible architectural structure. The glorification of VOC achievement was not materialized by the visible architecture but by the significance of the place name itself. Therefore, not all the place names derived from the Netherlands would achieve this goal. Only selected place names that denote much more significance to the contemporary VOC officials would do. It explains why the six place names where the VOC Chambers located were regarded as the glorification of VOC achievement.

A good example is the reclaimed agricultural fields around Fort Provintia or Sakkam (赤崁), today's downtown Tainan (臺南), Taiwan. In addition to the personal names of high-ranking VOC officials, only the place names where the six VOC Chambers were located were employed for the land titles. No other place names from the Netherlands were used here. Table 1 displays the six polders named after place names where the VOC Chambers were located.[5]

---

[5]The VOC names for the polders in Tables 1 and 2 were in both Dutch and Hokkien. The characters for the Hokkien language here follow Ang (2006: 70–85).

**Table 1** Polders named after place names in the Netherlands

| VOC toponyms | Hokkien | Characters | |
|---|---|---|---|
| **Amsterdam**s Polder ofte Orakan | *Hó-liâu-káng* | 禾寮港 | VOC Chamber |
| **Delft**s Polder ofte Leiseinhoin | *Jī-chàn-hâng* | 二層行 | VOC Chamber |
| **Hoorn**s Polders ofte Tonglouw | *Tiong-lâu* | 中樓 | VOC Chamber |
| **Enckhuysen**s Polder ofte Lockauw | *Lòa-kháu* | 瀨口 | VOC Chamber |
| **Rotterdam**s Polder ofte Hoenjouwa | *Hun-iû$^{n}$-á* | 薰洋仔 | VOC Chamber |
| **Middelburch**s Polder ofte Scholsiackhou | *Kū-siā-kháu* | 舊社口 | VOC Chamber |

## 4 Contemporary Prominent Figures

Places named after contemporary prominent figures ranged from water bodies and islands to reclaimed polders. Personal names used for place-naming mostly have no direct relationship with the designated places either.

Regarding the naming of the contemporary prominent figures for water bodies, for instance, modern Jiazi Gang (甲子港) or the Port of the Sixty-year Cycle in China was named after the aforementioned VOC fleet commander Cornelis Reijsersen: *Reiersebsbai* or the Bay of Reijsersen. The case of *Straat Diemen* or Diemen's Strait, the modern Osumi-kaikyo (大隅海峡) or the Osumi Strait in Japan, was named after the VOC Governor-general Anthony van Diemen (1635–1645). However, the hydro-geographical survey in Japan led by Maarten Gerritsz. Vries in 1643 produced another toponym named after Anthony van Diemen, the *Straat Antony van Diemen* (Leupe 1858: 313–314).

The aforementioned agricultural fields around Fort Provintia also produced eight polders named after personal names among all of the 14 baptized ones. Seven of the eight anthroponyms named after VOC officials were the names of Governors in Formosa, namely Sonck (Martinus Sonck, 1624–1625), De Witt (Gerard Frederikszoon de With, 1625–1627), Nuijts (Pieter Nuyts, 1627–1629), Putmans (Hans Putmans, 1629–1636), Van der Burch (Johan van der Burg, 1636–1640), Traudenius (Paulus Traudenius, 1640–1643), and Lemaires (Maximilian le Maire, 1643–1644). The only personal name who was not a Governor in Formosa was Koeckebacker (Nicolaes Coeckebacker), who was the Chief of the Dutch trading factory at Hirado (1637–1639) and also a commissioner who inspected Tonkin and Formosa in 1639. Table 2 shows the eight polders named after personal names of contemporary high-ranking VOC officials.

## 5 The Highlighted Nature of Incidents

The places named after the underlined nature of incidents include islands and water bodies. The emphasized events used for place-naming normally have the most direct relationship with the designated places, at least to the VOC personnel who

**Table 2** Polders named after personal names of the VOC

| VOC toponyms | Hokkien | Characters | |
|---|---|---|---|
| **Soncks** Polder ofte Tampoasiam | *Tháng-phôa$^n$-chàn* | 桶盤棧 | Governor (1624–1625) |
| **De Witts** Polder ofte Kaptwahoij | *Kak-tòa-ûi* | 角帶圍 | Governor (1625–1627) |
| **Nuijts** Polder ofte Sijsijnnangh | | | Governor (1627–1629) |
| **Putmans** Polder ofte Cekankon | *Siōng-káng-kong* | 上港崗 | Governor (1629–1636) |
| **Van der Burch** Polder ofte Tongsoija | *Tiong-chiu-á* | 中洲仔 | Governor (1636–1640) |
| **Traudenius** Polder ofte Tiokankon | *Tiong-káng-kong* | 中港崗 | Governor (1640–1643) |
| **Lemaires** Polder ofte Siamsiamticke | *Sam-sià-kah* | 三舍甲 | Governor (1643–1644) |
| **Koeckebackers** Polder ofte Heijsouga | *Hí-chhiū-á* | 喜樹仔 | Commissioner(1939) |

would like to commemorate the occasion that happened on the spot. Most of the commemorated events, nevertheless, are tragedies or unfortunate episode.

When the aforementioned Maarten Vries sailed to Japan in 1643, for instance, he christened Hachijō-shima (八丈島) or Hachijō Island as *Ongeluckich Eylant* or Unlucky Island, named after the incident of losing contact with the ship Breskens in a storm while off the said island (Leupe 1858: 63–64). *Verraders of Moordenaar Rivier* or Traitors or Murderer River on the southwestern Taiwan was to commemorate the killing of 63 VOC soldiers under the command of Governor Pieter Nuijts in 1629 by the local Austronesian warriors.

## 6 The Display of Place Names on VOC Cartographical Works

The way in which VOC place names were displayed on a small-scale map may also reveal a pattern for us to discern. For instance, we may generalize that the exhibition of VOC toponyms on the aforementioned "Chart of a part of the China Sea and Coast from Punta Lampacao to the Overveens Islands, the island of Formosa" into three types roughly in a northerly direction: the juxtaposition of Dutch commemorative toponyms named after ships and the spellings based on local languages, Dutch commemorative toponyms named after ships, and Dutch commemorative toponyms named after geographical names from the Netherlands, in addition to the Dutch landscape-descriptive toponyms scattered everywhere. Table 3 shows some examples of all the three types of VOC toponyms approximately based on the

**Table 3** VOC commemorative place names off the southeast coast of China

| VOC toponyms | Hokkien | Characters | Year of event |
|---|---|---|---|
| **Vissingen** | | 大门岛 | |
| **Haarlem** | | 鹿西岛 | |
| **Hoorn** | | 状元澳 | |
| **Delft** | | 霓屿 | |
| **Enchuissen** | | 洞头岛 | |
| **Durkerdam** | | 浮鹰岛 | |
| **Domburch** | | 黄湾屿 | 1629 |
| **Overveen** | | 高登島 | |
| **Zirik zee** | | 北竿 | |
| **Texel** | | 井尾 | |
| Tangsoa of **Victorie** bay | *Tâng-soaⁿ* | 铜山 | 1622 |
| **Oranje**bai of Zoatoupou | *Kiā-thâu-pho͘* | 崎头铺 | |
| **De Goede Hoop** of Pheio | | 钱澳岬 | |
| **Wieringen**s bai of Zinhaitou | *Chēng-hái-thâu* | 靖海头 | 1631 |
| **Nassou**en bai of onbekende Imbocht | | 神泉港 | |
| **Reiersen**sbai of Cathei | *Kah-chú* | 甲子湾 | |
| **Groeningen** / Oucham | | 田尾湾 | 1622 |
| **Haarlem**sbai of Pingaisou | *Pêng-hái-só͘* | 平海所 | 1623 |

Note: The words in *bold* type stand for commemorative purposes. Those in *black* are named after place names, in *red* after ship names, and *blue* after people

spatial sequence of the map. Figure 1 displays the VOC commemorative toponyms along the southeastern coast of China and the island of Taiwan.

The distribution of the three types of VOC toponyms actually reveals the language boundary (zone) in the southeastern coast of China. The places where we have the juxtaposition of Dutch commemorative toponyms named after ships and the spellings based on local language roughly correspond to the Hokkien (Southern Min)-speaking area. It is obvious that the VOC sailors tended to rely on the information provided by the Hokkien-speaking people when they conducted their hydrological survey, and the spellings based on local languages are no doubt Hokkien. The places where we only have the Dutch commemorative toponyms named after ships roughly correspond to the Fuzhou (Northern Min)-speaking area, whose language is unintelligible to the Hokkien-speaking people. Hence, there are no VOC toponyms with local spellings appearing in this area. As for the place where we only have the Dutch commemorative toponyms named after geographical names from the Netherlands, this only correspond to the Wu-speaking area. The Hokkien-speaking people were unfamiliar with not only the language there but also the coastal waterways, and thus, the area turned out to be where we have randomly assigned geographical names from the Netherlands used for place-naming.

**Fig. 1** Map showing distribution of VOC commemorative place names

# 7 The Nature of Commemorative Place Names

It can sometimes be difficult to judge whether a commemorative toponym is named after a ship or a geographical name from the Netherlands because the majority of ship names were named after place names in the Netherlands. For instance, *Haarlem* was used for both Luxi Dao (鹿西岛) near Wenzhou (温州), Zhejiang Province, and Pingaisou or Pinghaisuo (平海所) of Huizhou (惠州), Guangdong Province of China. The former was named after the place name *Haarlem* in the Netherlands whereas the latter, *Haarlemsbai*, after the ship *Haarlem*, as discussed in the earlier section. The case of *Haarlemsbai* here is usually called secondary nomination. Thus, the same *Haarlem* conveys two different meanings to the local designated spot. The *Haarlem* named after the ship commemorated the ship of the same name that was in the said bay, whereas another *Haarlem*, named after the geographical name of the Netherlands had nothing to do with the *Haarlem* in the Netherlands but was simply a randomly assigned symbolic landscape familiar to the contemporary Dutch and might also serve as a landscape of nostalgia for the Dutch sailors.

Although the nature of commemorative place names was either for memorializing events, celebrating achievement or soothing homesickness, the selection from among personal names, ship names, place names at home or the appellations of incidents themselves did follow some principles. Take the commemoration of tragedy as an example; to commemorate the killing of 63 VOC soldiers in 1629 by the local Austronesian warriors under the reckless command of Governor Pieter Nuijts, we had *Verraders of Moordenaars Rivier* or Murderers' River. Here we have the tag of an incident for the purpose of commemoration. However, to memorialize the killing of the sailors of the ship *Gouden Leeuw* by the inhabitants of Lamey island in 1622, we had *Gouden Leeuwseylandt* or Golden Lion Island instead. Here we had the ship name rather than the tag of incident for commemoration such as *Verraders Eiland* or Murderers' Island.

Nevertheless, we do have the case of *Verraders Eiland* in the Pacific Ocean. The VOC baptized Niuatoputapu, an island in today's Tonga, *Verraders Eylandt* or *Isle des Traitres*, meaning Murderers' Island or Traitors' Island, because the islanders attacked the VOC ship *Eendracht* in 1616 (Douglas 2014: 18–19). Thus, we may conclude that it is possible for the contemporary VOC personnel to have both *Gouden Leeuw* and *Verraders Eiland* being applied to the island of Lamey.

As for the case of *Verraders of Moordenaars Rivier*, it seems highly unlikely to have the river being called *Nuijts Rivier* to commemorate the tragedy since the contemporary VOC personal names were employed for commemorating glorious achievement rather than unfortunate incidents. It also explains why Lamey Island was called *Mattysen Eylant* only in *Caert vant Eijlant Packan* by Jacob Noordeloos in 1625 (Gommans and van Dissen 2010: 212), the first example of VOC cartography that depicts the said island. *Mattysen Eylant* was named after Mathijs Jacobsz., the highest ranking officer among all the victims killed on Lamey in 1622. The place name *Mattysen Eylant* never appeared in the subsequent VOC maps and was replaced by *Gouden Leeuwseylandt* as a commemorative place name since a place name after a prominent figure was supposed to celebrate his grand accomplishment rather than his calamity or faults.

Although the VOC might employ a ship name to commemorate a tragedy or a misfortunate, the majority of commemorative toponyms named after ships in East Asia were for providence or salvation, as we see along the southeast coast of China. For instance, the aforementioned yacht *Domburch*, along with junks Middelburch, Colomsou (Kó-lōng-sū 鼓浪嶼), Coissu (Gô-sū 浯嶼), Wankan (Ōng-káng 魍港), were captured and burnt by pirates led by Toutzailacq (鍾斌) on the Chincheo River (漳州河) in 4 June 1630 (Blussé et al. 1984: 31). Nevertheless, *Domburch* did not leave its name there but was only used for Huangwanyu to commemorate its hydrographical mission.

## 8 Conclusion

There are three points regarding VOC commemorative place names in seventeenth century East Asia. First, the VOC commemorative toponyms named after ships were mainly for memorializing providence or salvation, although in a few cases they may serve to commemorate tragedy. Second, the commemorative toponyms named after personal names were used for the purpose of commemorating achievements. Personal names used for memorializing misfortune at the initial stage would normally be replaced by other type of commemorative names later. Third, the commemorative toponyms named after geographical names from the Netherlands may function as a landscape of nostalgia to the sailors, but they also worked for the glorification of VOC achievement if the geographical names were restricted to the six place names where the VOC chambers were.

The way in which the VOC toponyms were displayed on the map also roughly reveals the linguistic geography in the southeastern coast of China. It is mostly due to the fact that the VOC sailors heavily relied on the information offered by one particular language group, the Hokkien-speaking people, who were the primary trading partner, and also competitor, in East Asian maritime trade.

**Acknowledgement** I am indebted to the valuable comments of Prof. Ferjan Ormeling and the cartographic work of Dr. Wee Chuen Pek. Any errors and mis-interpretations in the article remain only my own.

## References

Ang K (2006) Cóng jiù dìmíng yǔ gǔ dìtú kàn táiwān jìndài chūqí shǐ 從舊地名與古地圖看臺灣近代初期史 [Taiwan's early modern history based on old place names and ancient maps]. In: Táiwān shǐ shíyī jiǎng臺灣史十一講 [Lecture on history of Taiwan 11]. National Museum of History, Taipei.

Blussé L, van Opstall WE, Ts'ao YH (eds) (1984) De Dagregisters van het Kasteel Zeelandia, Taiwan, deel I, 1629–1641. M. Nijhoff, Gravenhage

Cheng WC (2014) Hélán dōng yìndù gōngsī rényuán zài táihǎi liǎng'àn jiān de shuǐwén tàncè huódòng荷蘭東印度公司人員在台海兩岸間的水文探測活動 (1622–1636) [The VOC hydrographic survey across the Taiwan Strait]. In: 2014 International symposium on Maritime history: transmission of information and mutual understanding across the Asian waters, Academia Sinica, Taipei

Dapper O (1670) Gedenkwaerdig bedryf der Nederlandsche Oost-Indische Maetschappye, op de kuste en in het keizerrijk van Taising of Sina: behelzende het tweede gezandschap aen den onder-koning Singlamong en veldheer Taising Lipoui; door Jan van Kampen en Konstantyn Nobel. Vervolgt met een verhael van het voorgevallen des jaers zestien hondert drie ein vier en zestig, op de kust van Sina, en ontrent d'eilanden Tayowan, Formosa, Ay en Quemuy, onder 't gezag van Balthasar Bort: en het derde gezandschap aen Konchy, Tartarsche keizer van Sina en Oost-Tartarye: onder beleit van Zijne Ed. Pieter van Hoorn. Beneffens een beschryving van geheel Sina. Verciert doorgaens met verscheide kopere platen. J. van Meurs, Amsterdam

Douglas B (2014) Naming places: voyagers, toponyms, and local presence in the fifth part of the world, 1500–1700. J Hist Geogr 45:12–24

Gommans JJL, van Dissen JR (2010) Grote Atlas van de Verenigde Oost-Indische Compagnie, deel VII: Oost-Azië, Birma tot Japan & Supplement. Asia Maior, Voorburg

Leupe PA (1858) Reize van Maarten Gerritsz. Vries in 1643 naar het noorden en oosten van Japan, volgens het journaal gehouden door C. J. Coen, op het schip Castricum. Frederik Muller, Amsterdam

Ormeling F (2006) Geographical names in the Atlas Isaak Graaf/Atlas Amsterdam. In: Schilder G et al (eds) Grote Atlas van de Verenigde Oost-Indische Compagnie, deel I. Asia Maior, Voorburg

Ormeling F (2007) Geographical names in the Atlas Isaak Graaf. Caert-Thresoor: Tijdschrift voor de geschiedenis van de cartografie [J Hist Cartogr Neth] 26(3):103–108

Schilder G, Moerman J, Ormeling F, van den Brink P, Ferwerda H (2006) Grote Atlas van de Verenigde Oost-Indische Compagnie, deel I. Asia Maior, Voorburg

**Peter Kang** is a professor of the Department of Taiwan and Regional Studies at National Donghwa University, Taiwan. He received a Ph.D. in geography from the University of Minnesota in 1996. His publications include "China's Island Frontier: Geographical Ideas on the Continent-based Nationalist Narratives on Taiwan," in the *Island Studies Journal*, 6(1), 2011; "Remapping National Geography in the Street-naming of Hồ Chí Minh City," *Journal of Geographical Sciences* 77, 2015; "Boulevardization, Commercial Speculation, and Regional Identity: renaming of urban arterial roads in the post-Cold War Taiwan," in Peter Jordan and Paul Woodman, eds., *Place-Name Changes*. Hamburg: Verlag Dr. Kovač, 2016; "Koxinga and his maritime regime in the popular historical writings of Post-Cold War Taiwan," in Tonio Andrade and Xing Hang, eds., *Sea Rovers, Silk, and Samurai*. Honolulu: University of Hawai'i Press, 2016.

# On the 'De-measuring' of Time and Space in the Models of the World of the Eighteenth and Nineteenth Centuries

**Andreas Christoph**

**Abstract** Terrestrial and celestial globes as well as planetaria, telluria and lunaria are representative models of the cosmos with high cultural and scientific importance. At a time when a global point of view is increasingly adopted by historians, the importance of these 'chronotopical' objects in the fields of history of science, cultural history, history of geography and arts necessitates the appreciation of the variety of globes, the involved protagonists, the history of manufacturing methods and technical innovations in globe production. Moreover, new technologies for visualizing old globes, will allow these physical objects to be transferred into virtual reality.

## 1 Introduction

> "The World (*Mundus*, or, universe) is the astonishing entirety that comprehends heaven and earth together with everything created within it." (Kleinsorg 1787: 4)

Models of the world represent complex space-time relationships. Addressed here are globes and globe-related instruments such as planetaria, telluria and lunaria. The terrestrial globe reproduces the shape of the Earth—its length, area and angularity—free of distortion. In terms of visualization this sets it apart from other cartographic representations. Maps, or map-related images project curved surfaces and must therefore make compromises with respect to the angles, area and equidistance. The two-dimensional methods of demonstration have distinct disadvantages, even if the technology of their production makes them a very easy medium to distribute, and their historical course of development still continues to unfold within the digital world.

I would like to look at globe objects from a history of science *and* knowledge background, as a physical model, as an image on paper, as an artistic abstraction and, from a contemporary perspective, as a virtual representation. I will also consider their pedagogical use, together with some biographical profiles of the

A. Christoph (✉)
Friedrich Schiller University, Jena, Germany
e-mail: andreas.christoph@uni-jena.de

 
M. Altić et al. (eds.), *Dissemination of Cartographic Knowledge*, Lecture Notes in Geoinformation and Cartography, DOI 10.1007/978-3-319-61515-8_15

major protagonists involved, and try to contextualize this in the knowledge landscape. In this way it is possible to obtain insights into the cultural history of models of the world that, whilst reaching back to antiquity, are still able to find some usage in current interpretations.

## 2 Historical Classification

If we look back to ancient times, it shows that the mathematical model which is the basis of the terrestrial globe is found much earlier in the design of so-called celestial globes (Künzl 1998). Thales of Miletus (*624*–547 BC) and Eudoxus of Cnidus (*400*–350 BC) used the calculations of the spheres of the universe to astronomically define certain celestial coordinates of stars and planets. From the ninth century AD, Islamic astronomers for the first time built and designed celestial globes made of metal. Some of the individual copies are still traceable today.

The verifiable tradition of the terrestrial globe carries features of Greek and Roman antiquity, steeped in Arabic cultural circles, and becomes important again only in the Middle Ages. The oldest terrestrial globe originated in 1492 and was designed by Martin Behaim (1459–1506). He recorded the geographical and cartographical knowledge of his time which of course omitted both of the Americas and Australia. It is almost certain that the first representation of America is found on Martin Waldseemüller's (1472–1522) world map and globe segments from 1507. At this time there were discussions about Christopher Columbus's search for an alternative sea route to Asia (Lester 2009: 358–370). After the Treaty of Tordesillas in 1494 the model of the globe was divided along a line separating the Earth into a Spanish and Portuguese half. Globes thus became strongly expressive symbols of domination or rule. With the increasing possibility of contextual experience globes became an object of demonstration. In the sixteenth and seventeenth century the celestial globe began to experience a renaissance in Europe. The Gottorf globe, which was created between 1656 and 1664 as a so called hollow globe, provided the opportunity to go into its interior and to look out at the firmament through small holes that had been introduced in its outer shell. This design anticipated the development of the modern projection planetarium. Much larger still were the terrestrial and celestial globes of Vincenzo Coronelli (1650–1718) with a diameter of about 3.80 m. In the seventeenth century various globe manufacturers in the Netherlands, England and France experienced their heyday. The models of the world had already obtained a general distribution and provided an important service in the systematization of contextual knowledge. The transition from the manufacturing of magnificent and lavish individual pieces to the much cheaper terrestrial and celestial globes in the German speaking areas is generally attributed to Johann Baptist Homann (1664–1724), Johann Ludwig Andreae (1667–1725) and Johann Gabriel Doppelmayr (1677–1750), all from Nuremberg. At the beginning of the nineteenth century, the "Landes-Industrie-Comptoir" and the "Geographical Institute" in Weimar, asserted themselves as Globe manufacturers (Christoph 2012) (Fig. 1).

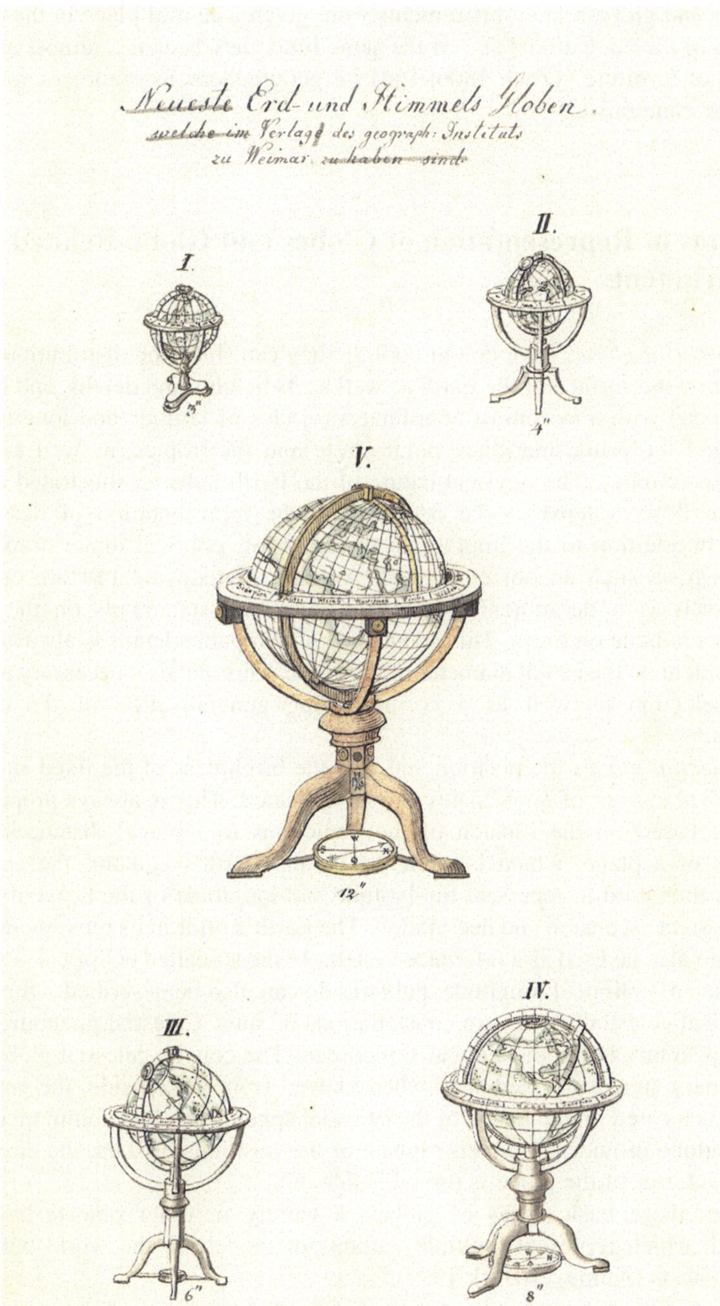


**Fig. 1** To improve the sales of the globes, which were relatively expensive, the "Geographical Institute" invested in vivid illustrations of the various globes from around 1850 (Stadtarchiv Schweinfurt, Sammlung Rückert A 89-16)

Globes and globe-related instruments were given a central place in the didactic portfolios of the educationalists. At the same time, they became "almost an essential piece of furniture" (Zach 1806: 154) for geographers, astronomers, mathematicians and cameralists.

## 3 Forms of Representation of Globes and Globe-Related Instruments

Once *terrestrial globes* have colour coding, they can show the distribution of land and water on the surface of the Earth as well as its heights and depths, and they can be referenced with a system of coordinates (circles of latitude and longitude, the equator and the prime meridian, polar circle and the tropics, as well as on old globes, the ecliptic). The physical nature of the Earth is better illustrated with the distinctive flow systems of the oceans, and the representation of deserts and glaciers. In addition to the impression of the relief, political dimensions can be seen shown on such an object as well. Also of thematic interest are economic aspects such as trade routes or industrial sites and statements on the natural resources available on Earth. This abundance of mappable details is always limited and dependent on the actual diameter of the globe. This makes it necessary to have a critical selection as well as a corresponding generalization of the depicted structures.

On *celestial globes* the position and also the brightness of the fixed stars in an astronomical system of coordinates can be indicated. This is always proportional, and itself based on the illusion of the depictions of the real distances. In the extension of a plane, which is stretched by the Earth's equator, the equatorial system is thus used to represent the latitude and longitude of the heavenly bodies in their correct ascension and declination. The Earth's orbit in its movement around the sun can also be used as a reference system. In the so-called ecliptical system the coordinates of ecliptical longitude and latitude can also be described. Often found on historical celestial globes are constellations of stars. Celestial phenomena such as the appearances of comets are also recorded. The convex celestial globe shows an imaginary sphere of fixed stars when viewed from the outside, the geocentric observer is located on the inside of the celestial sphere. The less common, concave celestial globe provides an inverse image of the positions of stars, the direct view from the interior of the globe is turned inside out.

Besides these basic types of globes, a variety of other objects have been preserved, which represent multiple variants of models of the world with many different facets (Allmeyer-Beck 1997).

### 3.1 *Excursus 1: Lunar Globes*

Since the beginning of the seventeenth century and the invention and technical development of the telescope, the success of lunar globes has been modest. The design of the lunar globe of Tobias Mayer (1723–1762) is one of the first demonstrable examples (Mayer 1750). The English painter John Russell (1745–1806) produced a small number of lunar globes in the late 1790s. He designed the globe frame in such a way that the phenomena of the wobbling motion of libration could also be shown (Russell 1797). Already in the mid-nineteenth century a first representation of the moon 'en relief' was available. It is attributed to Wilhelmine Witte (1777–1854) (Mädler 1855). On the basis of shadow lengths, they calculated certain heights on the lunar surface and then modelled a lunar globe with the details of craters, the seas and mountain ranges, which was presented to the public in 1839. When it was brought to the attention of the British astronomer John Herschel (1792–1871), he declared that the execution of this globe was "[...] of the most perfect kind, the result of incredible diligence and assiduity" (Herschel 1893: 289).

In 1849 Thomas Dickert (1801–1883), curator of the Natural History Museum at the University of Bonn, was persuaded by the astronomer Johann Friedrich Schmidt (1825–1884) to model the visible hemisphere of the moon as a relief. This was based on a lunar map published between 1834 and 1837. In 1854 an impressive plastic model of the front lunar hemisphere was completed with a diameter of about 580 cm (Schmidt 1854). In so doing the surface of the moon had been inflated by a factor of three. As many as 20,000 different objects (craters, seas, plains and mountain ranges) had all been highlighted moulded on 116 spherical gypsum plasterboards. Each board covered an area of 15° longitude and 15° latitude, the joints forming a coordinated grid. With appropriate illumination of the lunar hemisphere the corresponding heavenly constellations could be illustrated. This representative lunar relief which took up considerable space was exhibited at various national and international exhibitions. We lose track of it in the US from around 1900, here it was exhibited for the last time in Chicago at the "Field Columbian Museum" (Farrington 1899), (Fig. 2).

Only the front side of the moon is visible to us from the Earth due to synchronous rotation. In 1959 the Soviet spacecraft Luna 3 provided the first views of the far side of the moon. NASA photographs from 1965 onwards resulted in making more crucial details available of the planets and their moons visible.

### 3.2 *Excursus 2: Globe-Related Instruments*

Under this umbrella term we can include armillary spheres, planetaria, telluria and lunaria. The armillary spheres, also known as ring globes, represent the celestial globes' circles in the form of rings from a primarily geocentric perspective. They provided the determination of the positions and movements of the stars in a

**Fig. 2** The Schmidt-Dickert moon globe (City Library Braunschweig)

particular reference system. Hence equatorial armillary spheres differed from zodiacal armillary spheres. In the late eighteenth century 'Sphaerae armillares' had already lost their significance and were looked at and generally categorised as objects "from olden times" (Voigt 1799: 116). To convert (or transfer) celestial mechanics into the form of a model is the functional concern of a planetarium. The Earth, moon and planets were thereby presented in relation to one another and to the sun. In the salon culture of the eighteenth century complex celestial phenomena could really only be explained with presentable instruments and by means of a simple turn of a handle (King 1978). The first modern projection planetarium was opened in 1925 in Jena/Germany.

Telluria and lunaria were developed as teaching aids for the subjects of geography and astronomy, and could exhibit the Earth and its motion around the sun, as well as its synchronous rotation with the moon. The mechanism was fairly simple; the effect of the demonstration was impressive. A lunarium served to represent the movement of the moon around the Earth. If this apparatus moves around a central light source, which represents the sun, one speaks of a tellurium. In the second half of the nineteenth century the number of such appliances on the market soared, as did the number of its potential manufacturers. Mechanical telluria as models of the constellations of heavenly bodies functioned via cogwheels and belts and were set in motion by crank rotation or clockwork. Telluria which were dismountable also provided the possibility for separate demonstrations of cosmic processes. Telluria and lunaria were objects of demonstration in school lessons and were often provided with separate operating instructions so as to better illustrate the visible movements and to suggest possible activities.

## 4 On Globe Manufacturing

The most documented objects until the early sixteenth century are manuscript globes. The globe was formed by a spherical canvas or the design was engraved in metal or carved from stone. Globe segments or so-called spherical lunes in paper could eventually be produced in series when created by copperplate printing. This technique probably goes back to the Swiss polymath Glarenus (Heinrich Loriti, 1488–1563), who first described the mathematical construction method in 1527. A short time before that, Albrecht Dürer (1471–1528) also gave an account on the mathematical structure of spherical bodies in his "Underweysung der Meßung mit dem Zirckel und Richtscheyt" ("Instructions for Measuring with Compass and Ruler", Nuremberg 1525). Further adjustments to the design or construction process continued up until the early eighteenth century.

For a variety of globes contemporary explanatory notes appeared, which on the one hand described the presentation of the globe and on the other provided a portfolio of applications and didactic tasks. Completely departing from the conventional design of such generally available publications is a work by an anonymous author published in Königsberg in 1768: "Unterrichtende Briefe von denen Einwohnern der Planeten, Verfertigung und Verbesserung derer Erd- und Himmelskugeln, nebst einem Anhange die sonderbaren Begebenheiten eines grossen Weltmannes und R** in H*** betreffend" ("Letters of education on the inhabitants of planets, manufacture and improvement of terrestrial and celestial globes, together with a supplement concerning the strange events of a great *homme du monde* and R** in H***"] The author goes into detail about some of the mistakes of globe production, from graphic design through the printing process to installation, and he vividly describes the difficulties in this process chain. The eighteenth century *Encyclopédie* of Denis Diderot also included a comprehensive guide to building globes, together with detailed illustrations (*Construction méchanique des Globes*).

Individual actors did try alternative manufacturing methods, but although producing presentable results, these were ultimately not carried through due to their complexity. The astronomer Johann Hieronymus Schroeter (1745–1816), domiciled in Lilienthal, published further instructions on globe production in 1783. The basic framework of the globe was made out of brass and the spherical shape was formed by a wire mesh, in which flax was interwoven, impregnated with glue. The sphere was mounted on a lathe and then coated with a chalk base: "Henceforth I found my created round circular ball so solid and tough, that I was not afraid of it splitting, and despite its considerable weight, could boldly allow it to fall on the ground." ["Nunmehro fand ich meine geschaffene cirkulrunde Kugel so dauerhaft hart und zäh, dass ich sie ohne Spaltung zu befürchten, ihrer beträchtlichen Schwere ungeachtet, dreist zu Erde fallen lassen konnte"] (Schroeter 1783: 158).

In 1829 another very detailed explanation of globe production appeared which significantly wrestled with the various steps in the actual production, and thus combined old and new technologies. The sequence of operations, from the globe's design to the actual physical object, was of general interest and emerging industrial and commercial exhibitions in the nineteenth century demonstrated this (Fig. 3).

**Fig. 3** From the 9th to the 22nd of July 1861 the '2te Thüringer Gewerbeausstellung' (Second Thuringian industrial exhibition) took place in the Weimarer Schießhaus. The "Geographical Institute" was, on account of its products, rewarded with medals and a certificate. Visitors gained an insight into the variety of products as well as into the steps in their production—for example, by exhibiting copperplates (Stadtmuseum Weimar)

The occurrence of really only slight marginal transformations in globe production thus points towards a tradition of craftsmanship that lasted more than 400 years. Despite these long established serial production processes, there remained to be separate, now rather historical steps in the sequence of operation.

A lynchpin of the exact geographic imaging of the Earth on a globe or the positioning of a star was always the determination of corresponding coordinates. In addition, one of the outstanding fundamental problems within several fields of scientific research was the uniform use of a measuring system; especially in cartography, where the discussion revolved around the determination of the prime meridian—"The geographers have still not yet reached agreement where they should place the first meridian, and from where they should actually begin counting" ("Die Erdbeschreiber sind unter sich noch nicht einig, wo sie den ersten Meridianum sollen hinsetzen, und von demselben zu zählen anfangen", Anonymous 1768: 106)—The very heated discussions were eventually resolved in 1892.

This unavailability of a comprehensive determination of positions was a further basic problem of globe production until well into the nineteenth century. A number of different materials and tools were necessary for the serial production of globes. A globe was made with papier-mâché, paper, gypsum, glue, and an axis of wood or metal (iron, or brass) by using drawing instruments, a turning-lathe and a bench vice. The sphere was coated with a special glue, and a part of the previously engraved map grid or graticule, provided the orientation for the actual mounting of the segments. Of course, particular caution was necessary here, since it was only with the mounting of the very last segment to be glued on, that it was found out if it fitted exactly, or if a gap had become too large, or if the paper had overlapped on the edge. Any subsequent mechanical working of the globe segments created the possibility of additional error, as place names would often be trimmed, in such a way that, "[...] one sees at once that an unwashed hand had castrated them" ("dass [...] man gleich siehet, daß eine ungewaschene Hand dieselbe castriret hat." Anonymous 1768: 124). The longevity of the product was also thought about; wormwood was mixed into the glue so as to prevent any damage of insects feeding upon it. Manuals on how to subsequently colour the finished globe were widespread up until at least the 1850s.

## 5 Educational Applications: Globes for the Blind

The first German school for the blind was opened in Berlin in 1806. The founder was the pedagogue August Zeune (1778–1853), an eclectic thinker who addressed geographical issues and also held lectures at the Berlin University on geological topics. Around 1808 the first model of the so-called "Tasterdball" appeared, which was formed "from a wood pulp made of sawdust" ("aus einer Holzmasse von Sägespänen", Zeune 1820: III). The relief was applied with gypsum; cities were characterized by pieces of cloth. From this a plaster mould was taken, which in the following years became the template for a number of other relief globes. Overall there were probably 150 copies made of these wooden balls.

In addition to the Tasterdball's explicit use in education for the blind, Zeune was keen to position himself so that he could sell the globe cheaply on the scientific market. The special feature here was the standardized application of colour to bring into focus the different topographical features on the globe "so as to provide a living picture of the Earth's surface" ("um ein lebendiges Bild von der Erdoberfläche zu geben", Zeune 1815: 152). There was also the cost factor that gave the Berlin globe much higher sales opportunities in comparison to the products of competing manufacturers. Zeune's intention, to pursue at a cheap price a simple cartographic representation and to mirror the profile of the Earth's surface, made a significant contribution to the wider distribution of these models in geography lessons of the early nineteenth century. In addition to the relief, the eleven colors used for its illumination (amongst other colours: 'grass-green', 'gray-green' and 'peach blossom') were further examples of tactile assistance for teaching the blind: "The water and the ice to an even greater extent felt cold because of the mixture of the oils, [and] all naked surfaces without any green covering, such as deserts, salt steppes, llanos,[1] karoo,[2]

[1]In the areas of Amazonian rain forest there are fertile plains known as llanos.

[2]The karoo is a semi-desert natural region of South Africa.

because of the sand mixed in with the colour" ("So fühlt sich das Wasser und noch mehr das Eis kalt an, wegen eigener Mischung des Lackes, alle nackte von keiner Kräuterdecke überzogenen Flächen, als Sandwüsten, Salzsteppen, Llanos, Karoo, wegen des unter die Farbe gemischten Sandes", Zeune 1815: 153–154). There was also an additional possible use of the globe, for example in history lessons, by applying wax and watercolours for the presentation of historical and political boundaries.

Relief globes experienced great demand in the following years, not only as teaching aids for the blind, but also in the practical sense of the visualisation of geographical data. The unique selling point of the relief globe was the enhanced depiction of the Earth's surface that found great popularity with teachers. Somewhat later it was the, dumb 'induction globe' that took the place of the most recommended teaching aid in the portfolios of the teaching profession.

## 6 Summary and Outlook

Globes and globe-related instruments in all their various types and designs were, to some extent, predestined to be used in the explanation of 'abstract truths' in mathematical and geographical astronomy. Basic issues relating to orientation on earth and in the surrounding universe could be questioned and resolved with these models. The contemporary, widely distributed instruction manual for the use of terrestrial and celestial globes not only contained an introduction into the montage, the components and the installation of the globe, but also provided tasks to what were complex phenomena. The rising and setting of a celestial body could be deducted, as well as the origin of the seasons on Earth, the annual revolution of the Earth around the Sun, or the movement of the 'Perioecians' or 'Nebenwohner', those who lived on the same parallel of latitude but on opposite meridians, rotated in length by 180°, the 'Antoecians' or 'Gegenwohner', those people who lived under the same meridian, but on opposite parallels of latitude, north and south of the equator, as well as the 'Antipodeans', who could be found on diametrically opposite sides of the globe.

Based on the historical classification of the object in question, as well as actor-related derivation, it can be said in summing up, that the globe as a model is primarily both a visual object and a demonstration apparatus, and less of an instrument of scientific precision. And yet the image of the globe that has been passed down since antiquity, that what we perceive as a slight curvature on the horizon does in fact represent a sphere, is emphatic. A globe made the world significantly distinct. It is no wonder that in sixteenth and seventeenth century Europe the globe first began its career as an instrument of demonstration and then in the eighteenth and nineteenth century it became an educational tool. Digital modernity now allows innovative approaches in historical, political and socio-cultural themes that for over 500 years were chiefly reserved for globes and globe-related instruments, and these objects now appear to only have a marginal position in geography, cartography and astronomy.

**Fig. 4** 3D-Scan of a relief globe (Fraunhofer Institute for Applied Optics and Precision Engineering IOF Jena, Friedrich Schiller University Jena)

However, it is essential to note the potential that is hidden in 'real' physical and political globes, as well as from a geocentrical perspective, to learn of the shortcomings of celestial globes and models of planets for popular scientific usage, and the basis of their construction. In the twenty-first century, the virtual presentation of the global has opened up new prospects for the representation of the Earth. The classical globe is not simply placed on the Internet, but set in a completely new form, which is also not simply a depiction of 'world reality'. It is true that with Google Earth and Open Street Map the global is made visible, but satellite data are also transferred into a spherical shape by means of standardized mathematical models. Based on historical globes and globe related instruments, new visualization methods allow using 3D scans and to virtually superimpose the historical objects. Thus, a variety of new notations is possible: historical maps and globes are linked via corresponding algorithms with modern satellite imaging. These projections permit a new global positioning. Finally, however, the basic form of the globe remains for orientation. It raises the question whether this all leads to a new renaissance of historical 'models of the world' that will become virtual? (Fig. 4)

## References

Allmeyer-Beck PE (1997) Modelle der Welt. Erd- und Himmelsgloben. Kulturerbe aus österreichischen Sammlungen. Brandstätter, Wien

Anonymous (1768) Unterrichtende Briefe von denen Einwohnern der Planeten, Verfertigung und Verbesserung derer Erd- und Himmelskugeln, nebst einem Anhange die sonderbaren Begebenheiten eines grossen Weltmannes und R** in H*** betreffend. Hartung, Königsberg

Christoph A (2012) Geographica und Cartographica aus dem Hause Bertuch. Zur Ökonomisierung des Naturwissens um 1800. Fink, München

Farrington OC (1899) The Schmidt-Dickert moon model. Science 9(210):35–36
Herschel J (1893) Outlines of astronomy. New edn. Longmans, Green and Co., London
King HC (1978) Geared to the stars. The evolution of planetariums, orreries, and astronomical clocks. University of Toronto Press, Toronto
Kleinsorg R (1787) Abriß der Geographie, zum Gebrauche in und ausser Schulen, nebst der besondern Geographie des Erzstifts Salzburg, und einem Unterricht vom Weltgebäude, und vom Gebrauche der Globen. Hochfürstliche Waisenhausbuchhandlung, Salzburg
Künzl E (1998) Der Globus im Römisch-Germanischen Zentralmuseum Mainz. Der bisher einzige komplette Himmelsglobus aus dem griechisch-römischen Altertum. Der Globusfreund 45 (46):7–80
Lester T (2009) The fourth part of the world. The epic story of history's greatest maps. Profile Books, London
Mädler J (1855) Der Witte'sche Reliefglobus der Mondkugel. Magazin für die Kunde des geistigen und sittlichen Lebens in Russland 3:11–17
Mayer T (1750) Bericht von den Mondskugeln, welche bey der kosmographischen Gesellschaft in Nürnberg, aus neuen Beobachtungen verfertiget werden [...]. Nürnberg.
Russell J (1797) A description of the Selenographia. An apparatus for exhibiting the phenomena of the moon. Together with an account of some of the purposes which it may be applied to. Faden, London
Schmidt JF (1854) Das Relief der sichtbaren Halbkugel des Mondes. Georgi, Bonn
Schroeter JH (1783) Anweisung zur genauen Verfertigung einer künstlichen Himmels- oder Erdkugel. Astronomisches Jahrbuch für das Jahr 1786:155–160
Voigt JH (1799) Lehrbuch einer populären Sternkunde nach dem gegenwärtigen Zustande der Wissenschaft; für Schul- und akademischen Unterricht, auch Selbststudium der Liebhaber; mit Beziehung auf einen vierzolligen Himmels-Globus, und einen Stern-Atlas mit doppelten schwarzen Charten. Landes-Industrie-Comptoir, Weimar
Zach FX (1806) Die Erdkugel, worauf alle Entdeckungen, welche Jacob Cook auf seinen drey grossen Weltreisen, vom Monat April 1768 bis zum 14. Febr. 1779, und nach ihm die englischen Schiffs-Capitaine Portlock und Dixon in den Jahren 1785 bis 1788 gemacht haben, befindlich sind [...]. Monatliche Correspondenz zur Beförderung der Erd- und Himmelskunde 13:152–185
Zeune A (1815) Erdansichten oder Abriss einer Geschichte der Erdkunde vorzüglich der neuesten Fortschritte in dieser Wissenschaft. Maurersche Buchhandlung, Berlin
Zeune A (1820) Erdansichten oder Abriss einer Geschichte der Erdkunde vorzüglich der neuesten Fortschritte in dieser Wissenschaft. Fortgesetzte Ausgabe. Maurersche Buchhandlung, Berlin

**Andreas Christoph,** born in 1982, studied biology, anthropology and history of science at the Friedrich-Schiller-University, Jena. In 2011 he obtained his PhD in the history of science with a work on 'Geographica and Cartographica' in Friedrich Justin Bertuch's Geographical Institute at Weimar. Since 2013 he is a member of the digiCULT project group in Thuringia; he was responsible for the design and implementation of an exhibition for the German National Academy of Sciences, Leopoldina: "Weltansichten. Vom Globus zur Globalisierung" ("World views. From Globe to Globalisation"). Since 2014 he is chief editor of "Der Globusfreund"/"Globe Studies". He is a founding member of the DHnet|Jena (The Digital Humanities Network at the University of Jena) and a member of the research group "Kartenarchiv Plus".

# Obliterating Historical Complexity as Academic Practice: Historiographical Maps of 7th c. BCE Egypt

**Melanie Wasmuth**

**Abstract** This contribution focuses on the interdependence between map production and map usage including adaption and re-circulation, which can be witnessed by a historiographical approach to mapping. Case study are the maps displaying the history of 7th c. BCE Egypt, which date from the later 20th and early 21st c. AD. Key observation is the loss of specialised cartographic and even historiographical knowledge in the process of its dissemination. Though the case study presents a very specific example and, in addition, one that is based on a rather small corpus of sources, the author assumes that the detectable mechanisms are in operation on a much larger scale, though maybe less visible.

## 1 Introduction

Historiographical mapping is a rather recent phenomenon in the studies of the ancient Near East and Egypt: major projects did not come into being before the late 1960s. Within these, 7th c. BCE Egypt provides a very stimulating case study for discussing the dissemination of cartographic knowledge: Atlases, which aim at telling ancient history by a series of historiographical maps (in contrast to maps illustrating historical issues presented in text format) do exist, but not specifically for Egypt. Nevertheless, Egypt is included in several such historiographical atlases on the wider Eastern Mediterranean and Near East. All of them cover a large time span in addition to their (often very broad) geographical scope. Therefore, they are heavily dependent on close cooperation within a wide field of specialisations for the basic data as well as their display in thematic maps.

Regarding the visualisation of 7th c. BCE Egypt, all these works are characterised by a strong focus on political maps and a condensation of the first half of the 1st millennium BCE to very few maps. More specifically, they tend to visualise the perspective of the most powerful ruler in the area of academic specialisation,

M. Wasmuth (✉)
Basel University, Basel, Switzerland

Allard Pierson Museum, Amsterdam, Netherlands
e-mail: m.wasmuth@gmx.net



M. Altić et al. (eds.), *Dissemination of Cartographic Knowledge*, Lecture Notes in Geoinformation and Cartography, DOI 10.1007/978-3-319-61515-8_16

thereby obliterating not only the political diversity underlying the (claimed) official unity, but also the impact of the subsequent regional and cross-regional strategies on the social and material history. On an institutional level, this is paralleled by a marginalisation of the social and transcultural history of ancient Egypt and the Near East in the wider Eastern Mediterranean Area of Connectivity (see Wasmuth 2016: vi–ix).

On the basis of a dissection of the existing historiographical maps on 7th c. BCE Egypt and their reception, I will focus on the following questions: Why is it easier to detect the academic background of the cartographer than information on ancient historical complexity? Why is the widest reception achieved by the most oversimplifying maps? And how may historiographical mapping contribute to shifting the paradigm from obliterating historical complexity to opening up major research foci in a transdisciplinary environment?

### *1.1 Outline of the Historical Setting*

The example of 7th c. BCE Egypt is exceptionally productive for a study of the dissemination of cartographic knowledge due to its historical events and their reconstruction within the current academic structure. While Egypt is widely associated with cultural stability and a strong regionally focused kingship, this was much more fluctuating in ancient reality. One of the periods, in which struggles for local, regional and cross-regional power become easily visible in the ancient records, is the later 8th and 7th c. BCE: In the 8th c., Egypt—the Nile delta and valley up to the 1st cataract near modern Assuan—is ruled by a number of rather local dynasties (22nd to 24th dynasties; see Moje 2014; Kitchen 2015/1986). This changes in the late 8th c., when Egypt becomes part of the political agenda of the Kushite kingdom with its roots in the Gebel Barkal region around the 4th Nile cataract. At least the southern Nile valley and temporarily even parts of the eastern delta are actually controlled by the Kushite kings, who are accepted as Egyptian pharaohs (of the 25th dynasty) between c. 740 and 656 (for an introduction to chronological issues regarding the 25th dynasty see Zibelius-Chen 2006; Gautschy 2015: especially 90). The other cross-regional power with expansion politics towards Egypt is the Assyrian empire with its centre in the northern Tigris region in the wider area around modern-day Mossul. Whether Assyria was actually aiming at including Egypt into their empire or to establish it as a buffer state between themselves and the Kushites, is open to discussion and needs some further cross-disciplinary research (for an introduction see Kahn 2006). In any case they strengthened the local kingdom of Sais in the western delta, whose kings were allies and became vassals under the Assyrian kings Esarhaddon (681–669) and Assurbanipal (669–c. 630), who led successful military campaigns into Egypt up to the Kushite-ruled centre of Thebes (671, 667, 664/663). This Assyrian interference results in the 670–650s in a major power shift in Egypt towards the Saitic dynasty, which will end up controlling all of Egypt with military forays into the remaining Kushite area of power in the south and into the Levantine territories claimed by the Assyrian and Babylonian kings. Whether the Saitic king Psammetichus I, who had become an Assyrian vassal under Esarhaddon and Assurbanipal was at some time officially released from this vassaldom, cannot be ascertained: currently, no actual source stating the release or the upholding is known. Given his political actions, one has to assume that he actually ruled Egypt at least from 656, possibly even from 664/663, without Assyrian interference [see also Kahn 2006; Manley 1996: 121 (see Fig. 2b); Kessler and Schlaich 1991 (see Fig. 2a)].

## *1.2 The Corpus of Sources*

The partial lack of preserved sources and the tendency of all cultural traditions involved to display history as successful events controlled by their kings with the help of their gods is occasionally vexing for the historian. However, it triggers an exceptionally powerful corpus of sources to study the academic minds responsible for the creation of historiographical maps of these times. Cartographic choices become more easily detectable, when history has obviously been a highly complex affair, but the available sources and the academic structure are deficient to reveal this complexity.

This I will illustrate by an examination of the available historiographical atlases, which focus on the history of Egypt or of a wider area including Egypt. The venture which aimed at the highest degree of complexity is (still) the Tübinger Atlas des Vorderen Orients (TAVO). In a large interdisciplinary project running from the early 1970s to the mid 1990s, both natural and cultural history were plotted for the area of connectivity contextualising the Levant and Mesopotamia. With its four maps featuring Egypt in the 7th c. BCE (B IV 1, 3, 8, 13; Gamer-Wallert and Schefter 1993; Zibelius and Haas 1981; Kessler and Schlaich 1991; Wittke et al. 1993), it provides the pivotal point for this contribution. The maps quite clearly represent four different perspectives, which can be shown to reflect the academic background of the authors.

The examination of this prominent case study will be embedded into an example (Guthe 1911) of earlier historiographical atlases, which are common in Bible Studies at least since the early twentieth century. The second major part of this contribution analyses to which extent and why the cartographic as well as some of its underlying historiographical knowledge displayed in the TAVO got lost in the later, more widely circulated atlases published between 1990 and 2016 (Roaf 1998/1990; Manley 1996; McEvedy 2002; Birken 2004; Haywood 2005; Wittke et al. 2007; Bryce and Birkett-Rees 2016).

## 2 Forerunners

While Ancient Near Eastern Studies and Egyptology discovered the potential of historiographical atlases only in the mid- to later 20th c., they have been used long before in Bible Studies for illustrating the historical events to be gleaned from the Old Testament. As Assyria and Egypt are prominent players in the cross-regional history of the times, featuring as enemies and as allies, Egypt tends to be included in maps visualising the history of the 7th c. BCE.

### *2.1 Guthe Atlas*

A very revealing example for studying the background of cartographic choices is the so-called Guthe Atlas, the *Bibelatlas in 20 Haupt- und 26 Nebenkarten* by Guthe, which was first published in 1911. Characteristically, the underlying historical sources to be deduced from the cartographic display are the Old Testament and the Assyrian royal inscriptions: Accordingly, Egypt is visualised as absolutely ruled by the Assyrians and Babylonians in the 7th c. BCE (see Fig. 1a). Unsurprisingly, these sources do not provide information on the borders of the supposedly ruled territory: In fact, the actual extent of Assyrian control over Egypt was probably limited to some military campaigns against the Kushites reaching into southern Egypt (up to Thebes), and a certain amount of influence over the rulers of the principality of Sais in the western delta.

Apart from these cartographic characteristics, which reveal the academic background of the author, the Guthe Atlas shows a general feature of the contemporary *zeitgeist*. Though the western border of Egypt as part of the Assyrian Empire is left undefined, the southern border is given. In accordance with political maps of contemporary Africa, which are still reflected in various borderlines of modern states including Egypt, the borders were drawn with the ruler without reference to geophysical features or locally perceived boundaries. Quite obviously, this provided the input for drawing the southern border of 7th c. BCE Egypt: the straight line of contemporary Egypt was plotted through the southern-most border post traditionally associated with Egypt and featuring a garrison including a Jewish community, i.e. Elephantine.

## 3 The Complex Venture of the Tübinger Atlas der Vorderen Orients (TAVO)

As already indicated, historiographical mapping of the wider region, was only launched in the 1960s. The interdisciplinary research project *Tübinger Atlas des Vorderen Orients* (financed by the Deutsche Forschungsgemeinschaft as

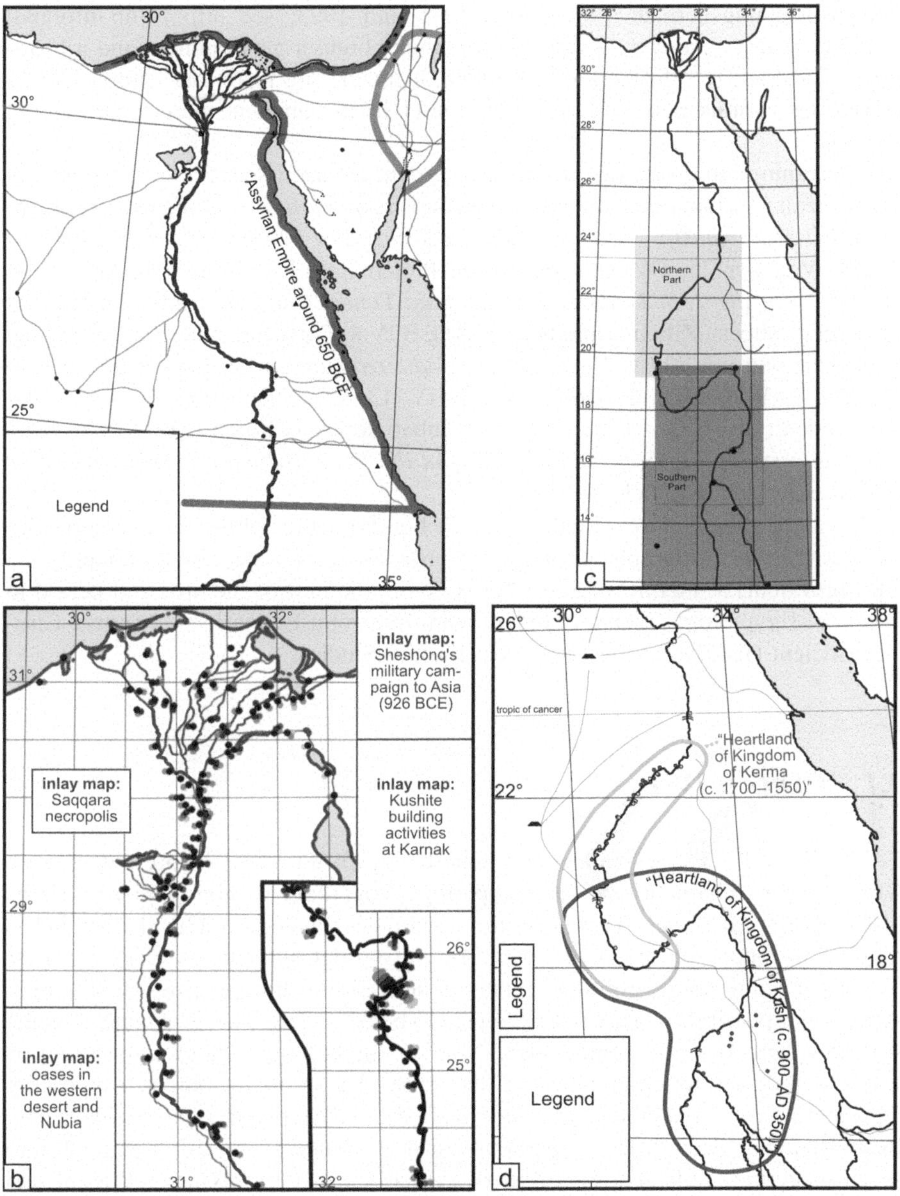


**Fig. 1** The history of 1st millennium BCE Egypt respectively Nubia in a nutshell according to (**a**) Guthe 1911: 5 [8th–7th c. BCE], (**b**) Gamer-Wallert and Schefter 1993 [1089–332 BCE], (**c**) Zibelius and Haas 1981 [c. 1st millennium BCE to c. 350 AD; location maps], (**d**) Haywood 2005: 67 [c. 1700 BCE to 350 AD] (schematized drawings by the author)

Sonderforschungsbereich 19 between 1975 and 1993; see http://d-nb.info/gnd/2116341-8, last accessed 14 Jan 2017) features both, a geographical and a historiographical section. In addition, it aimed at illustrating historical complexity by presenting various perspectives to be found in the ancient sources throughout the area.

Concerning 7th c. BCE Egypt four maps are to be discussed: B IV 1 'Ägypten in der Spätzeit (21. bis sogenannte 31. Dynastie)/*Egypt in the end of the epoch (21st to so-called 31st dynasty)*' (Gamer-Wallert and Schefter 1993), B IV 3 'Nubien und Sudan. Von der 25. Dynastie bis in die Ptolemäisch-Römische Zeit/*Nubia and Sudan. From the 25th Dynasty to Ptolemaic Times*' (northern and southern part on separate sheets; Zibelius and Haas 1981), B IV 8 'Östlicher Mittelmeerraum und Mesopotamien um 700 v. Chr./*Eastern Mediterranean and Mesopotamia around 700 B.C.*' (Wittke et al. 1993), and B IV 13 'Das Neuassyrische Reich der Sargoniden (720–612 v. Chr.) und das Neubabylonische Reich (612–539 v. Chr.) /*The Neoassyrian Empire (720–612 BC) and the Neobabylonian Empire (612–539 BC)*' (Kessler and Schlaich 1991).

As can be shown, they provide so far the highest degree of historical complexity visualised in historiographical maps. Nevertheless, they reflect predominantly the (still prevalent) academic structure for studying the past of the area and period in question: Egyptology, Nubian Studies (with its strong Egyptological roots), Classics/Ancient History, and Ancient Near Eastern Studies.

## *3.1 TAVO B IV 1*

TAVO B IV 1 (Gamer-Wallert and Schefter 1993; see Fig. 1b) presents a map, which is very characteristic for Egyptology, regarding cartographic as well as historiographical issues. Typically, the potential of historiographical mapping is reduced to topographical issues, based on very rough indications of geophysical features. Demographic, political, economic and administrative issues (to cite only some major possibilities) are not attempted to be displayed. This is enhanced by the still prevalent focus on sources dealt with in Egyptology, where archaeology is strongly reduced to epigraphics of the elite. The plotted sources supposedly representing all available archaeological data ('Archäologisch nachweisbare Reste/Remains according to archaeological evidence') are differentiated into seven categories indicated by different symbols (cited from Gamer-Wallert and Schefter 1993: legend): 'royal palace, temporarily royal residence', 'fort-like construction, fortified city', 'sacral building activity', 'royal necropolis, remains of royal burial', 'private necropolis', 'animal cemetery' and 'stele'.

The legend features a quite diversified overview of the dynasties ruling parts of Egypt as kings/pharaohs. It clearly indicates the partial overlap of rule of the 22nd to 25th dynasties as well as of the end of the 25th and the beginning of the 26th dynasties by providing their dates of rulership and their power centres.

Unfortunately, a substantial part of this information is lost in the cartographic display, for which the information provided is reduced to four colour-coded time periods (dates in brackets taken from additional chronological overview): the local 21st to 24th dynasties (1089–715 BCE), the Kushite 25th dynasty (760–656 BCE), the local 26th and 28th to 30th dynasties [664–656 (this is a mistake: this is only the time of overlap with the 25th dynasty and potential direct influence by the Assyrians on the Saitic rulers; the 26th dynasty is in place until 525, when Cambyses conquers Egypt and integrates it into the Persian Empire); 404–343 BCE], and the Persian 27th and 31st dynasties, under which also the 7th c. evidence for Assyrian conquests are subsumed according to the more detailed chronological overview (671, 667–665, 525–404, 343–332 BCE).

Key issue is obviously a differentiation of perceived indigenous versus foreign rule, which obliterates the social and political complexity of the period depicted, which includes 7th c. Egypt. Similarly, the complexity of 8th c. delta and lower Nile valley politics could have been indicated—if wished for—by separating the contemporary dynasties in the cartographic display. Nevertheless, the map still provides the most complex cartographic visualisation of the available sources as well as indications in inlay maps on special issues, e.g. an outlook of the major sites of 8th and 7th c. Nubia.

## *3.2 TAVO B IV 3*

For the 8th and 7th c. BCE, the TAVO maps B IV 1 (Gamer-Wallert and Schefter 1993; see Fig. 1b) and 3 (Zibelius and Haas 1981; equally focused upon topography; see Fig. 1c) have to be merged into one, as one prominent feature of the later 8th and earlier 7th c. BCE is the Kushite rule (and probable actual control) over the vast area between at least the 5th Nile cataract and the Mediterranean coast along the central and eastern delta. For this specific time period a differentiation into Egypt (north of the 1st Nile cataract; B IV 1) and Nubia/Sudan (south of the 1st Nile cataract; B IV 3) is therefore highly artificial and misleading, though adequate concerning the much longer time span visualized in both maps.

Nevertheless, the cartographic choice closely reflects academic practice, which is still prevalent in the cartographic display of the Kushite kingdom (see below). Characteristically, Kushite rule over the lower Nile area is not displayed as part of Kush (this could easily have been indicated by a small inlay map similar to the indication on the map of Egypt, see Fig. 1b). It also reflects that the subject area of Nubian Studies is largely defined by negation, i.e. what is not part of Egyptology. The two other principal roots, which can equally be gleaned from the set-up of the legend, are the rescue excavations in the wake of damming the Nile, most prominently due to the Assuan High Dam, and the study of the Meroitic language.

In contrast to the epigraphic focus of B IV 1 reflecting the Egyptological perspective, B IV 3 (Zibelius and Haas 1981: legend) categorises the archaeological sources more generally: into 'settlement', 'fort/fortified settlement', 'sanctuary', (unspecified) 'building', 'single tomb/cemetery' and 'important single find(s)'. The archaeological focus is also present in the colour-coded categories, which are not defined as periodisation, but as 'Kulturkreise/Classification (cultural stage)'. The earlier phase is called 'Napatan' indicating its centre around Napata at the Gebel Barkal in the 4th Nile cataract region. This category includes the Kushite kingdom, whose kings rule over (parts of) Egypt as Egyptian pharaohs of the 25th dynasty (c. 740–656 BCE). The sources grouped into this category may date to any time between c. 1000 and 300 BCE. The second category marked by colour-coding is 'Meroitic' indicating the new centre at Meroe north of the 6th Nile cataract and the emergence of the Meroitic script. The kingdom of Meroe is dated roughly between 400 BCE and 350 AD. The last category is 'Ptolemaic-Roman', which indicates the contemporary Mediterranean-influenced finds.

The origin in Meroitic language studies is presented in the additional section of 'Philologischer Beleg/*Epigraphical record*', which categorises place names according to their degree of certainty and whether they are known from sources in Egyptian or in Meroitic.

As in the case of B IV 1, B IV 3 still provides the most comprehensive plotting of known sources (though newer excavation results are to be added by now). Nevertheless, for a study of the political and social history of the 7th c. BCE, it is primarily an indicator of potential source locations—as any solely topographical map with a large chronological frame, which is grouped into only a few sections. As indicated by the contemporary categories of Meroitic and Ptolemaic-Roman finds, an attempt at visualising historical complexity is possible, if wished for.

## 3.3 *TAVO B IV 8*

While B IV 1 and 3 concentrate on topographical issues, though with different focus as the legends illustrate, B IV 8 and B IV 13 add a thematic layer of political history to their maps. Concerning political history, the map including the highest degree of complexity of actual rule and local power is B IV 8 (Wittke et al. 1993; see Fig. 2c). Though hardly visible when looking at the map sheet as a whole, three different areas of control are indicated by slight hatching: according to the authors, in 700 BCE the Nile delta is divided into the principality of Sais in the far west (mainly west of the still existing Bolbitine/Saitic/Rosette branch), the smaller principalities of Mendes, Sebennytos and Bousiris along the (still existing) Bucolic/Phatnitic/Damiette branch, and (partially interrupted by those smaller principalities) the major part of the delta east of the Bolbitine branch, which is under Kushite control.

Despite this (in the printed map nearly invisible, but nevertheless) indicated degree of local political complexity, the map is clearly created from a Classics perspective. The Eastern Mediterranean is defined rather widely in order to include the whole of the Black Sea region with its exceedingly close contacts with the Aegean. On the other hand, Egypt is reduced to the Nile delta. According to this

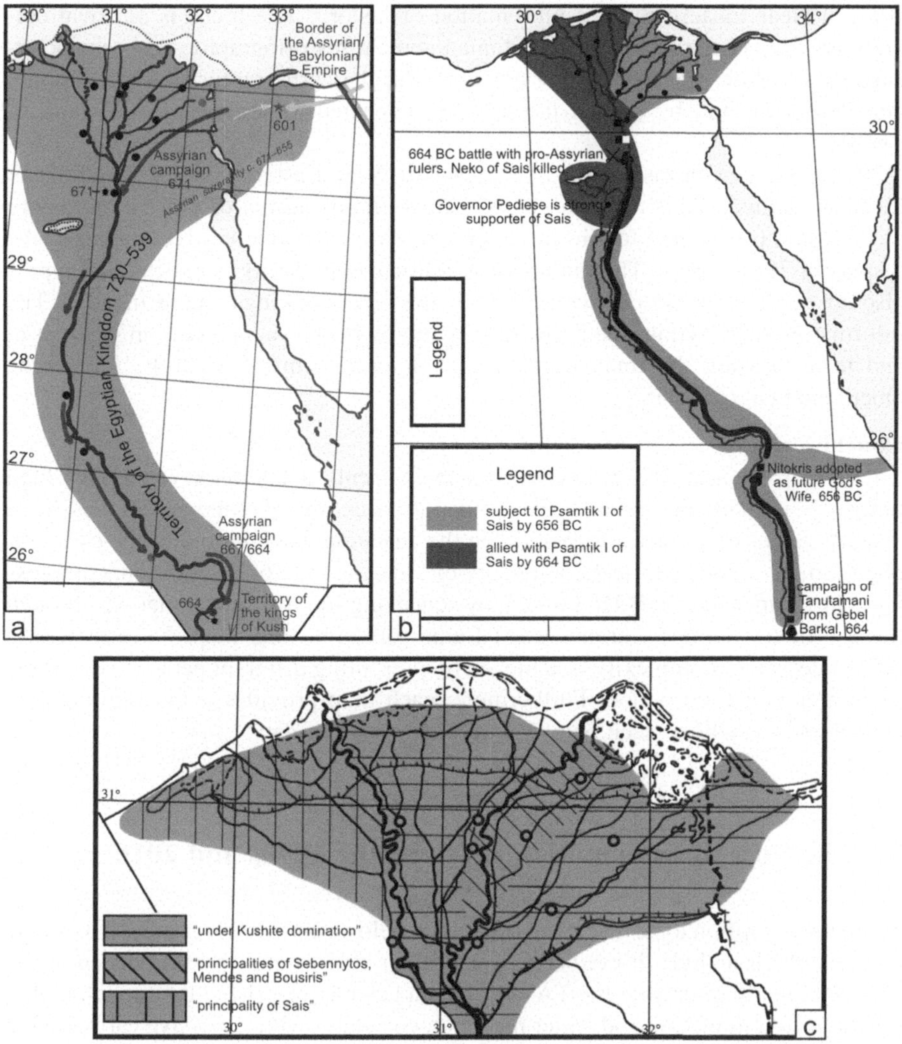


**Fig. 2** 7th c. Egypt in a nutshell, display of (partial) Egyptian rule according to (**a**) Kessler and Schlaich 1991, (**b**) Manley 1996: 121, (**c**) Wittke et al. 1993 (schematized drawings by the author)

cartographic visualisation, Egypt is rather obviously perceived as the area of Greek settlers in the delta region.

## 3.4 *TAVO B IV 13*

The fourth map in the TAVO providing information on 7th c. Egypt is B IV 13 (Kessler and Schlaich 1991; see Fig. 2a). In many respects, it is a very typical

Ancient Near Eastern Studies presentation of history. The focus is on royal epigraphics and on Assyrian and Babylonian ancient historiography. Also characteristically, Kushite rule over Egypt is negated in the cartographic display. It is not depicted at all and by textual means even relegated to somewhere far south of Thebes.

However, in contrast to the later Ancient Near Eastern Studies creations of historiographical maps to be discussed below, the paradigm of presenting history as indicated (or wished for) in the Assyrian royal inscriptions is countered in the display of 7th c. Egypt. Though no local political complexity is indicated, Egypt is also not depicted as actually controlled by the Assyrian kings, but as invaded. The information of Assyrian rule between 671 and 655 BCE is added only in small print and in a phrasing that may imply actual or only nominal control: ‘Assyrische Oberhoheit ca. 671–655’.

**Conclusions**

To sum up: though the TAVO obviously presents a complex venture of high potential, it markedly reflects the academic structure of Ancient Studies. It is actually easier to extract information on the academic background of the cartographer(s) than on the historical complexity of 7th c. BCE Egypt. In addition, complexity is lost, especially in B IV 1 and 3, by squeezing too much information (through the use of too large a chronological framework) into one map. Nevertheless, a certain degree of complexity is achieved by presenting different perspectives of the same area and time period. Furthermore, each map provides a specific unprecedented and (as will be seen below) unrivalled scope of data.

## 4 The Loss of Information Between the 1990s and 2016

With one exception to be discussed separately below, the historiographical thematic atlases, which include 7th c. BCE Egypt in their studies and were produced in the very final phase of or after the TAVO, share a key characteristic: their much higher potential for circulation and wider reception correlates with a massive cartographic oversimplification of (even the political) history of the period in question. Different versions of the presentation of oversimplified, one-sided, or conflicting historical information can be traced through the process of dissemination, going as far as the complete negation of its relevance of display.

### *4.1 Presentation of One-sided Oversimplified Information*

The first and probably most influential example for the display of exceedingly oversimplified history regarding 7th c. Egypt is to be found in the *Cultural Atlas of Mesopotamia and the Ancient Near East* by Roaf (1998/1990: 191; see Fig. 3a).

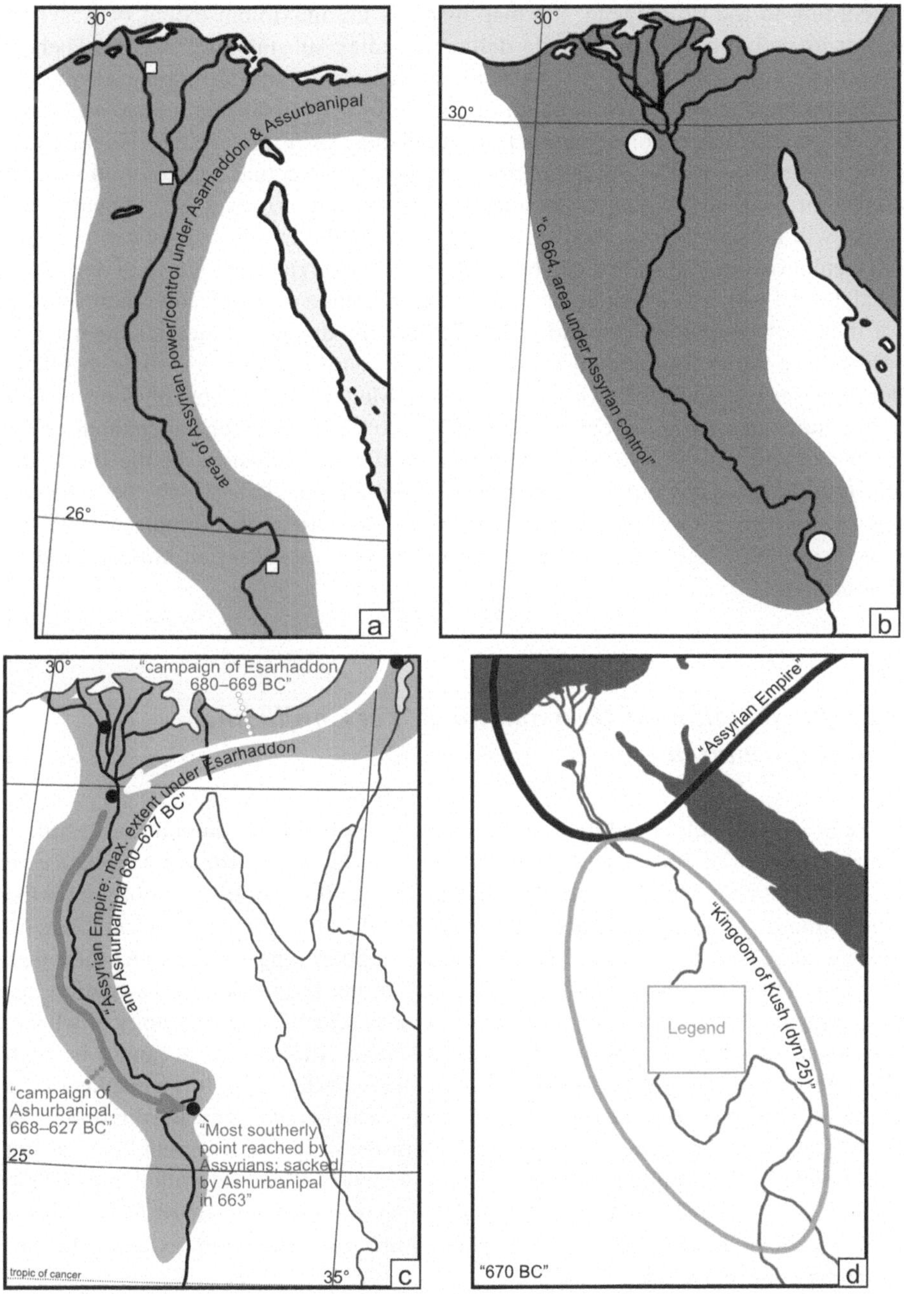


**Fig. 3** 7th c. Egypt in a nutshell, display of Assyrian control according to (**a**) Roaf 1998/1990: 191, (**b**) Manley 1996: 119, (**c**) Haywood 2005: 47, (**d**) McEvedy 2002: 53 (schematized drawings by the author)

According to the title/legend, the map displays the maximum extent of Assyrian power as including the wider Nile delta and valley substantially beyond Thebes. The presentation of a monolithic block of absolute Assyrian control may adequately represent the ancient Assyrian claim of control in their historiographical accounts, i.e. Assyrian royal history construction. Its display as a modern historiographical view of the history of Egypt at the time, presenting the actual degree of power and control of Esarhaddon and Assurbanipal, is misleading however (see above).

As will be seen below, this had fatal impact on the map's use for cartographic re-circulation (an impact that can be assumed by the obvious similarity of display to Roaf 1998/1990: 191 in the maps to be discussed, none of which state their sources for their cartographic information). In addition, the display obliterates the need of researching the cross-regional impact on local history from the various available perspectives (for an introduction to the Eastern Mediterranean Area of Connectivity at the times and some urgent research issues to be tackled for understanding the impact of this cross-regional connectivity on the social history of the times see Wasmuth 2016). Nevertheless, it achieved something vital, which the complex venture of the TAVO never did: a wide circulation of the importance of the cross-regional history of that period and of the role the Assyrian kings played in that history.

## *4.2 Presentation of Oversimplified, Yet Conflicting Information*

Quite obviously influenced by Roaf and equally problematic, though successful, is the presentation of 7th c BCE Egypt in the *Penguin Historical Atlas of Ancient Egypt* by Manley (1996: 119, 121; see Figs. 2b and 3b). Without offering a solution for the conflicting display, Egypt is depicted as absolutely ruled by the Assyrian kings on the one side (see Fig. 3b) and by the local Egyptian king on the other side (see Fig. 2b) of two adjoining double pages. Once more, the mode of presentation has high potential. What is missing however is a more sensitised terminology of what is displayed, which is not the history of Egypt at the time, since it cannot be both absolutely controlled by the Assyrian and by the Egyptian king. It is a display of the Assyrian and the Saitic history constructions, which both deny the limitation of power due to the presence of the other. In all probability, the Assyrian king has not much more than nominal power over Egypt, which is effectively ruled by 664/656 by the Saitic king Psammetichus I, though he is nominally a vassal of Assyria since 671 (see above; for a suggestion of displaying this historical complexity and further references see Wasmuth 2017).

## 4.3 Re-circulation of Oversimplification: Copy of Display

As already indicated above, the cartographic display of the Assyrian empire presented in Roaf 1998/1990: 191 achieved substantial circulation and re-circulation. Copies or at least exceedingly similar modes of display can be seen e.g. in Manley (1996: 119; see above and Fig. 3b), Birken (2004) and Haywood (2005: 47). The map of 7th c. Egypt in the digital atlas on the history of the ancient Near East by Birken (2004) provides one of the few displays of Kushite rule over Egypt. This is interrupted by an exceedingly Roaf-like display of the area of (doubtful) Assyrian absolute control up to the Theban area. Similarly, a 'copy' of the display of absolute Assyrian control can be found, e.g., in the *Penguin Historical Atlas of Ancient Civilisations* by Haywood (2005: 47, see Fig. 3c). To this territorial depiction, two routes of military campaigns are added, the earlier to Memphis, and the later one to Thebes.

## 4.4 Re-circulation of Oversimplification: Copy of Concept

Though the borders are drawn much more schematically, also the display in the *New Penguin Atlas of Ancient History* by McEvedy (2002: 53; see Fig. 3d) is likely to be influenced by the cartographic presentation in Roaf 1998/1990: 191 or its re-circulations. Once more, Egypt up to Thebes is depicted as absolutely controlled by Assyria, south of Thebes under control of the kings of Kush. No indication is provided, that both cross-regional super-powers actually fight in Egypt against each other and try to involve the local Egyptian powers in their strategies of keeping the major cross-regional foe at bay. Though there is no actual copy of display, we can observe a 'copy' of concept.

In addition, McEvedy (2002: 53) is a good example of the much wider practice of displaying (political) entities as adjoining each other without indications of potential overlaps, conflicting claims or fuzzy borders.

## 4.5 Negation of Relevance: Partial Omission

Similarly effective as the oversimplified display of Assyrian power is the lack of presenting Kushite rule, even though (substantial parts of) Egypt were under actual political control of Kushite kings for at least 80 years in the later 8th and earlier 7th c. BCE. Though included in the topographic map on late period Egypt (TAVO B IV 1, Gamer-Wallert and Schefter 1993, Fig. 1b), this political control is visualised in none of the regionally specialised atlases of Egypt and the Ancient Near East. In TAVO B IV 13 (Kessler and Schlaich 1991, Fig. 2a) it is even explicitly negated, by relegating Kushite rule to south of the area displayed on the map. By not explicating the depiction as reflecting Assyrian and Saitic history constructions, the depictions in Roaf (1998/1990: 191; Fig. 3a) and Manley (1996: 119, 121; Figs. 2b and 3b) equally negate Kushite rule.

It is therefore not surprising that it does not enter cartographic display in works with a wider chronological and geographical frame as for example Haywood (2005). Though Haywood (2005: 67; see Fig. 1d) indicates geographically overlapping areas of Nubian kingship in two different periods and could easily have added a similar line to depict, e.g., the maximal extent of Kushite power, this is not done.

### *4.6 Negation of Relevance: Complete Omission*

Most drastically, the very recent *Atlas of the Ancient Near East* by Bryce and Birkett-Rees (2016) exemplifies the potential outcome of mapping history either only topographically or in a highly oversimplified way, yet presenting conflicting information. Though Egypt is displayed as part of the maximum extent of the Assyrian empire in the chapter on Assyria (Bryce and Birkett-Rees 2016: 168–169), the section on 'Egypt in the Third Intermediate and Saite Periods (1069–525 BC)' (ibidem: 195–196) does not include a single thematic map, only a very rough physical/topographical one for the whole period (ibidem: 196). The Kushite 25th dynasty is not mentioned at all in the historiographical overview of the section.

## 5 The Potential of the Complex Venture of NP Suppl. 3

A team of researchers overlapping with that of TAVO map B IV 8 (Wittke et al. 1993; see Fig. 2c) completed a second complex venture of mapping in 2007: the *Historischer Atlas der antiken Welt*, which was published as Supplement 3 of *Der Neue Pauly* (Wittke et al. 2007). Instead of a large box of loose map sheets like the TAVO, the maps are bound into a book and each map page is accompanied by a page of historiographical comment. Once more, the complex history of 7th c. BCE Mesopotamia and adjoining areas including Egypt is spread over four maps, which cover the area up to the first Nile cataract (see Fig. 4a–d): On p. 53 the maps on c. 710–661 and c. 660–631 BCE are grouped together, p. 55 shows the same areas in 631–606 and 605–539/525 BCE.

Though the southern centre of the Kushite kingdom is not displayed due to the choice of area to be mapped, at least for what is traditionally perceived as Egypt (also in ancient times), some complexity is achieved. The choice of periodisation allows to start the series with a map predominantly visualising Egypt as Kushite-ruled territory. By marking the domain of power of the local Saitic rulers, which are specified in the legend as the ruling kings of Egypt in the latest phase displayed on the map, a certain amount of power struggle also in the earlier phase is implied. The

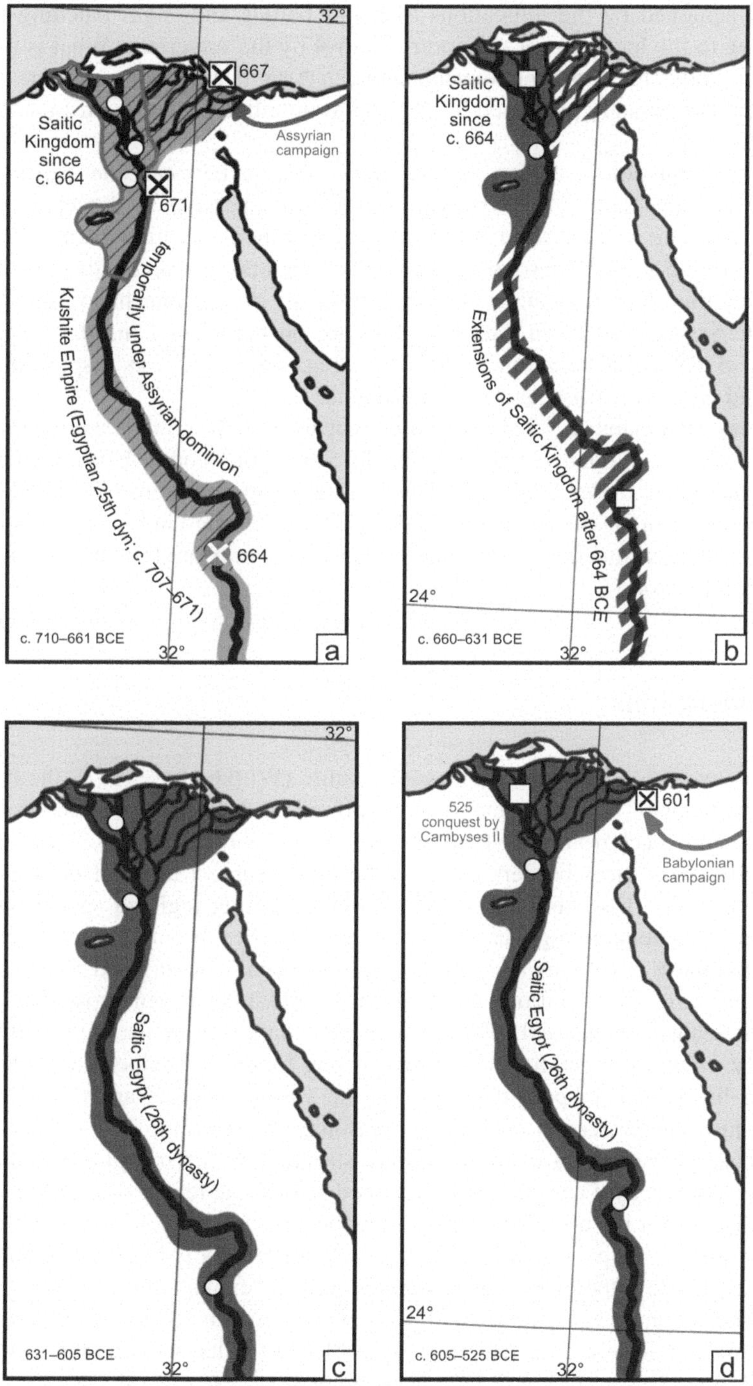


**Fig. 4** The history of 7th c. Egypt in four maps according to Wittke et al. 2007 (**a**) 53 above, (**b**) 53 below, (**c**) 55 above, (**d**) 55 below (schematized drawings by the author)

same is achieved by the indications of major battles and slight hatching used—according to the legend—for temporary control by the Assyrians. What is unfortunately not indicated is the relationship between Assyria and the local rulers of Sais, who were deliberately promoted by Assyria as an allied buffer against the Kushite kingdom.

By different colour-coding, the dynastic change between the Kushite 25th dynasty and the Saitic 26th dynasty of Egyptian pharaohs is depicted. For the second map, the periodisation is obviously due to non-Egyptian events. The shift of power to the Saitic dynasty was likely to be much faster than the display implies. However, this distortion allows the depiction of, on the one hand, the struggle between Assyria and Kush (see Fig. 4a) and, on the other hand, the expanding politics of the delta kingdom of Sais (see Fig. 4b), which, as displayed, likely happened without further Assyrian interference.

The last two maps (p. 55 above and below; see Fig. 4c–d) indicate the on-going stability of the Egyptian kingdom under the Saitic rulers until their defeat against the Persian army under Cambyses. Once more, temporary control is indicated by slight hatching in the third map (beyond the section selected in Fig. 4c), this time of Egyptian thrusts into the Levant and beyond, i.e., into predominantly Assyrian-controlled territory.

## 6 Conclusions

With the exception of the last discussed venture (Wittke et al. 2007), the historiographical maps of Egypt presented above provide at least as much information on the academic background of their authors as on ancient historical complexity. Given the rather clear borders between the major subject areas of the traditional academic setup, this is not surprising. Though a strong regional specialisation is required for understanding local developments, it defies the study of cross-regional impact on these developments. For this, an additional teaching and research focus with a cross-regional outlook as well as closely interwoven cross-disciplinary research teams are needed. This is a major challenge in the humanities, and certainly in antiquity studies, due to the strong focus on one-person publications and the financial *cum* personal setup of the various subject areas, which often feature just one permanent position per regional specialisation.

Though the reflection of the academic structure in the mapping ventures and their display of Egypt can easily be explained, this is—at least to the author—much more difficult for the question of why the widest reception is achieved by the most oversimplifying maps. As life is a complex venture, this is to be assumed for the past as well. How therefore, can a simple or linear explanation of events or their display be deemed satisfactory? As this is not the place for a philosophic discussion, the question will be reduced to (a) why the TAVO failed its potential to open up research questions and to push harder for diversified research on 7th c. BCE Egypt, and (b) what can be learned from the observed processes for the dissemination of knowledge.

In the author's experience, the TAVO does not play any significant role in teaching either in Egyptology or in Ancient Near Studies. Most students encountered up to PhD level did not even know of its existence. Having been introduced to the atlas very early on in my studies in Tübingen this was at first surprising, but considering its format, it may be something to be expected. The format of the sheets is much too small to use as a teaching instrument for more than four to six students. The only way to achieve this would be through digitalisation, but various maps are still too densely marked and lacking contrast for successful beamer presentation. The format of rather large single sheets is obviously meant for poring on it by a single viewer or to hang on a wall for closer inspection. In addition, the TAVO is mainly found in university libraries to be used under special conditions due to the format. They are therefore never likely to achieve high circulation in their current print format.

Considering the fact that also the much more easily accessible *Neue Pauly Suppl.* 3 atlas (Wittke et al. 2007) did not find reception e.g. in Bryce and Birkett-Rees (2016), it has to be asked whether this is only due to the format (e.g., printed maps, works in German) or also due to the aim of visualising historical complexity. If we as scientists analysing the past wish to uncover and present (ancient) historical complexity, we obviously have to change our strategies. Our staff resources do not allow for letting already achieved historical and cartographic knowledge go lost in the process of dissemination. An important step would be to make the already existing detailed maps more easily accessible, ideally digitally and with the possibility to interlink information from various mapping projects. In addition, the above survey clearly indicates that we have to strive for avoiding undue oversimplification though still keeping the maps readable. The perception of what is undue or necessary simplification undoubtedly depends on the context, the author and the reader. However, the survey clearly indicates the danger as well as the potential of mapping as a tool for opening up or obliterating research questions. In order to avoid subsequent loss of cartographic and historical knowledge in the process of dissemination, much more attention should be paid to explicate what is actually visualised, especially in the case of easy accessibility (which is to be aimed for).

Given these caveats, the combination of much smaller chronological scopes and the creation of more diversified thematic maps reflecting the different ancient (and/or modern) perceptions of history as well as different historical aspects (in addition to cross-regional politics) could be a powerful tool to open up new research questions, to illustrate their importance, and to make historical complexity palpable.

## References

Birken A (2004) Neuer Atlas zur Geschichte des alten Orients. Mantis, Gräfelfing

Bryce T, Birkett-Rees J (2016) Atlas of the ancient Near East: from prehistoric times to the Roman imperial period. Routledge, New York

Gamer-Wallert I, Schefter A (1993) Ägypten in der Spätzeit (21. bis sogenannte 31. Dynastie) [Egypt in the end of the epoch (21st to so-called 31st dynasty)]. Tübinger Atlas der Vorderen Orients B IV 1. Ludwig Reichert, Wiesbaden

Gautschy R (2015) The tepi shemu feast: a tool for testing chronologies of dynasty 21 to 25? J Egypt Hist (JEH) 8:81–114

Guthe H (1911) Bibelatlas in 20 Haupt- und 28 Nebenkarten. Wagner & Debes, Leipzig

Haywood J (2005) The Penguin historical atlas of ancient civilizations. Penguin Books, London et al

Kahn D (2006) The Assyrian invasions of Egypt (673–663 B.C.) and the final expulsion of the Kushites. Studien zur altägyptischen Kultur (SAK) 34:251–267

Kessler D, Schlaich X (1991) Das Neuassyrische Reich der Sargoniden (720-612 v. Chr.) und das Neubabylonische Reich (612-539 v. Chr.) [The Neoassyrian Empire (720-612 B. C.) and the Neobabylonian Empire (612-539 B. C.)]. Tübinger Atlas der Vorderen Orients B IV 13. Ludwig Reichert, Wiesbaden

Kitchen KA (2015) The Third Intermediate Period in Egypt (1100–650 BC), reprint of 2nd edn (1986) including preface of 1996 edn. Arts & Phillips, Warminster

Manley B (1996) The Penguin historical atlas of ancient Egypt. Penguin Books, London et al

McEvedy C (2002) The new Penguin atlas of ancient history. Penguin Books, London et al

Moje J (2014) Herrschaftsräume und Herrschaftswissen ägyptischer Lokalregenten. Soziokulturelle Interaktionen zur Machtkonsolidierung vom 8. bis zum 4. Jahrhundert v. Chr. Topoi. Berlin Studies of the Ancient World [BSAW] 21. De Gruyter, Berlin and Boston

Roaf M (1998) Mesopotamien: Kunst, Geschichte und Lebensformen. Bildatlas der Weltkulturen, Lizenzausgabe. Bechtermünz/Weltbild, Augsburg. Original: (1990) Cultural atlas of Mesopotamia and the ancient Near East. Facts on File, New York

Wasmuth M (2016) Introduction. The eastern Mediterranean area of connectivity in the 8th–6th c. BCE – setting an agenda. J Anc Egypt Interconnect (JAEI) 12:vi–xvi

Wasmuth M (2017) Mapping political diversity: some thoughts on devising a historiographical map of 7th c. BC Egypt. In: Grunwald S, Hofmann KP, Werning DA, Wiedemann F (eds) Mapping ancient identities – Kartographische Identitätskonstruktionen in den Altertumswissenschaften. Berlin (forthcoming)

Wittke A-M, Olshausen E, Szydlak R (2007) Historischer Atlas der antiken Welt. Der Neue Pauly Supplement 3. Metzler'sche Verlagsbuchhandlung and Carl Ernst Poeschel, Stuttgart

Wittke A-M, Prayon F, Röllig W, von Bredow I, Wank S (1993) Östlicher Mittelmeerraum und Mesopotamien um 700 v. Chr. [Eastern Mediterranean and Mesopotamia around 700 B. C.]. Tübinger Atlas der Vorderen Orients B IV 8. Ludwig Reichert, Wiesbaden

Zibelius K, Haas M (1981) Nubien und Sudan. Von der 25. Dynastie bis in die Ptolemäisch-Römische Zeit [Nubia and Sudan. From the 25th Dynasty to Ptolemaic Times]. Tübinger Atlas der Vorderen Orients B IV 8. Ludwig Reichert, Wiesbaden

Zibelius-Chen K (2006) The chronology of Nubian kingdoms from dyn. 25 to the end of the kingdom of Meroe. In: Hornung E, Krauss R, Warburton DA (eds) Ancient Egyptian chronology. Handbuch der Orientalistik [HdO] 1/83. Brill, Leiden

**Melanie Wasmuth** has studied Egyptology, Ancient Near Eastern Studies, and to a lesser degree Prehistory, Archaeological Theory and Ancient Philosophy at the universities of Tübingen (D), Munich (D), Cambridge (UK) and Vienna (A). Since her PhD studies at Vienna, Munich and Basel (CH) her research focuses on cross-regional mobility and cultural contacts in the wider eastern Mediterranean area of connectivity. Special interests include identity constructions and strategies of living in culturally diverse societies in the 1st millennium BCE. A second focus concerns the question of how ancient historical complexity can be rediscovered and made tangible—e.g. with the help of historiographic maps. Currently, Melanie Wasmuth is postdoctoral researcher at the University of Basel: Egyptology and Guest Curator (Near East/Egypt) at the Allard Pierson Museum Amsterdam.

# Part IV
# Libraries, Accessibility and Specific Cartographic Resources

# Identifying Metadata on Soviet Military Maps: An Illustrated Guide

Martin Davis and Alexander J. Kent

**Abstract** In the aftermath of the dissolution of the Soviet Union, the commercial availability of previously classified Soviet military mapping has revealed the vastness of a global mapping project, previously unknown to the rest of the world, comprising detailed topographic maps and city plans at several scales. Although numerous libraries and archives around the world have acquired these maps, and continue to do so, difficulties in interpreting the metadata of each sheet have led to errors and inconsistencies in the cataloguing of the maps, potentially hindering their accessibility. This paper focuses on approaches to identifying metadata on the maps and will present some illustrated examples as a guide. It is hoped that such assistance in the interpretation of these maps in general will also be of use to cataloguers.

## 1 Introduction

Throughout the Cold War, the General Staff of the Soviet military undertook an extensive, standardised and highly detailed mapping programme covering the globe. Although produced in secrecy, the years following the collapse of the USSR saw maps from this previously inaccessible series become commercially available, principally from commercial enterprises within former Soviet republics. The maps range in scale from reference maps at 1:15,000,000 to regional maps at 1:2,500,000, 1:1,000,000, 1:500,000 and 1:200,000, with selected areas (including much of Europe) also mapped at 1:100,000 and 1:50,000 (Watt 2005; Kent and Davies 2013). Topographic maps at larger scales were reserved for areas within the Soviet Union, and were produced by a sister civilian organisation, the Chief Administration of Geodesy and Cartography (GUGK). However, urban areas, both within and outside the Soviet Union, were the subject of a separate series of military city plans, predominantly at the scales of 1:25,000 and 1:10,000, although some examples are known at 1:15,000 and 1:5,000. Over 2000 such city plans are known to exist (Davies and Kent 2017) and, along with maps from the topographic

M. Davis (✉) • A.J. Kent
Canterbury Christ Church University, Canterbury, UK
e-mail: martin.davis@canterbury.ac.uk; alexander.kent@canterbury.ac.uk



M. Altić et al. (eds.), *Dissemination of Cartographic Knowledge*, Lecture Notes in Geoinformation and Cartography, DOI 10.1007/978-3-319-61515-8_17

series, have become very popular acquisitions for numerous libraries and public collections around the world.

Although the procurement of Soviet mapping by libraries is a promising first step in facilitating the accessibility of this relatively untapped source of geospatial data, identifying the metadata on the maps presents cataloguers with several difficulties, some of which become particularly problematic when such cataloguing is not carried out by those familiar with the Russian language. Toponyms on Soviet maps, including those used in the titles of city plans, are transliterated phonetically from the native language of the locality being mapped. Therefore, cataloguers need a grasp of the Cyrillic alphabet if a further conversion (e.g. into the Latin alphabet) is required in the cataloguing process. Identifying the production date of the plans can also be problematic, with edition, printing, source material and coordinate system dates all frequently appearing in marginalia of the map sheets, and can be open to misinterpretation if cataloguers are unfamiliar with their layout. Even basic metadata, such as the name of the organisation responsible for producing the map, can also be subject to confusion, with terms such as Soviet Union, USSR, Soviet Army, General Staff and Military Topographic Directorate all providing accurate descriptions of the responsible entity, albeit with varying precision. The fact that each of these names can be also presented in Russian or phonetic forms, or in any other language, adds complexity to the cataloguing process.

Perhaps the most significant barrier to accessibility is the use of varying methods for identifying and presenting this metadata, not only between the catalogues of different institutions, but also within catalogues. It is hoped that by highlighting these issues and outlining the layout of various types of Soviet map sheets, cataloguers will be better equipped to accurately and consistently identify the relevant metadata.

## 2 Key Elements of Metadata on Soviet Military Maps

### *2.1 Nomenclature and Titling*

The nomenclature used on Soviet topographic maps is directly derived from the system used by the International Map of the World (IMW) at the scale of 1:1,000,000 (see Fig. 1). This system is explained in detail by Soviet topographic mapping manuals (e.g. Military Topographic Directorate 1966), Psarev (2005), and by Davies and Kent (2017), all of which form the basis of the following summary.

Bands of latitude, spanning 4°, are expressed by (Latin) letters from A–U in each hemisphere, while 6° longitudinal zones are numbered 1–60, beginning at the 180th meridian. For example, the 1:1,000,000 Soviet topographic sheet covering the central part of Great Britain is numbered N-30. Each 1:1,000,000 sheet is divided into four sheets at 1:500,000 (each of which covers an area of 2° latitude by 3° longitude), with one of the first four letters of the Cyrillic alphabet (А, Б, В, Г)

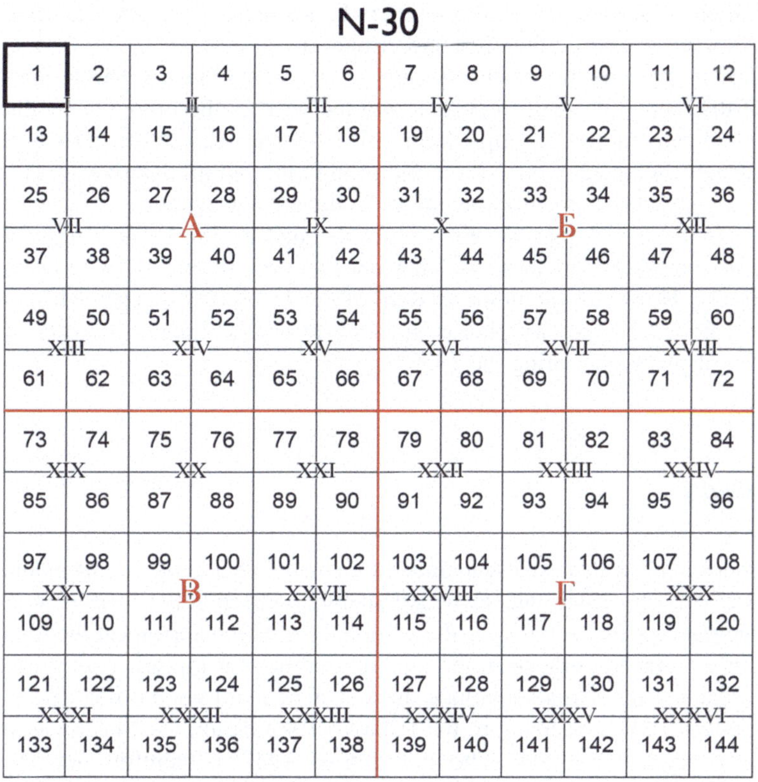


**Fig. 1** Diagram showing the gridding of Soviet map sheets according to the system adopted by the International Map of the World and illustrating the division of 1:1,000,000 sheets into four 1:500,000 sheets and their further sub-division into 36 1:200,000 and 144 1:100,000 sheets

appended to the 1:1,000,000 nomenclature to denote the appropriate part of that sheet (in left-to-right, top-to-bottom order). For example, N-30-Б, refers to the top-right quarter of N-30. Each 1:1,000,000 sheet is also divided into 36 sheets at 1:200,000 (each of which covers an area of 40′ latitude by 1° longitude), which are numbered (from top-left to bottom right) using Roman numerals (I to XXXVI). These Roman numerals are appended to the relevant 1:1,000,000 sheet number, for example N-30-XXIV. Similarly, 1:1,000,000 sheets are also divided into 144 sheets at 1:100,000 (numbered 1–144 in the same order), with this number appended to the relevant 1:1,000,000 sheet number (for example, N-30-105). Each 1:100,000 covers an area of 20′ latitude by 30′ longitude.

Upper-case Cyrillic letters are appended to 1:100,000 nomenclature to denote 1:50,000 sheets (А, Б, В, Г), for example N-30-105-В. A further lower-case letter is added to this to denote 1:25,000 sheets (а, б, в, г), for example, N-30-105-В-г. Deviations are only made from this system north of the 60th parallel; between the 60th and 76th parallels, laterally adjacent sheets are merged to form a double sheet, whereas quadruple sheets are produced for areas north of the 76th parallel. The sheet reference for locations in the southern hemisphere are given the suffix 'ЮП' for Южное полушарие (southern hemisphere).

All Soviet topographic sheets also have a toponymical title. This is usually the name of the largest or most important city included in the map. On smaller scale maps, which may include multiple large cities, the one nearest the centre of the map area is usually selected. Countries, regions and other administrative areas covered by each topographic map sheet are stated in the top-left corner.

The extents of city plan sheets are not dictated by the IMW grid system but instead are centred on urban areas. They are therefore titled only with name of the relevant city. However, the nomenclature of the 1:100,000 sheet(s) which include the area covered by the city plan is included in brackets under the title, with the exception of plans pre-dating the 1950s.

## 2.2 Dates

Soviet maps at all scales usually include more than 1 year in their marginalia, which can cause issues for cataloguers who are unable to interpret the accompanying text. The principal date on each map is the edition date, which appears at the top of each map, either under the title or in the top-right corner (or top-left corner before the 1950s). Many maps will also include the text 'Система координат 1942' (Coordinate system 1942), referring to the Pulkovo 1942 datum used in Soviet military mapping from 1946 onwards. The inclusion of '1942' is therefore not relevant to the age of the map or its source materials. A print code is included in the bottom-right corner of map sheets, indicating the relevant map series (and scale), job number, month and year of printing (which may sometimes differ from the edition date) and the factory in which the map was produced. The map series (and scale) are indicated in the print code by a prefix letter as follows:

Ж 1:1,000,000 topographic map
Е 1:500,000 topographic map
Д 1:200,000 topographic map
Г 1:100,000 topographic map
В 1:50,000 topographic map
Б 1:25,000 topographic map
А 1:10,000 topographic map
З 1:2,000,000 and 1:4,000,000 aeronavigation chart
И 1:10,000 and 1:25,000 city plan

The factory in which the map was printed is indicated by a suffix letter as follows:

Б (unknown)
В (unknown)
Д Dunayev (Moscow)

Е (unknown)
И Irkutsk
К Kiev
Л Leningrad (now St Petersburg)
Ср Sverdlovsk (now Ekaterinburg)
Срт Saratov
Т Tashkent
Тб Tblisi
Хб Khabarovsk

All other dates on the plans are preceded by explanatory text, including 'Составлено' (Compiled), 'Повторно отпечатано в' (Reprinted in) or 'Составлено по карте издания' (Compiled from maps published in). On earlier sheets, particularly before the 1970s, the compiler(s) and/or editor of the sheet are named e.g. 'Составили ст техники' (Technical compiler) and 'Редактор' (Editor). Where present, this information is located below the print code at the bottom-right of the sheet.

From 1963 onwards, a separate coordinate system, SK-63, was used for maps produced by GUGK for non-military purposes. The origin of SK-63 coordinates is slightly offset from the 1942 datum, resulting in a different arrangement of sheet lines. Maps were produced at 1:100,000, 1:25,000 and 1:10,000 using the SK-63 datum but contained no coordinates, removing the possibility of precisely locating a point on the map. The nomenclature of the SK-63 system also differed, with 1:100,000 sheets numbered by row and column (e.g. 45–24) within discrete zones, rather than the IMW system. Objects of military sensitivity were removed from maps using the SK-63 datum (E. Šleiterytė-Vilūnienė, *pers. comm.*, March 27, 2015).

## 2.3 Author

Although general terms, such as 'Soviet Union', 'USSR' and 'Red Army', can be used to accurately describe the author or producer of Soviet maps, marginalia typically include a more precise indication of the map's origin. On military maps at scales larger than 1:1,000,000 and on city plans, the term 'ГЕНЕРАЛЬНЫЙ ШТАБ' (General Staff) is usually placed above the title at the centre-top of the sheet. More precisely still, it was the Military Topographic Directorate of the General Staff which was responsible for cartographic outputs, although this full term is not given on the sheets themselves.

Civil maps and plans of Soviet territory were produced under the jurisdiction of the Chief Administration of Geodesy and Cartography (GUGK or Главное управление геодезии и картографии) and therefore do not include the 'General Staff' label. Though its role largely remained the same, the civil body was renamed several times early in the Soviet era, known originally as the Higher

Geodetic Administration (1919–1926) and latterly the Geodetic Committee (1926–1928), the Chief Geodetic Committee (1928–1930), the Chief Geodetic Administration (1930–1932), the Chief Geological-Hydrological-Geodetic Administration (1932–1935), the Chief Administration of State Survey and Cartography (GUGSK) (1935–1938); finally settling on the Chief Administration of Geodesy and Cartography (GUGK) (1938–1991) (Postnikov 2002).

### 2.4 Other Metadata and Marginalia

Scales on Soviet military maps are expressed using a ratio, a statement, and a scale bar, all of which follow conventional, recognised formats, and are usually placed in the centre of the bottom of each sheet. Maps at smaller scales (e.g. 1:1,000,000) also incorporate diagrams indicating the adjacent sheets at the same scale and a schematic map denoting national boundaries of the area covered. Magnetic declination is also depicted with a diagram and on larger-scale topographic maps, a profile indicating gradients of slope is also included. Each 1:200,000 map has a written description on the back of each sheet explaining the military geology of the area and incorporates a schematic map of the surface geology. The city plans include a 'справка' (*spravka*), a description of the urban subject of the map and its environs. Where space does not permit on the map itself, the *spravka* is printed in a separate booklet, along with a street index and a list of important sites.

It is possible to determine the number of colour plates used in the printing of the Soviet map sheets by examining the vertical colour bar (if it is intact) which is usually placed in the lower half of the left-hand margin. A typical topographic map sheet may use from six to ten colours, while city plans often utilize up to 12 colours. The accuracy of registration of the printing plates can be determined by examining the ticks around the outer edges of the map sheet, if they have not been trimmed off.

## 3 Examples of Soviet Maps

The following examples serve to illustrate the locations of useful metadata found on various Soviet sheets, with a view to facilitating their cataloguing and improving accessibility for users. The locations of metadata on each map sheet within a Soviet map series were standardised by production manuals which thoroughly and precisely set out where each item should be placed. Although these layouts vary by series, scale and the age of the map as the map production manuals were periodically updated, the consistency with which the key metadata are stated and positioned on the maps is helpful to those tasked with cataloguing or interpreting the maps. One need only be familiar with common layouts and a few commonly used Russian terms to reliably extract metadata with accuracy and it is possible to identify sheets should some of the metadata be missing. Nevertheless, variations do currently exist between, and even within, some library catalogues and this lack of consistency poses a potential barrier to the accessibility of the maps which, in turn, limits the scope of potential applications of the rich geographic data they provide. It is hoped that, by clarifying the nature of Soviet cartographic metadata in the examples shown here (Figs. 2–6), the scope for misinterpretation and inaccurate cataloguing can be reduced.

1. **Title—**Murmansk—Мурманск
2. **Countries shown on the map**—СССР. РСФСР. Норвегия. Финляндия—USSR. RSFSR. Norway. Finland.
3. **Nomenclature—**Based on IMW Sheets
   **Edition date—**Издание 1968 г.—Edition 1968
4. **Print Code—**Series, Job number, month of printing (Roman numerals), year of printing (last 2 digits), factory (Ж 40 I 82 Л—1:1,000,000 Topographic Series, Job 40, January 1982, Leningrad)
   **Compilation and drafting dates—**1967 and 1968 respectively
5. **Schematic diagram of adjacent sheets—**showing names and nomenclature
6. **Schematic diagram of political-administrative boundaries—**including a list of the relevant countries and regions
7. **Scale—**1:1,000,000
8. **Legend**

1. **Producer**—Soviet Union. General Staff (ГЕНЕРАЛЬНЫЙ ШТАБ, *General'nyy shtab*)
   **Title—**Thimphu—Tkhimpkhu—Тхимпху
2. **Coordinate System—**Система координат 1942—Coordinate System 1942 (Pulkovo 1942)
   **Countries and regions shown on the map**—Индия. Бутан (Инд). Сикким (Инд). Китай. Непал. Пакистан.—India. Bhutan (Ind.). Sikkim (Ind.). China. Nepal. Pakistan.
3. **Former classification**—секретно—secret
   **Nomenclature**—Based on IMW Sheets
   **Edition date—**Издание 1971 г.—Edition 1971
4. **Reprint date**—Повторно отпечатано в 1976 г.—Re-printed in 1976
5. **Print Code—**Series, Job number, month of printing (Roman numerals), year of printing (last 2 digits), factory (E 64 XII 76 И—1:500,000 Topographic Series, Job 64, December 1976, Irkutsk)
   **Compilation, source material and drafting dates—**1969, 1965–1969 and 1970 respectively
6. **Schematic diagram of adjacent sheets—**showing names and nomenclature
7. **Schematic diagram of political-administrative boundaries—**including a list of the relevant countries and regions
8. **Scale—**1:500,000
9. **Legend**

**Fig. 2** Soviet 1:1,000,000 topographic map of Murmansk, Russia (1968), annotated to show the locations of key metadata

1. **Producer**—Soviet Union. General Staff (ГЕНЕРАЛЬНЫЙ ШТАБ, *General'nyy shtab*)

   **Title—**Alma-Ata—Алма-Ата
2. **Coordinate System—**Система координат 1942—Coordinate System 1942 (Pulkovo 1942)

   **Countries and regions shown on the map**—СССР. Казахская ССР Алма-Атинская область. Киргиская ССР Иссык-Кульская область.—USSR. Kazakh SSR Alma-Ata region. Kyrgyz SSR Issyk-Kul region.
3. **Former classification**—секретно—secret

   **Nomenclature**—Based on IMW Sheets

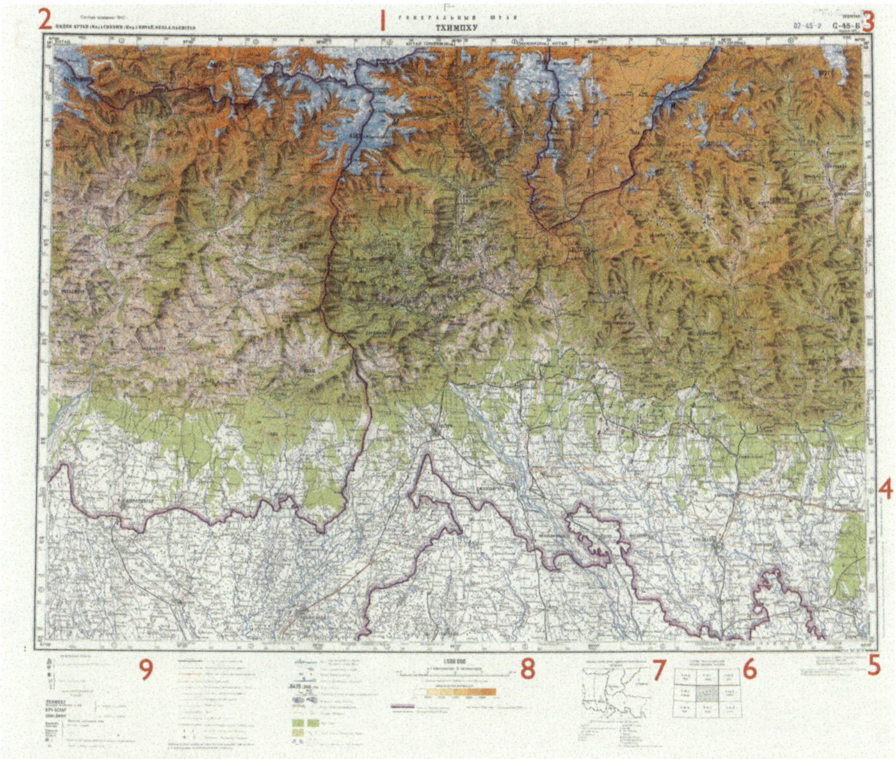


**Fig. 3** Soviet 1:500,000 topographic map of Thimphu, Bhutan (1971), annotated to show the locations of key metadata

**Date of information shown on the map**—Состояние местности на 1979–1984 гг.—The state of the area between 1979 and 1984

**Edition date**—Издание 1987 г.—Edition 1987

4. **Print Code**—Series, Job number, month of printing (Roman numerals), year of printing (last 2 digits), factory (Д 223 IX 87 T—1:200,000 Topographic Series, Job 223, September 1987, Tashkent)

   **Compilation and source material dates**—with diagram showing which dates apply to which parts of the map
5. **Scale**—1:200,000
6. **Legend**
7. **Magnetic variation notes**—with diagram

1. **Producer**—Soviet Union. General Staff (ГЕНЕРАЛЬНЫЙ ШТАБ, *General'nyy shtab*)

   **Title**—London—Лондон

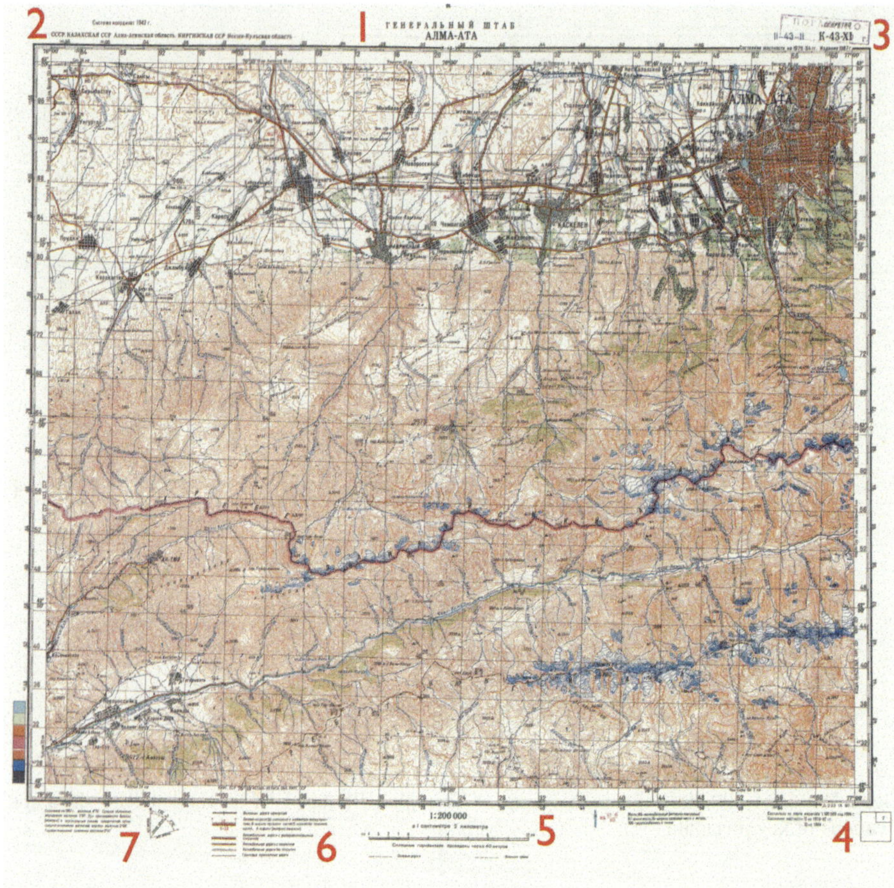


**Fig. 4** Soviet 1:200,000 topographic map of Alma-Ata (Almaty), Kazakhstan (1987), annotated to show the locations of key metadata

2. **Coordinate System—**Система координат 1942—Coordinate System 1942 (Pulkovo 1942)

   **Countries and regions shown on the map**—Великобритания. Англия графства Бакингемшир, Боль. Лондон, Суррей и Хартфордшир—Great Britain. England county Buckinghamshire, Greater London, Surrey and Hertfordshire.

3. **Former classification**—для служебного пользования—for official use

   **Nomenclature**—Based on IMW Sheets

   **Date of information shown on the map—**Состояние местности на 1974–1979 гг.—The state of the area between 1974 and 1989

   **Edition date—**Издание 1983 г.—Edition 1983

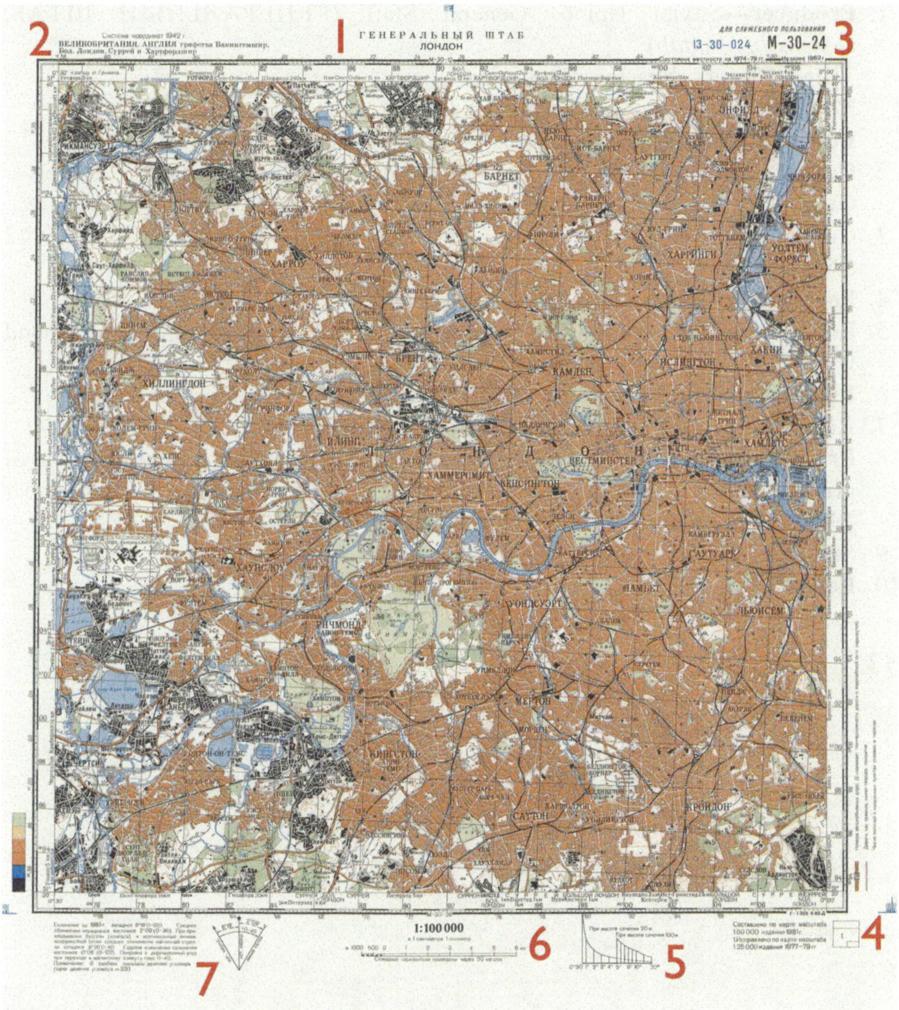


**Fig. 5** Soviet 1:100,000 topographic map of London, UK (1983), annotated to show the locations of key metadata

4. **Print Code—**Series, Job number, month of printing (Roman numerals), year of printing (last 2 digits), factory (Г 1325 II 83 Д—1:100,000 Topographic Series, Job 1325, February 1983, Dunayev, Moscow)
   **Compilation and source material dates—**with diagram showing which dates apply to which parts of the map
5. **Cross section graph**
6. **Scale—**1:100,000
7. **Magnetic variation notes—**with diagram

1. **Producer—**Soviet Union. General Staff (ГЕНЕРАЛЬНЫЙ ШТАБ, *General'nyy shtab*)
   **Title—**Exeter—Ekseter—Эксетер
   **Nomenclature—**Based on IMW Sheets
   **Edition date—**Издание 1982 г.—Edition 1982
2. **Country and Region**—Великобритания. Англия графство Девон—Great Britain. England county Devon
3. **Coordinate System—**Система координат 1942—Coordinate System 1942 (Pulkovo 1942)
4. **Former classification—**секретно—secret
5. **List of important sites—**Перечень важных объектов (numbers correspond with numbered buildings on the plan)
6. **List of street names—**Перечень названий улиц (with grid references)
7. **Reference**—Справка (description of the settlement)
8. **Print Code—**Series, Job number, month of printing (Roman numerals), year of printing (last 2 digits), factory (И 23 IX 82 Л—City Plan, Job 23, September 1982, Leningrad)
9. **Scale—**1:10,000
10. **Legend**

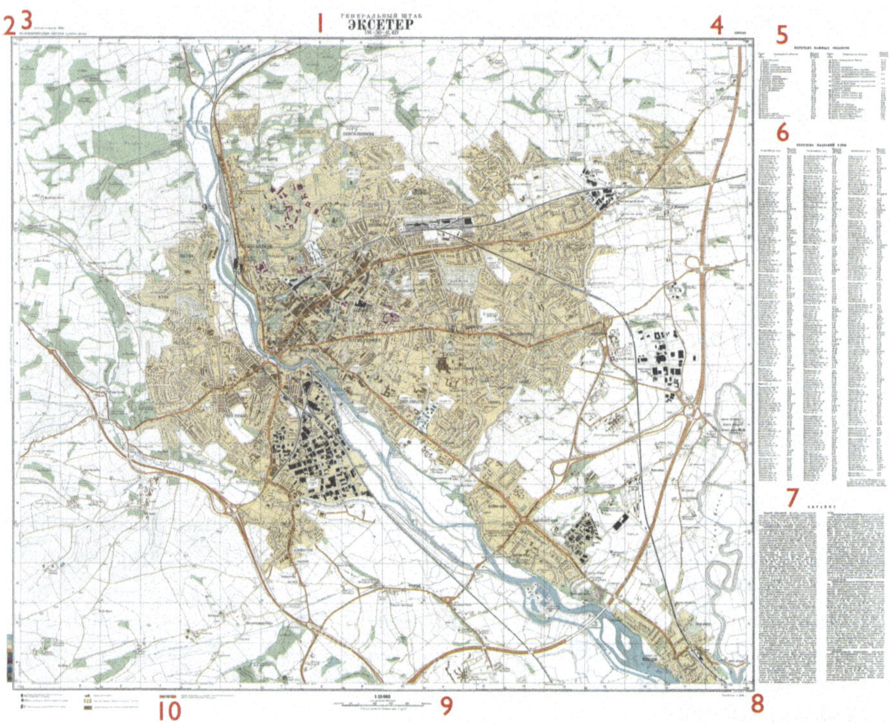


**Fig. 6** Soviet 1:10,000 plan of Exeter, UK (1982), annotated to show the locations of key metadata

# 4 Conclusion

The increasing availability of Soviet military mapping over the last two decades has seen their unprecedented use as sources of geospatial information, particularly as their acquisition by libraries around the globe continues to grow. Nevertheless, the difficulties posed by their unfamiliar format and language (especially for non--Russian-speakers) presents a significant obstacle to their accessibility and application. By offering an illustrated guide to locating and identifying key metadata on typical Soviet map sheets of different series and scales, this paper has focused on knowing what to look for when encountering or cataloguing this valuable and continually emerging resource.

# References

Davies J and Kent AJ (2017) The Red Atlas: how the USSR secretly mapped the world. The University of Chicago Press, Chicago. See http://www.redatlasbook.com for associated resources. Last accessed 23 Dec 2016

Kent AJ, Davies JM (2013) Hot geospatial intelligence from a cold war: the Soviet military mapping of towns and cities. Cartogr Geogr Inf Sci 40(3):248–253

Military Topographic Directorate of the General Staff (1966) Conventional signs for topographic maps of the USSR. Translated by Martin Davis (unpublished). Military Topographic Service, Moscow

Postnikov AV (2002) Maps for ordinary consumers versus maps for the military: double standards of map accuracy in Soviet cartography, 1917-1991. Cartogr Geogr Inf Sci 29(3):243–260

Psarev AA (2005) Russian military mapping: a guide to using the most comprehensive source of global geospatial intelligence. East View Cartographic, Minneapolis

Watt D (2005) Soviet military mapping. Sheetlines 74:1–4

**Martin Davis** has worked as an Instructor in Geography at Canterbury Christ Church University since 2014, alongside his ongoing PhD research into Soviet military cartography. In 2015, Martin was awarded the British Cartographic Society's Ian Mumford Award for excellence in original cartographic research. Martin is a member of the British Cartographic Society and is Reviews Editor and Editorial Assistant of *The Cartographic Journal*.

**Alexander J. Kent** is Reader in Cartography and Geographic Information Science at Canterbury Christ Church University. He is also the current President of the British Cartographic Society, Chair of the ICA Commission on Topographic Mapping, Editor of *The Cartographic Journal*, and is a committee member of UK Cartography, the Charles Close Society, and the Society of Cartographers. Prior to joining CCCU, Alex was Head of the Cartographic Unit at the University of Southampton and since graduating from Cambridge he has spent 18 years lecturing in the associated fields of cartography, GIS and remote sensing.

# Discovering Hidden Maps: Cartographic Representations as Arguments for Historical Narratives

**Tillmann Tegeler**

**Abstract** The national revivals of the nineteenth century contributed to the nationalization of historiography in Eastern and Southeastern Europe. Maps often illustrated history textbooks of the time and served as an argument for the territorial ambitions of the arising nations. Maps on military operations, boundary lines, medieval statehood or ethnography aimed to prove that a nation had occupied a certain territory for centuries; they functioned as a tool for the legitimation of historical claims. In contrast to national topographic map series, maps printed in books are always related to the surrounding narrative. The contextualization of both text and cartographic representation helps us to understand the function of these maps within a national narrative. Another way to capture the intention of the map is to embed it into the geographic reality. Once maps have been digitized and georeferenced, their projection within a GIS can offer clues about their accuracy. Ultimately, the cartographic representation in relation to the real world gives researchers an analysis tool to resituate maps stating an argument about a discourse [Harley (Cartographica 26(2): 1–20, 1989)] within a measurable framework. This method may provide new evidence about the intention of the author's statement.

## 1 A Brief History of Thematic Cartography

When does thematic mapping begin? In research there different times for the publication of the first thematic map are mentioned: Jürgen Schweikart and Immelyn Domnick date the publication of "one of the first known thematic representations" back to the early seventeenth century. According to them, this map is the *Designatio orbis Christiani* showing the dispersion of the major religions, produced in 1607 by Jodocus Hondius (Schweikart and Domnick 2013: 140). In contrast, historians of data visualization suggest August Friedrich Wilhelm Crome's statistical *Producten-Karte von Europa*, self-published in Dessau in 1782, might be classified as the first thematic map. For them, it is "possibly the

T. Tegeler (✉)
Leibniz Institute for East and Southeast European Studies, Regensburg, Germany
e-mail: tegeler@ios-regensburg.de

 
M. Altić et al. (eds.), *Dissemination of Cartographic Knowledge*, Lecture Notes in Geoinformation and Cartography, DOI 10.1007/978-3-319-61515-8_18

first economic and thematic map, [showing] geographic distribution of 56 commodities produced in Europe" (Friendly and Denis 2001). In the 19th century a new kind of thematic maps appeared: the infographic. The visualization of data—often in a geographical context—served a growing demand for information. So in 1869 Charles Joseph Minard for example created the *Carte figurative des pertes successives en hommes de l'Armée Française dans la campagne de Russie 1812–1813*, which was according to Edward R. Tufte the "best statistical graphic ever drawn" (Tufte 2001: 40), showing the losses suffered by Napoleon's army in the Russian campaign. This was the time when thematic mapping reached a stage of matured quality. One reason for this was the emergence of "more specialized fields of study" like geography which became a modern geospatial science and required a "new kind of mapping" (Robinson 1982: x). Others were the improvement of printing—even in newspapers—and the establishment of modern statistical administrations, which let the amount of data increase. Furthermore, land surveys created the topographical basis for the thematic contents (Kohlstock 2014: 123).

Besides the professionalization of science, another development shaping the nineteenth century was nationalism. The growing self-consciousness of small and bigger nations manifested in the use of maps as an instrument within the fight for the prerogative of historical interpretation. At that time historical maps were not only used to show a historical event, but also to claim territories for a certain nation. Many maps on military operations or boundary lines came into being, others showed medieval statehood or the ethnic distribution of people.[1] They were published in historical atlases[2] or in monographs. In contrast—and complementary—to the national topographic map series, these thematic maps were published together with a narrative, meaning their purpose was to substantiate the arguments which were put forward in the surrounding text. Finally these publications served as weapons in a propaganda war between the emerging nations.

The information contained in these maps was undoubtedly prejudiced. A new type of maps occurred, described by Haslinger and Oswalt as a "geopolitical cartography of planning" (Haslinger and Oswalt 2012: 5–6), which visualizes options for action and spatial utopias. With their expertise cartographers influence political decisions and contribute to the pursuit of geopolitical interests—always arguing with alleged facts, which are reflected "objectively" in visualized geographical data. The detection of this manipulative characteristic resulted in *critical cartography* (see Harley 1989; Crampton 2010). According to Crampton this criticism arose in the 1980s and early 1990s, even there were forerunners in questioning our knowledge of maps (Crampton 2010: 16–21). It was the work of Brian Harley which contributed to the development of a critical theory the most. He "brought into the discipline the ideas of power, ideology, and surveillance, arguing that no understanding of mapping was complete without them" (Crampton 2010:

[1]For Hungarian ethnic mapping, see Horváth (2010: 187–199).

[2]For the development of historical atlases in Germany, Austria, England and the USA during the period from 1860 to 1960, see Schraut (2011).

20). Especially in the data driven thematic mapping, the cartographer has the opportunity to mislead the recipient. Peter Jordan distinguishes between three main representations for ethnic mapping: choropleth mapping, dot distribution mapping and proportional symbols (Jordan 2010: 178). This means that on choropleth maps the majority in an ethnically mixed area occupies the whole territory, while minor nationalities are not represented. In contrast the other both cartographical representation types (dot distribution and proportional symbols) cover all nationalities, regardless their quantities. This shows, that the cartographic representation of quantitative data always risks to be fragmentary or false; whereas the current example of the "Atlas of knowledge" illustrates, that cartographic distortions can help to visualize quantitative distribution of data (Börner 2015). Here, areas with a bigger volume of data are drawn disproportionately larger. In his book "How to lie with maps" (Monmonier 1991), Mark Monmonier has described other reasons for errors and falsifications in cartography. He distinguishes between white lies, which are necessary to make a map readable, and deliberate manipulations. These latter are used to serve political and military propaganda or just to help companies to increase their profits; like estate agents who falsify maps to make the neighbourhood of an offered property appear more beautiful.

## 2 Historizing Thematic Maps

As we have seen, the aim of thematic mapping is to visualize data in a geographical context, to spatialize non-territorial elements (Haslinger and Oswalt 2012: 4). This data may be statistical material like results from surveys or census data. The manner of interviewing the probands already influences the results and drawing the maps further increases the scope for interpretation. In times when open data makes science transparent and good scientific practice is the guideline of research funding, the disclosure of data is necessary to clarify the motivation of a certain interpretation. With regard to research data management funders ask for the publication of data to verify the results and to maintain the integrity of the research. This applies all the more for publications based on historical data. Several databases already provide statistical data from printed publications to be used in current research or to verify older research results.[3] This is also true for thematic maps. If a cartographer publishes the data on which a map is based, one can bring to light his motivation. However, the relevant sources might be lacking or the data collection might not be comprehensible. Another way of verifying maps is therefore to integrate them in a Geographical Information System (GIS) digitizing and georeferencing them to generate the needed geographical data.

[3]For example: projects like IPUMS by the Minnesota Population Center (see https://www.ipums.org/, last accessed: 16 Jan 2017) or MOSAIC (see http://www.censusmosaic.org/, last accessed: 16 Jan 2017).

# 3 The Geoportost Project

These considerations led to a project conducted at the Leibniz Institute for East and Southeast European Studies (IOS) in Regensburg, Germany. Researchers at this institute work on questions regarding the political, social and economic conditions of Eastern and Southeastern Europe in a historical and comparative perspective. The library is responsible for the electronic research infrastructure and serves as a department for research support. In this capacity the library has developed a project to provide digitized and georeferenced maps in open access.

## *3.1 The Catalogue*

The starting point for this project was a library catalogue containing 16,000 index cards. This catalogue was built up since 1952, when one of the predecessor institutes of the IOS started to index books, journals and even their content. This way catalogues of so-called *hidden* bibliographies, pictures and maps came into being. *Hidden* describes the indexed material as enclosed in—and in this manner depending on—books, journals or conference proceedings. Whereas the catalogue of these latter, independent publications had already been transferred to an electronic Online Public Access Catalog (OPAC), searches on the depending material could only be performed in the card catalogues. For researchers in area studies like the East and Southeast European ones, maps are always of large interest to examine and understand the spatial concepts of a certain area. So the idea was born to retro-convert the hidden map catalogue. To receive funding by the German Research Foundation (DFG), the digitizing of those maps which are free of copyright was included in the application as an added value.

After the project had been approved, the retro-conversion started and resulted in a surprise. A company specialized in converting analogue catalogues scanned the cards and transcribed their content into the union catalogue of the Bavarian Library Network (Bibliotheksverbund Bayern—BVB). First the record of the source work was searched, then a record for the map was created and both were linked. The new record of the depending map included the title, the reference, the year and the subject headings. In the end the catalogue cards described nearly 27,000 maps, instead of the supposed 16,000 maps. When the bibliographical records were entered into the catalogue database, those free of copyright were indexed by using the German authority file (GND) and the Dewey Decimal Classification (DDC). These very valuable descriptions are accessible and searchable via the catalogue.[4] As MarcXML

[4]https://www.regensburger-katalog.de/TouchPoint/start.do?View=ubrwios&Language=de&SearchProfile=GeoPortOst. Last accessed: 23 Jun 2017.

files the records are public domain under a creative commons license CC0 on the website of the Bavarian library network.[5]

The regional focus of the collection was the former Soviet Union and Poland, because of the research area of one predecessor institute. It was therefore decided to add maps from the other research area of the IOS, Southeastern Europe, to increase the number of copyright-free maps. Altogether 912 maps were identified as public domain. These maps date from 1717 to 1922,[6] while the whole catalogue includes records up to 2011. About two thirds of the 912 maps have a temporal reference (from the ancient migration period to the ethnographic conditions of Hungary in the 1920s). More than a third are history maps; others are street or city maps (each about 100). On the whole thematic maps predominate. Their subjects are heterogeneous: the majority are military (especially of the Russo-Japanese War or the Balkan Wars), statistic and directory maps, but there are also maps on infrastructure, agriculture or ethnicity.

## 3.2 Digitizing and Georeferencing

In cooperation with the University library of Regensburg, the IOS then started to digitize the 500 maps from the catalogue and the newly selected maps regarding Southeastern Europe. As hidden maps are always related to the context of the surrounding text and their message is not comprehensible without this, it was decided to digitize not only the single maps, but all of the books and volumes as well. Four maps dated from the 18th, 279 from the 19th and 629 from the early 20th century, printed in about 100 source works. The digitization was conducted at a high resolution (400 dpi) in the TIFF format.

To fulfil the objective of verification, geographical data of the maps had to be generated. Therefore, after the digitizing process the georeferencing was prepared. Because of the heterogeneity of the collection it was decided to implement an application that allowed to open up the georeferencing to a maximum number of experts on these maps. Consequently the IOS licenced the proprietary tool *Georeferencer* by Klokan Technologies which makes it possible to generate geographical data by crowdsourcing. This application was already used by bigger libraries and archives such as the British Library,[7] National Library of Scotland,[8] Czech libraries within StaréMapy.cz[9] or Leiden University.[10]

[5]http://www.bib-bvb.de/web/b3kat/open-data. Last accessed: 8 Dec 2016.

[6]The maps from 1922 were created by Pál Teleki, who died in 1941, which is why they are free of copyright.

[7]http://www.bl.uk/georeferencer/georeferencingdata.html. Last accessed: 12 Dec 2016.

[8]http://maps.nls.uk/projects/georeferencer/. Last accessed: 12 Dec 2016.

[9]http://www.staremapy.cz/. Last accessed: 12 Dec 2016.

[10]http://www.library.leiden.edu/special-collections/special/maps-in-the-crowd-eng.html. Last accessed: 12 Dec 2016.

One precondition for starting the georeferencing process was having a metadata sheet[11] which included the following mandatory fields: identity number, filename, image format and image URL. To improve the description on the georeferencer site several optional columns were offered within the csv file: links to the library website, the viewer and the catalogue entry, title of the map, temporal coverage, publication date, person names (for creator or contributor), publisher and—important for the expected date—width, height, scale and scan resolution. Some of the requested data were already part of the catalogue dataset, others were created additionally. Another preparatory work was to cut the images into separate tiles. This measure allows to zoom into the images, which is very important in order to take the details of the maps into account during the georeferencing.

Although the images are not provided on the company's but on the project's server, every necessary step for georeferencing can be taken within the web application. After registration every volunteer can take part in the project. There is a tool for clipping the map to mark only those areas whose geographical data should be generated. After doing so, one can set three control points for a basic georeferencing job. Another tool makes it possible to visualize the accuracy of the map, so that any distortion resulting from the mapping process or any mistake caused by the georeferencing will become visible. To identify these kinds of errors one has to identify at least five control points. Finally the most interesting result of the whole process is the visualization of the map image as an overlay over a modern map in a GIS. By adapting the transparency of the overlay one can compare the historical map with its present-day counterpart so changes can be spotted at once.

### *3.3 The Portal*

During the preparation of the georeferencing process it was decided to not only provide images of maps on the IOS website and to georeference them within *Georeferencer*, but to set up an individual portal that goes further than a site for images and their geographical data. It should offer at least three components: an explanation of georeferencing, the geoportal with all metadata and the description of the catalogue.

The first part of the website answers the questions *what georeferencing is* and *why it is useful*. Then it calls for participation and explains how one can support the project. On this site there is also a list of the top contributors, a progress bar of the project and finally a map on which every dot represents the centre of a georeferenced map, so one can see the collection's regional focus. A special button leads to a random map not georeferenced yet. The success of the crowdsourcing was overwhelming. Participation was promoted in a conventional way by posts on mailing lists and announcements on websites. Within the first 2 weeks, 33 persons

[11]http://help.georeferencer.com/metadata. Last accessed: 12 Dec 2016.

had georeferenced about 80% of the 912 maps, while after 4 weeks the quota was already 99%. Only one of these 899 maps was georeferenced wrongly. This one belongs to the part of the collection which up until now has proven impossible to georeference. It concerns maps printed in Russian and German books dealing with the Russo-Japanese war. Here the problem is that images of battles and skirmishes are located in China, but the location names on the maps are in Russian, Japanese or a German transcription of one of these languages. To get the remaining maps georeferenced IOS tried to contact experts on this subject from Germany and Israel. The future will show, if the project can recruit researchers, whose work is based on written sources, but who are willing to contribute to generating geographical data.

As a special service for users the geoportal provides a viewer and a retrieval system for all metadata of the maps, including information about their creation, content and electronic accessibility:

- ID of the map
- title of the map
- URL of the geoportal representation
- URL of the IOS viewer (zoomable map)
- URL of the map image (TIFF download)
- URL of the georeferenced representation
- URL of the catalogue entry
- name of the cartographer
- date of publication
- language
- width
- height
- scale
- type of map
- collection
- geographical subject heading
- topical subject heading
- personal subject heading
- temporal coverage
- additional information
- author of the source
- GND of the author of the source
- title of the source
- year of publication of the source
- URL of the source within the catalogue
- signature of the source within the catalogue
- reference of the map in the source
- PURL of the map image
- information on license.

This information is very important for the reuse of the maps. On the one hand it is possible to download metadata or images, on the other hand legal information for the use of an image is provided. Since most of the images are free of copyright, the project contributes to the open access of geographical and cartographical data. Due to the high resolution of the digital reproductions, another added value is the zoomable viewer that allows to obtain a closer look at the map. But also for the retrieval of historical maps GeoPortOst helps to get better results through its very detailed metadata. These include intellectually created subject headings, so one can search for a person, a historical event or even a subject in combination with other keywords.

After finishing the georeferencing process geographical searches are possible by using the *GeoBlacklight* discovery service. With this aim the generated geographical data were imported into the bibliographical metadata of the BVB union catalogue. By harvesting the bibliographical records from the OAI repository, the metadata including the coordinates can be downloaded in a MarcXML format. To index data for *GeoBlacklight* Solr, they have to be transformed from XML to a Dublin Core based JSON format. Then the data can be integrated into the geoportal. In *GeoBlacklight* the images can be downloaded in GeoTIFF format, so that finally overlays of the map images or bounding boxes with the maps' size can be visualized in a GIS and the user has the opportunity to search not only semantically, but also geographically.

The end result is a catalogue which provides access to the bibliographical data of almost 27,000 hidden maps from three centuries. Only a fifth of this number dates from non-socialist times: 1028 are from the time before the end of World War I, 1472 from the interwar period till the end of World War II, and 1893 come from books published after 1990. Over 80% of the indexed maps are from the years between 1946 and 1990. They cover all areas of research from history and economics to engineering and natural sciences like geology. Among them are of course many topographical maps and city plans, but all of them represent the scientific knowledge of the second half of the twentieth century and its political implications. Once they will become free of copyright in the future, their digital representations will also help to uncover political thinking and spatial concepts of divided Cold War Europe.

## *3.4 Examples*

**Ethnic Composition of Macedonia**

But back to the time before World War I. This historical event was a result of the growing nationalism and national chauvinism in all of Europe. However, even before the worldwide firestorm there were already regional armed conflicts that let one imagine, what was to follow. One of these events were the Balkan Wars. To examine war crimes, an international commission was set up by the Carnegie Endowment for International Peace. Their work resulted in the *Report of the International Commission on the Balkan Wars*, which was published in 1914 and which also included maps. Two of them show the ethnographical composition of

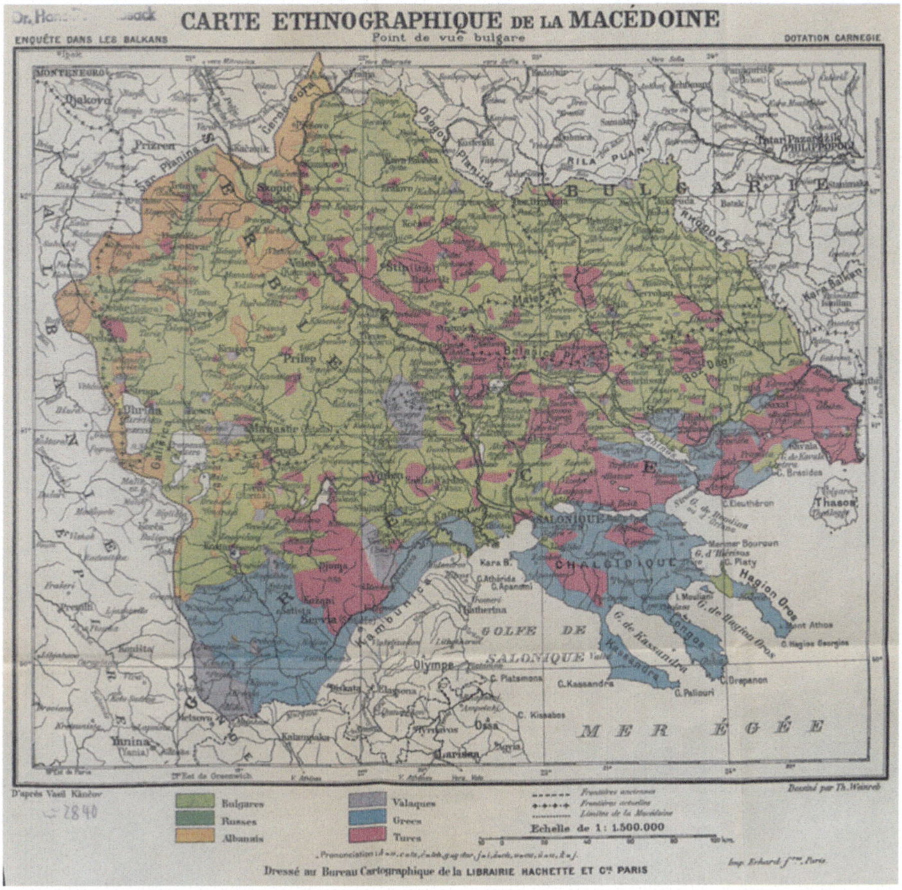


**Fig. 1** Ethnographical map of Macedonia: Bulgarian point of view (Source: Carnegie Endowment for International Peace 1914: Supplement)

Macedonia—once from the Bulgarian point of view, and once from the Serbian side. Not very surprisingly, large parts of the north and the centre of the country were controversial (see Figs. 1 and 2).

While the Serbs claimed this area for Serbo-Croats, the Bulgarians denied the existence of Serbs there completely. Macedonia stayed disputed at least until the creation of the Socialist Republic in 1944. The georeferenced maps[12] now provide the information on which particular areas were contested.

[12]For Fig. 1, see http://geoportost.ios-regensburg.de/map/BV042760754, for Fig. 2, see http://geoportost.ios-regensburg.de/map/BV042760767. Both last accessed: 12 Jan 2017.

**Fig. 2** Ethnographical map of Macedonia: Serbian point of view (Source: Carnegie Endowment for International Peace 1914: Supplement)

## Nationalities Map of Austria-Hungary

While in the case of Macedonia it is obvious, why the opposing parties used different data to support their views, another example does not at a first glance clarify the motivation for changing a map in different editions of its source work. In 1915, Rudolf Kjellén, who coined the term *Geopolitik*, edited a book in Sweden (Kjellén 1915), which was translated into German and published in Germany (Kjellén 1916) in the following year. It dealt with the *political problems of the World War* (title of Kjellén's book). In the Swedish edition the map shows that the Habsburg Monarchy was inhabited by Germans, Italians, Romanians, Hungarians, Serbs, Slovenes, Czechs, Poles, and Ukrainians (Fig. 3).

This was adopted in the German edition, although all South Slavs were named Serbs here, with no mention of Croats, Bosnians or others. Furthermore the German

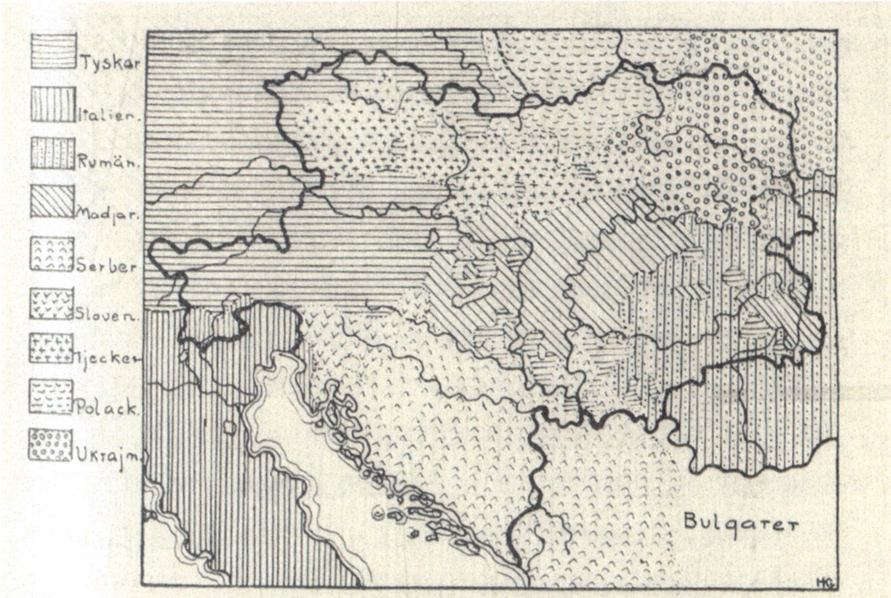


Karta IV. Nationalkarta öfver Österrike-Ungern.

**Fig. 3** Ethnographic map of Habsburg Monarchy: Swedish edition (Kjellen 1915: 99)

editors did not simply copy the map, but changed the dashes, which were showing the ethnicity in a certain area (Fig. 4).

The territory east of Prague was never inhabited by a sizeable German population. Nevertheless, the Germans marked this part of Bohemia as an ethnically mixed area of Czechs and Germans. One can only assume that this re-drawing of the map is related to the simultaneously occurring World War I.

## Coalition Wars

In 1799, during the Napoleonic wars, Field Marshal Alexander Suvorov led a Russian–Austrian coalition army to victory against French troops in the Battle of Cassano. In the same year a book was published in Germany describing the *life and campaigns of field marshal general Alexander Suvorov Rymniski Italski in Italy* including maps of this battle (Mxy 1799). In the introduction the author stated:

> Several meritorious writers have already taken the brush of history, and their works could undoubtedly be no other than satisfactory to most. But since the works published on this subject are exposed to drawings, they are usually incomprehensible to a large number of readers. The present can serve to supplement them; and as we succeeded in preserving the plans of the sieges and meetings of this campaign, we hasten to reach the purpose by their public notice, having freed them from all superfluous expedients. The whole, as well as the order of institutions for attack and defense, can be better overlooked. (Mxy 1799: 2).

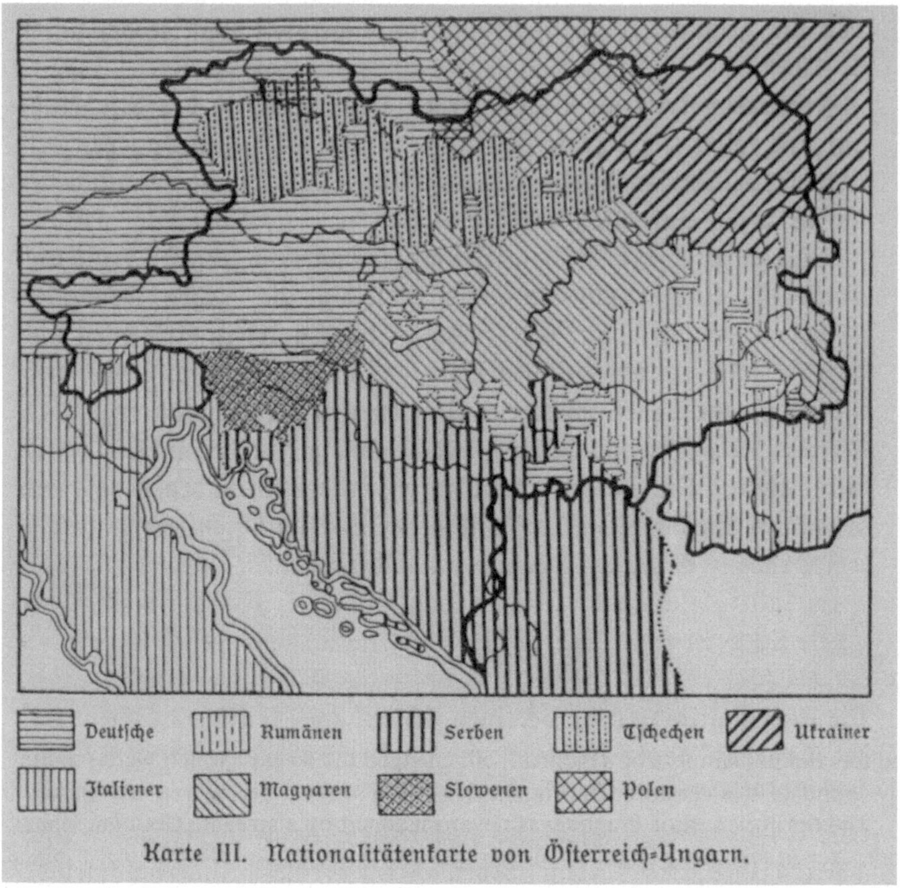


**Fig. 4** Ethnographic map of Habsburg Monarchy: German edition (Kjellen 1916: 78)

The reader thus expects to get exact information on the course of the battle. And indeed, the drawing[13] shows in great detail where the troops were situated and how many military units were deployed. It seems like a stroke of luck that data on a historical event were published so close after its occurrence. However, when current techniques try to verify the data, it is astonishing to see how the author could have raised expectations in this manner. By georeferencing one of these maps, the author's information is proven to be absolutely false. Because the settlements are wrongly located, the georeferenced map seems to be twisted.[14]

An analysis tool of Georeferencer illustrates the lack of accuracy.[15]

---

[13]http://geoportost.ios-regensburg.de/map/BV042512931. Last accessed: 13 Jan 2017.

[14]For Fig. 5, see http://geoportost.georeferencer.com/id/BV042512931/visualize. Last accessed: 13 Jan 2017.

[15]For Fig. 6, see http://geoportost.georeferencer.com/id/BV042512931/accuracy. Last accessed: 13 Jan 2017.

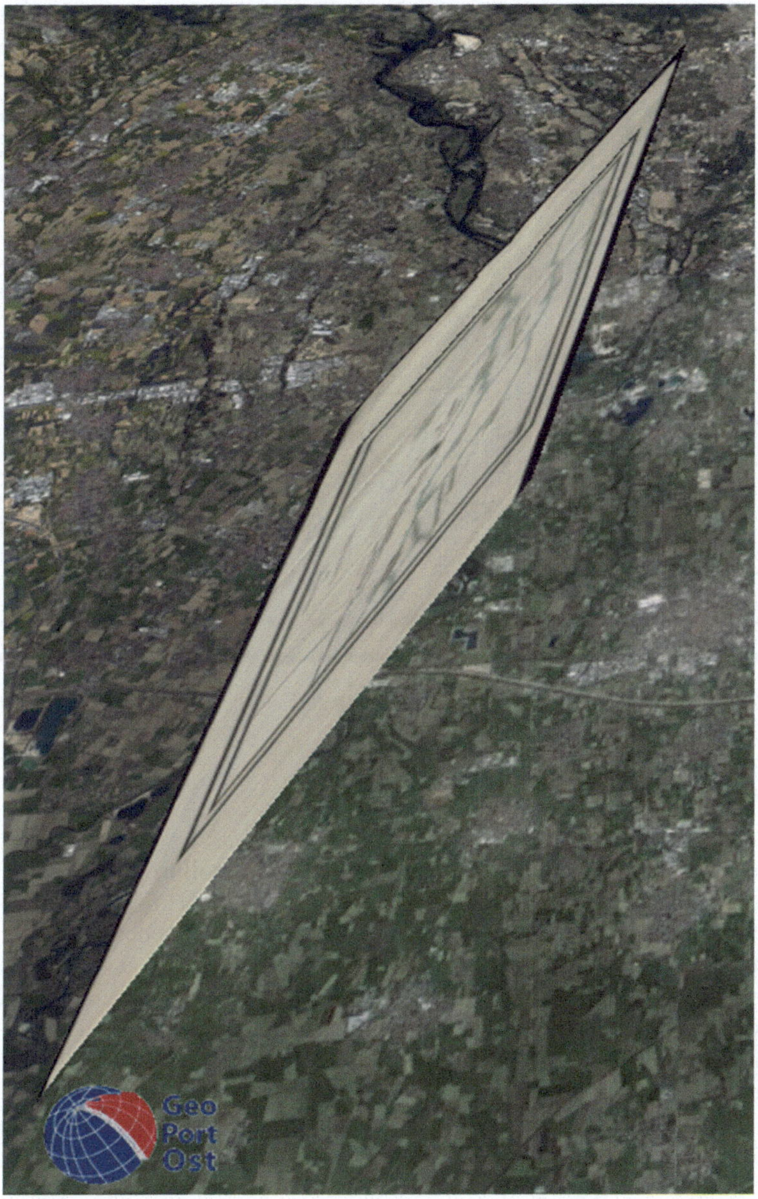

**Fig. 5** Visualization of georeferenced map, showing the Battle of Cassano (Source: Klokan Technologies GmbH 2017)

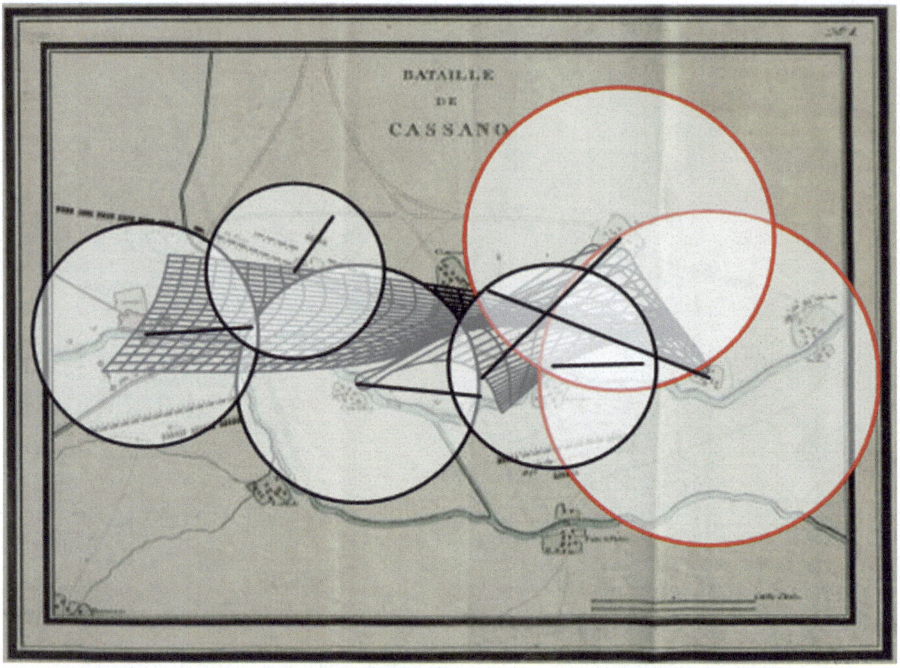


**Fig. 6** Accuracy analysis of map, showing the Battle of Cassano (Source: Klokan Technologies GmbH 2017)

## 4 Conclusion

As these examples have shown, *hidden* maps are used to illustrate the surrounding text of the source work. On occasion they inform about different views on a disputed issue (see the map of Macedonia). In another context we see maps changing their statement when the source work addresses a new audience (see Kjellén). They can also falsify the facts (see Cassano), perhaps intentionally to mislead the recipient or to mask the maker's ignorance about an issue.

In any case, the key characteristic of hidden maps is their close relation to the surrounding text. To *dis*close this, modern technologies can support the researcher in his work. Digitizing and georeferencing the maps can help to validate the author's statement. In the future it is therefore crucial to also publish the data of thematic maps to verify not only their intention, but also the used sources.

## References

Börner K (2015) Atlas of knowledge. Anyone can map. MIT Press, Cambridge, MA

Carnegie Endowment for International Peace (1914) Enquête dans les Balkans: Rapport présenté aux directeurs de la dotation par les membres de la commission d'enquête. Centre Européen de la Dotation Carnegie, Paris

Crampton JW (2010) Mapping. A critical introduction to cartography and GIS. Wiley-Blackwell, Chichester
Friendly M, Denis DJ (2001) Milestones in the history of thematic cartography, statistical graphics, and data visualization. Web document, http://www.datavis.ca/milestones/. Last accessed 16 Nov 2016
Harley JB (1989) Deconstructing the map. Cartographica 26(2):1–20
Haslinger P, Oswalt V (2012) Raumkonzepte, Wahrnehmungsdispositionen und die Karte als Medium von Politik und Geschichtskultur. In: Haslinger P, Oswalt V (eds) Kampf der Karten. Propaganda- und Geschichtskarten als politische Instrumente und Identitätstexte in Europa seit 1918. Herder Verlag, Marburg
Horváth FS (2010) Karten als Fortsetzung der Politik mit anderen Mitteln. András Rónai und sein Mitteleuropa-Atlas. In: Haslinger P, Oswalt V (eds) Kampf der Karten. Propaganda- und Geschichtskarten als politische Instrumente und Identitätstexte in Europa seit 1918. Herder Verlag, Marburg
Jordan P (2010) Methodik und Objektivität von Karten des nationalen/ethnischen Bewusstseins. In: Happel J, Von Werdt C (eds) Osteuropa kartiert – Mapping Eastern Europe. LIT, Zürich
Kjellén R (1915) Världskrigets politiska problem. Bonnier, Stockholm
Kjellén R (1916) Die politischen Probleme des Weltkrieges, 3rd edn. Teubner, Leipzig
Klokan Technologies GmbH (2017) Georeferencer. http://geoportost.georeferencer.com/. Retrieved 16 Jan 2017
Kohlstock P (2014) Kartographie. Eine Einführung, 3rd edn. Schöningh, Paderborn
Monmonier M (1991) How to lie with maps. University of Chicago Press, Chicago
Mxy (1799) Leben und Feldzüge des Generalfeldmarschals Alexander Suworow Rymniskoi Italskoi in Italien. Reinicke und Hinrichs, Leipzig
Robinson AH (1982) Early thematic mapping in the history of cartography. University of Chicago Press, Chicago
Schraut S (2011) Kartierte Nationalgeschichte. Geschichtsatlanten im internationalen Vergleich 1860–1960. Campus, Frankfurt am Main
Schweikart J, Domnick I (2013) Thematic mapping between individual issue and mass media. Kartogr Nachr 63(special issue):140–147
Tufte ER (2001) The visual display of quantitative information, 2nd edn. Graphics Press, Cheshire, CT

**Tillmann Tegeler** studied East and Southeast European History, Medieval History and Slavic Studies in Munich and Moscow (M.A. 2000); additionally he gained research experiences abroad (Moscow, Vilnius, London). In 2000 he started working at the library of the Osteuropa-Institut Munich, in 2004 also at the Südost-Institut. With the relocation of the two institutes to Regensburg in 2007, he became head of the libraries; since 2012 he is Head of the Library and Electronic Research Infrastructure Division at the Leibniz Institute for East and Southeast European Studies (IOS) in Regensburg, Germany. With its collection of more than 330,000 media items, the library ranks internationally among the largest libraries for East and Southeast European studies. The department also conducts projects for research support. So Tegeler managed a project to promote open access in the humanities; within the framework of another project half a million pages of German language periodicals from Eastern Europe were digitized. After the approval of funding, currently the second phase of GeoPortOst started, where tools will be developed to support the annotation of thematic maps. Tegeler is member of the Working Group of Libraries and Documentation Units for Research in East, East Central and South Eastern Europe (ABDOS) and of the Advisory Council of the Working Group of Special Libraries (ASpB).

# Revealing the Past: How Science Is Unlocking Cartographic Secrets

**Nick Millea and David Howell**

**Abstract** Although advanced imaging and analytical techniques have long been available, it is only over the last 3–5 years that they have been more generally applied to serious study of library materials. This has been driven by incredible advances in technology which allow for non-invasive, non-sampling, non-contact methods to be designed specifically with the curators' concerns for the safety of the material in mind. At the same time the curators' increasing awareness of what is possible has increased the demand for such services.

For maps, one of the outstanding developments has been hyperspectral imaging. This is a technique where an item is scanned and each 'pixel' represents a full colour spectrum, commonly 400–1000 nm. The large files produced in this way can be interrogated by sophisticated software to reveal erasures and obscured areas of detail invisible to the naked eye. Hyperspectral imaging can also be used for identifying pigments, especially organic materials. This technique is completely safe if applied correctly and with continuous dialogue between curators, conservators and the people carrying out the analysis/imaging.

It is now possible to employ science to date cartographic material, and to infer where maps might have been made. We can also use hyperspectral imagery to reveal underlying cartographic content previously obscured from view. This paper will examine how these techniques have been employed with varying degrees of success on the late medieval Gough Map of Great Britain, and two manuscript English estate maps, all items held within the Bodleian Library's collections.

## 1 Hyperspectral Imaging at the Bodleian: An Introduction

The way we experience the world is not unique. Other living creatures have different sensory receptors to humans; different abilities to hear high pitched sounds and the ability to smell very low concentrations of odour for example. The same is true for vision. We have just three colour receptors: red, green and blue, whilst other species have different numbers. An extreme example is the mantis

N. Millea (✉) • D. Howell
Bodleian Libraries, University of Oxford, Oxford, UK
e-mail: nick.millea@bodleian.ox.ac.uk; david.howell@bodleian.ox.ac.uk

 
M. Altić et al. (eds.), *Dissemination of Cartographic Knowledge*, Lecture Notes in Geoinformation and Cartography, DOI 10.1007/978-3-319-61515-8_19

shrimp. Being able to see four or five wavelengths in the ultra-violet region and also into the infra-red region of the electromagnetic spectrum as well as seven different sensors across the visible wavelengths will give the shrimp a very different version of the world to us (plus a very different appreciation of art!).

Most imaging, or digitisation, has as the main aim of creating as accurate a reconstruction or representation as possible for the viewer and is focussed on visual aesthetic. This means concentrating on the way the human eye responds to colour. Other imaging can show things invisible to the human eye, the most commonly known being in medical imaging such as X-radiography and Magnetic Resonance Imaging (MRI).

At Easter 2014 the Humanities Division within the University of Oxford was lucky enough to receive a grant from the internal University Fell Fund to purchase imaging equipment more akin to the shrimp's 'eye' than the human eye. Hyperspectral imaging is the process in which image data is captured at many frequencies across the electromagnetic spectrum to produce a complete reflectance spectrum for each pixel. The Headwall imaging system captures all the spatial (1600 pixels) and spectral data (972 bands from 400 to 1000 nm) that can be seen within the field of view of the imaging device. This new set of virtual eyes gives users a more complete view of any type of reflecting material than the human eye alone. It can be used to reveal information about both the materiality of objects (e.g., pigments, lichen, and binder) and hidden text or pictorial information in built heritage as well as objects in a museum, a library, or a university.

The equipment produces a great deal of data, effectively a stack of 972 high resolution images, each at a specific wavelength. While scanning through each of the images can sometimes reveal hidden information it is often necessary to use computational analysis to tease out detail. The Library is lucky to have access to ENVI software, and extremely powerful computer program designed specifically for analysing hyperspectral data. The software carries out a number of processes on the data to try to group similar spectra together. One of the most powerful techniques is called Principal Component Analysis (PCA) but ENVI has many algorithms which can be applied to tease out understandable information from the vast amount of data.

The equipment has been used most generally on codices; bound books. Sometimes the data can be analysed very quickly and a result delivered to the researcher within hours. In most cases people are looking for hidden texts or under-drawings and if these become visible one has an immediate and positive result. But the first object ever scanned in the Bodleian, even before the equipment was purchased, was the Gough Map. This object was chosen as a challenge for the technology as the map is very large and it is not entirely clear whether there is information to be uncovered. The data collection was extremely successful but even after two years we are still exploring the data to reveal more information (Fig. 1).

To place the research in its institutional context, the Bodleian Library, the library of the University of Oxford, was founded by Sir Thomas Bodley in 1602. It has become Europe's largest university library, holding 12 million books, and well in excess of 1.25 million maps. The Library's maps fall within the Department of

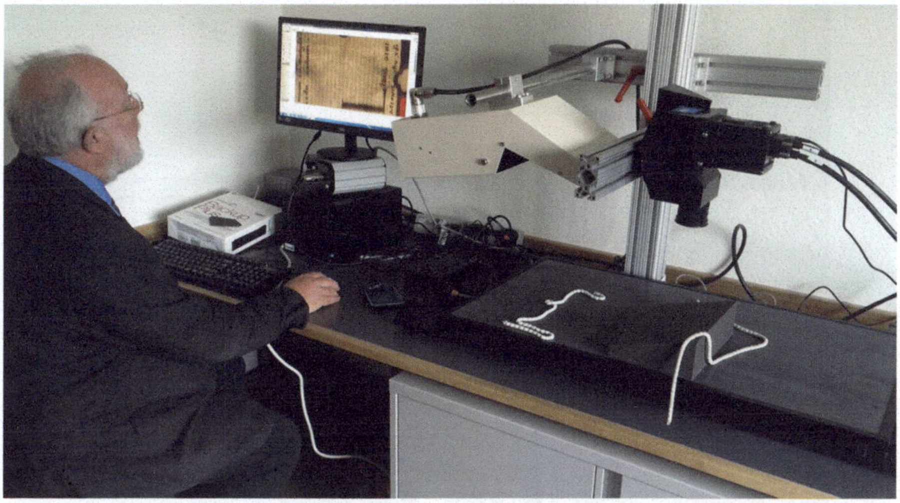

**Fig. 1** Here is the equipment. In essence the sensor looks at a very narrow band 1600 pixels long and one pixel in width. The sensor measures colour from 400 nm (near the UV end of the spectrum) up to 1000 nm (in the IR region) at something around 1 nm intervals. This means that each pixel is represented by 972 'measurements' and gives a huge amount of calibrated colour information (© Nick Millea)

Special Collections, administered from the recently-refurbished Weston Library, located in Oxford city centre. Both authors are based in this building.

## 2 The Maps

To prepare for the ICA History of Cartography gathering in Dubrovnik, the authors decided to select three manuscript maps from the Library's collections for hyperspectral treatment. The basic premise being to reveal detail unseen to the naked eye. As alluded to above, some preliminary work had already been undertaken on the Gough Map, but in order to broaden the research pool, two unheralded and virtually unknown manuscript maps were set aside for data capture.

### *2.1 The Ascott Park Map*

Firstly, the Ascott Park map, featuring a long-disappeared estate in South Oxfordshire (about 15 km south-east of Oxford). This map dates from the early eighteenth century, and was made by the prolific local land surveyor William Burgess. All that remains from the estate is an ornamental gateway bearing the inscription *si bonus es intres: si nequam nequaquam*, or *if good come in: if wicked, by no means*, and this

**Fig. 2** David Howell standing beneath the Ascott Park gateway (© Nick Millea)

gateway has been reconstructed within the Weston Library, opened to the public in March 2015 (Fig. 2). The gateway and its message are particularly apposite, representing the point at which only individuals bearing a Bodleian Library reader's ticket can move from the public area of the building to those areas set aside for Library readers. The map itself was also purchased by the Bodleian from a private individual in 2015, thus reuniting both manuscript and masonry. The map measures 133 × 84 cm (Fig. 3).

The intention of the hyperspectral work was to help identify the location of the gateway on the map. This objective has yet to be met, however, although the map's text is now clearly legible, and as can be demonstrated by Fig. 4 below, the map's

**Fig. 3** Ascott Park estate map by William Burgess (Bodleian Libraries, MS. C17:49 (326), © Fiona McLees, Bodleian Libraries)

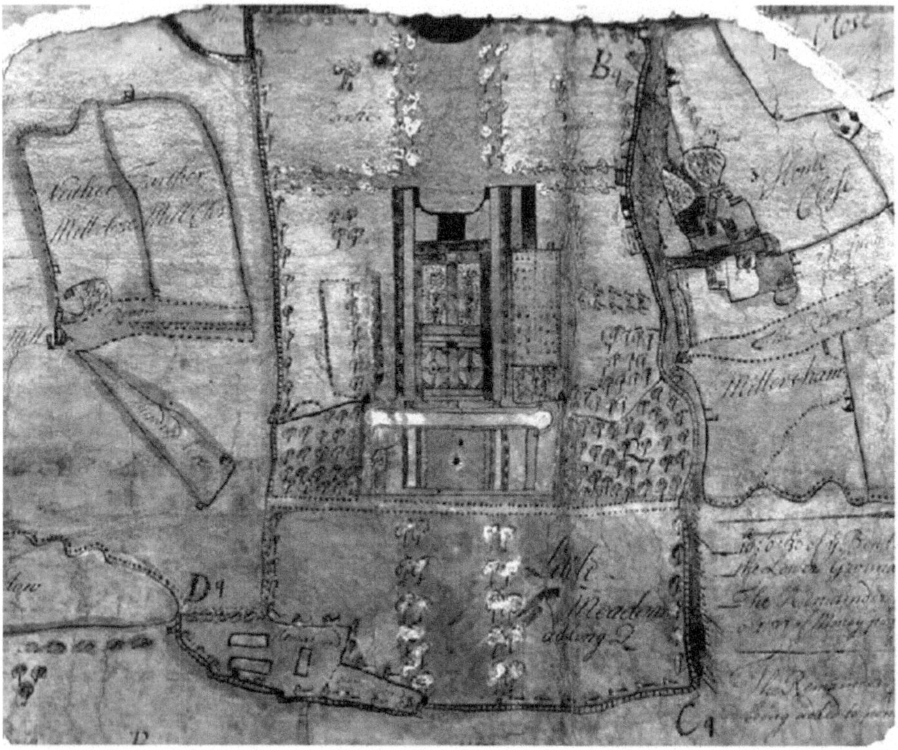

**Fig. 4** Hyperspectral image of the area around Ascott Park's main house (Bodleian Libraries, MS. C17:49 (326), © Bodleian Libraries, University of Oxford)

detail has become considerably clearer. The 'go to' series of algorithms, essentially ways that the computer searches for information in all this data, within the ENVI software are grouped together in one semi-automated procedure called the Spectral Hourglass Wizard. The hourglass processing flow uses the highly detailed nature of hyperspectral data to find the most spectrally pure, or spectrally unique, pixels (called *endmembers*) within the dataset and to map them. This processing flow begins with reflectance input data. Then the 'wizard' looks for 'endmembers' that are similar across the whole spectrum (including areas not visible to the human eye) and maps these, often giving several different outputs. It is then up to a human to look at these outputs to decide which ones give the most useful information.

## 2.2 *The "Cheshire Acres" Map*

The second manuscript selected was an altogether different challenge. This map has yet to be identified geographically. The text is in English, and it is believed to represent an English estate. But where? Numerous attempts have been made to associate the cartography with a precise geographical location, but without success. All that is known is that the map includes measurements in "Cheshire acres", there is a North arrow to give a sense of direction, and the text that is legible with the naked eye suggests the manuscript is an update of a 1641 original. The map measures 68 × 84 cm.

The hyperspectral imagery has made clearer some of the map's detail, but the exact location of this estate remains unknown. The darker, much-damaged left-hand side of the map features a large house or hall, but as yet, the accompanying text cannot be read, and so the hall continues to be nameless. The clearer, right-hand section does however give up some clues. Numerous field names are revealed, always presented with the rider equating to "formerly known as . . . in 1641 but now called . . ." whereupon the text ends and a frustrating blank space can be seen. There is also a house clearly named "Scotland", but alas such a house no longer appears to exist. One clue, however, is the appearance of a "Mr Harrop", and Harrop is a surname which can confidently be pinpointed to a small geographical area, principally to the south of Manchester, in other words north-east Cheshire, a convenient coincidence given the presence of Cheshire acres on the map (see Fig. 5).

**Fig. 5** The "Cheshire Acres" map, copied from a 1641 original (Bodleian Libraries, no call number, © Bodleian Libraries, University of Oxford)

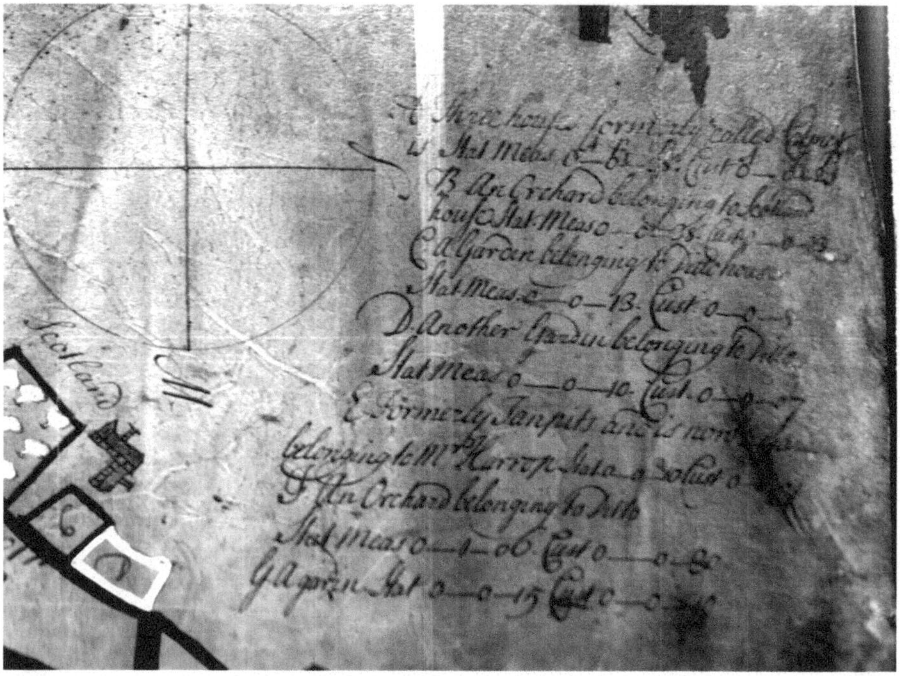

**Fig. 6** Extract from the "Cheshire Acres" map, showing the house known as Scotland and naming Mr. Harrop (Bodleian Libraries, no call number, © Bodleian Libraries, University of Oxford)

Figure 6 below has been derived from the ENVI Spectral Hourglass Wizard, and the descriptive text identifies a number of the small parcels of land depicted on the map.

Figures 7 and 8 place the successful research to date into geographical context. Cheshire is located in the north-west of England, and University College London's investigation into surname distribution focuses on the 1881 census and the whereabouts of those named Harrop entered onto that census. Clearly there is a degree of correlation which gives grounds for optimism for future exploration.

The principal house on the map, however, remains unidentified, despite experimenting with a variety of spectral combinations (see Fig. 9). This image was derived from the ENVI Spectral Hourglass Wizard and using 'false colour' to highlight differences.

## 2.3 The Gough Map

Much has been written about the late medieval Gough Map of Great Britain, described by Helen Wallis as "The most remarkable cartographic monument of

**Fig. 7** Location of Cheshire (© Michael Athanson, Bodleian Libraries)

medieval England" (Wallis 1994: 13–14). The map has its own website (Linguistic geographies 2011), which offers the opportunity to explore the manuscript in much greater detail.

The map measures 56 × 117 cm and has been held in the Bodleian since 1809 when it was bequeathed to the Library by Richard Gough, hence its name. The map is widely considered to be the earliest surviving map to show the island of Great Britain in a geographically recognisable form. It consists of around 600 place names and over 200 rivers, and clearly depicts late medieval England and Wales with a certain degree of familiarity. Scotland is somewhat less convincing. This iconic cartographic design was very much the template for maps of Britain until the latter part of the sixteenth century.

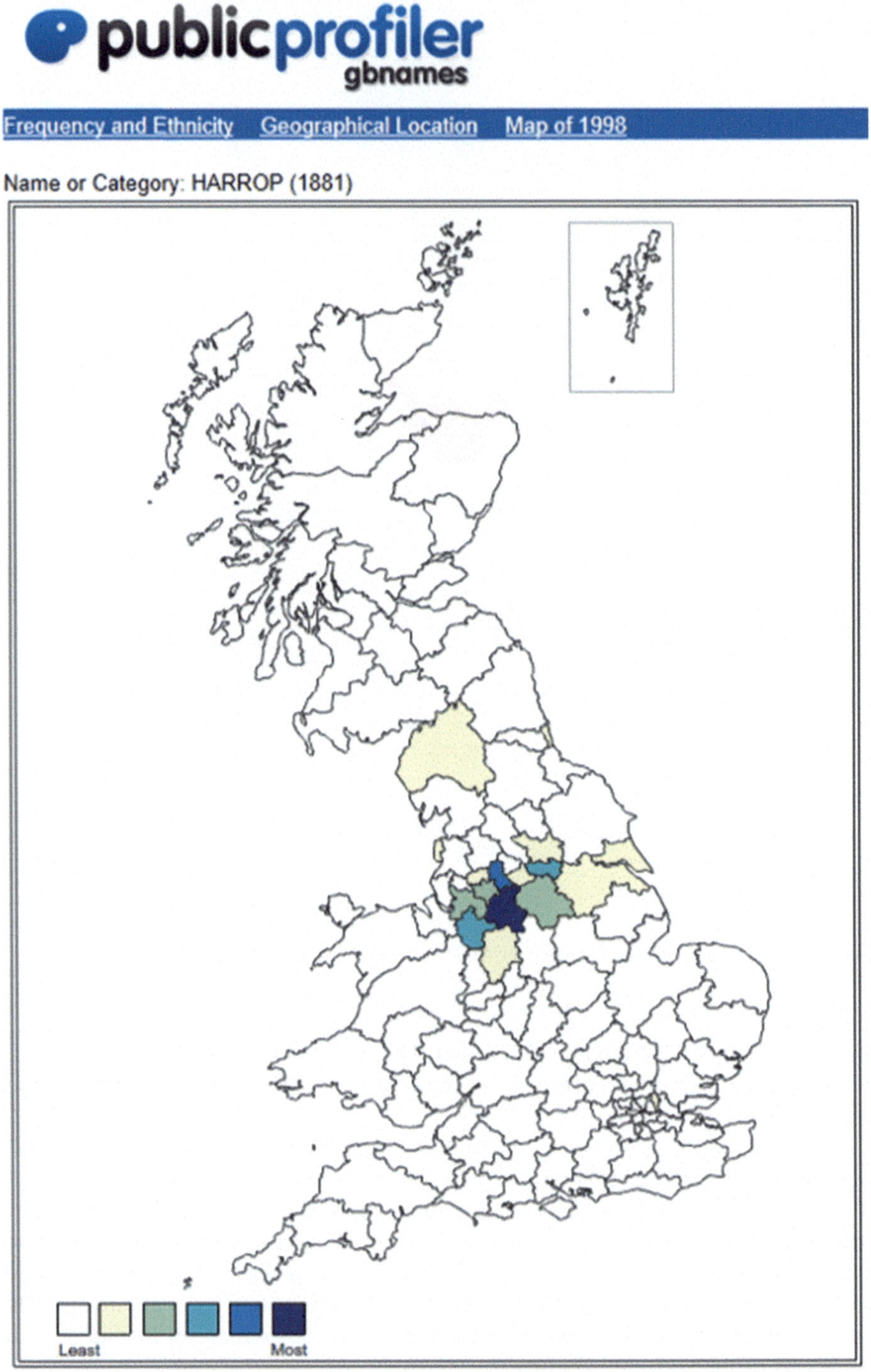


**Fig. 8** Surname locations from the 1881 census of Great Britain: distribution of the surname 'Harrop'. (Reproduced by kind permission of gbnames.publicprofiler.org and University College London)

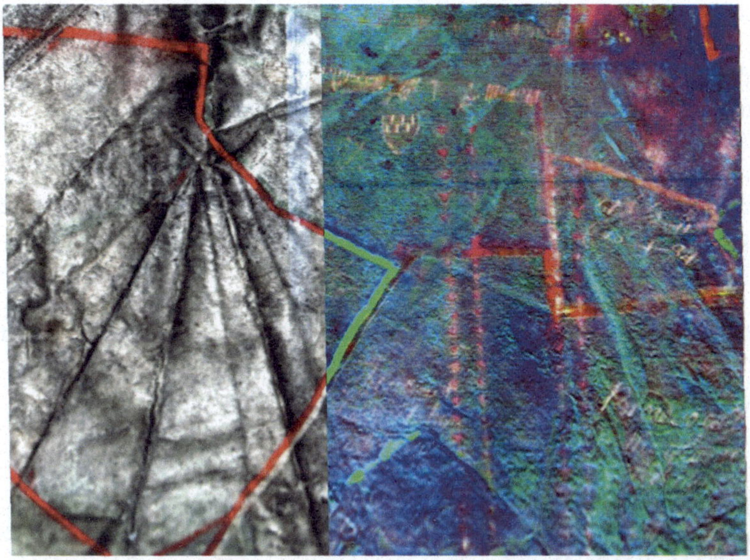

**Fig. 9** Combining spectra around the unidentified main house on the "Cheshire Acres" map (Bodleian Libraries, no call number, © Bodleian Libraries, University of Oxford)

Since high resolution imagery became available in the first decade of the twenty-first century, research interested has abounded, hence the reason for the map to be subjected to hyperspectral treatment.

In 2012, a "Gough Map panel" was set up by Catherine Delano-Smith (Delano-Smith et al. 2017: 1) with a view to developing research into one of the history of cartography's more enigmatic maps. Delano-Smith's team included Peter Barber, Damien Bove, Matthew Champion, the late Christopher Clarkson, Paul Harvey, Nick Millea, Nigel Saul, William (Bill) Shannon, Christopher Whittick and James Willoughby. This initial group has subsequently reached out to the scientific community, and in particular David Howell; the University of Durham's Raman spectroscopy team of Andrew Beeby, Kate Nicholson and Tony Parker; the 3-D scanning outfit from Madrid's Factum Arte—Adam Lowe, Carlos Bayod and Arthur Prior; and more recently the hyperspectral research undertaken by David Messinger and his colleagues at the Rochester Institute of Technology (RIT), New York. The first manifestation of this group's work can be found in an article published in 'Imago mundi' (Delano-Smith et al. 2017: 1–36).

Hyperspectral imagery of the map was initially captured in January 2015, but early results proved inconclusive. However, once the striking discoveries made by the Lazarus Project on the Beinecke's 1491 Martellus map of the World were reported (Van Duzer 2015, 2016), the Bodleian was able to open communications with hyperspectral colleagues in the United States, and a partnership with RIT was born.

On 30 June 2016, Oxford's Wolfson College hosted the Science and Engineering in Arts Heritage and Archaeology (SEAHA) special seminar in Multispectral and Hyperspectral Imaging, an event attended by members of the Rochester team, which included a paper by David Messinger, Director, Chester F. Carlson Center for Imaging Science, RIT. This presentation was partly illustrated with slides featuring the Gough Map.

Demonstrating how flexible hyperspectral imaging can be, Messinger presented an image showing the centre of the Gough Map with north at the top. Using Principal Component Analysis (PCA) he was able to demonstrate very clearly how two different reds have been used on the map, seen below. One red shows up as Hadrian's Wall and as the urban vignettes dotted across the map (red in the image); whilst the purple writing running from north to south on the image below, also shows up as red to the naked eye (Fig. 10).

**Fig. 10** The Gough Map after PCA—note the two different rendering of *reds* (© David Messinger, Chester F. Carlson Center for Imaging Science, Rochester Institute of Technology) The suggestion is that different pigments were used in the *red inks*

**Fig. 11** Note the different *greens* in the River Humber—the PCA image is on the *right* (© David Messinger, Chester F. Carlson Center for Imaging Science, Rochester Institute of Technology)

A further manifestation of the value of PCA can be seen when comparing a conventional Red/Green/Blue image of the Humber estuary with a PCA image. Where the naked eye cannot differentiate between different pigments, hyperspectral techniques clearly demonstrate that different inks must have been used for the greens employed to colour the map's watercourses (Fig. 11).

What may be at play here is that the mapmaker(s) are differentiating between tidal and non-tidal waters, although the Humber's tidal range does extend much further inland. Might it refer to river depth? Or a crossing point? All of these options may be worth further research.

Another hyperspectral technique is Spectral Angle Signature matching (SAM), and the illustration below shows how this can work, examining an area of north-east Wiltshire, just below the "albus equus" label, or (Vale of) White Horse (Fig. 12).

Because the text is more similar to the reference spectrum than the "blotch", it stands out more on the right-hand image, so some writing can be seen to the top right of the rooftop.

Much of what Messinger presented had been previously unseen by the authors of this paper, and with Messinger's permission, his slides were shared with the Gough Map panel, Delano-Smith and Bove enthusiastically engaging with the challenges and opportunities revealed. In the intervening six months, progress has been made with the Rochester team responding to queries from the UK-based panellists.

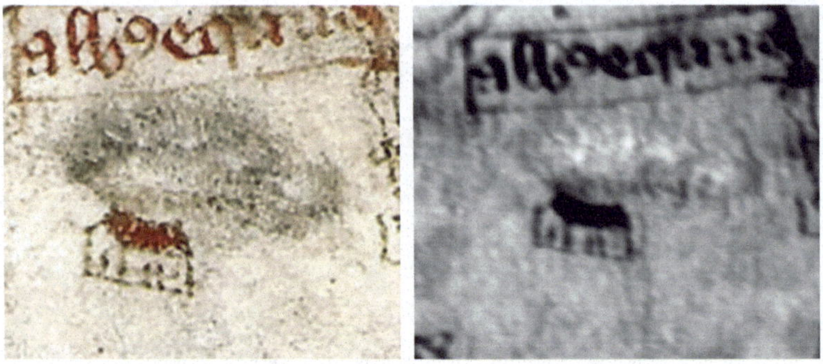

**Fig. 12** Note the emerging text using the SAM image is on the *right* (© David Messinger, Chester F. Carlson Center for Imaging Science, Rochester Institute of Technology)

**Fig. 13** Labelling the south-east corner of the Gough Map (© Di Bai, Chester F. Carlson Center for Imaging Science, Rochester Institute of Technology)

Since the Bove/Messinger communication channel has opened, the RIT team has been looking towards the application of novel image processing algorithms to the map image, with the intention of identifying both clustering of materials such as substrate, inks and pigments. There is also the option to "sharpen" features on the map.

Latest developments have seen a labelling exercise initiated, whereby all features on the map are colour-coded, and different colours have been allocated to: buildings or rooftops; rivers; Hadrian's Wall; red text; and "older" ink. To date, about one third of the map has been completed, as exemplified below in Fig. 13.

These labels establish what are referred to as "ground truths" by the RIT team, which can be employed to assess the performance of what are referred to as "pattern recognition algorithms". In turn, this permits RIT specialists to interrogate inks and their pigments, as well as the material on which the map is made. The intention is to make all this data publicly available for further testing, safely secured by those ground truths which give the data a set of solid starting reference points.

## 3 Conclusion

The hyperspectral imaging of these maps has been successful in showing that additional ‘knowledge’ can be gained by looking at the images created in ways which are simply not possible to see with the naked eye. This is a fairly alien concept to those who have historically wanted to produce images that are as close to the original as possible. As such it will take time and experience to fully appreciate the value of these techniques.

The aspiration that hyperspectral would answer all our queries has not materialised. But it may do in the fullness of time. We have not spotted the gateway on the Ascott Park map; and we have not definitively identified the location of the “Cheshire Acres” map. The Gough Map, however, has started to give up some of its secrets, and the more the RIT consider the imagery they have, the more will be revealed. Those nuggets of information coming back across the Atlantic are starting to excite the Gough Map panel. Those at RIT are beginning to benefit from input and questions delivered by the Gough Map specialists, and gradually more data informs us of the map’s compilation and content. Neither team can work in isolation—RIT need input from the panellists, the panellists need to know what the hyperspectral technology is capable of delivering. This is all part of the cross-fertilisation of ideas between the scientists and the curators and historians. We have seen that the Martellus map has produced some stunning results. Perhaps we can hope for something similar with the Gough Map?

We know it is now possible to employ science to date cartographic material, and to infer where maps might have been made. We know we can also use hyperspectral imagery to reveal underlying cartographic content previously obscured from view. We need to press on with the research to extract as much new information as possible from these manuscripts, and embrace the science that is now available. A decade ago, a partnership such as this would have been beyond the academic community’s research aspirations.

## References

Cheshire Archives & Local Studies. e-Mapping Victorian Cheshire http://maps.cheshire.gov.uk/tithemaps/Default.aspx. Last accessed 9 Dec 2016

Delano-Smith C et al (2017) New light on the medieval Gough Map of Britain. Imago Mundi 69 (1):1–36

Linguistic geographies: the Gough Map of Great Britain (2011) http://www.goughmap.org/. Last accessed 9 Dec 2016

University College London (2008) Great Britain family names profiling. http://gbnames.publicprofiler.org/. Last accessed 9 Dec 2016

Van Duzer C (2015) Graphic record of a lost wall map of the World (c. 1490) by Henricus Martellus. Peregrinations J Mediev Art Archit 5(2):23–47

Van Duzer C (2016) Multispectral imaging for the study of historic maps: the example of Henricus Martellus’s World map at Yale. Imago Mundi 68(1):62–66

Wallis H (1994) Historians’ guide to early British maps. Royal Historical Society, London

**Nick Millea** has been Map Librarian at the Bodleian Library, University of Oxford, since 1992; he was previously Map Curator at the University of Sussex. Nick is the Bibliographer for 'Imago Mundi', and a director of Imago Mundi Limited. He is also Chair of the British and Irish Committee for Map Information and Catalogue Systems (BRICMICS), and convenor of The Oxford Seminars in Cartography (or TOSCA). Nick's previous publications include numerous articles on both map librarianship and the history of cartography, as well as:

- The Gough Map: the earliest road map of Great Britain? Oxford: Bodleian Library, 2007.
- Street mapping: an A to Z of urban cartography: an exhibition in the Bodleian Library, February–April 2003. Oxford: Bodleian Library, University of Oxford, 2003.
- A celebration of 50 years of the British Cartographic Society/co-edited with Seppe Cassettari ... [et al.]. Ilkeston: British Cartographic Society, 2013.

**David Howell** has been at the Bodleian Library, University of Oxford since 2005 as Head of Conservation and Collection Care, and since 2012 Head of Heritage Science; he was previously Head of Conservation Research for Historic Royal Palaces. David is a Trustee of Icon and The National Heritage Science Forum and is an accredited conservator. As Head of Conservation he oversaw the financing and conservation of the Gough Map of Great Britain and the Selden Map of China. His current research is applying imaging and analytical techniques to heritage objects.

# Leiden and the Dissemination of Asian Cartography

**Martijn Storms**

**Abstract** Since the foundation of Leiden University in 1575 much attention was paid on Oriental cultures. Nowadays, vast collections on the South-, East- and Southeast-Asia are kept in the collections of Leiden University Libraries. An important, and for sure an attractive, part of these collections consists of cartographic material. On the one hand these are European, especially Dutch, products, largely made by the VOC (Dutch East India Company) and later by the Topographic Service in Batavia, the capital of the Dutch East Indies. On the other hand, documents and maps of oriental origin were collected by scholars in Leiden. In this paper the role of Leiden, and Leiden University in particular, in the production, trade, consumption and preservation of Asian cartography is examined.

Who collected these maps and atlases? What aims defined their collecting strategies? How ended these maps up at Leiden University? Various personal and institutional collectors from the late sixteenth to the early twenty-first century will pass in review. Among them famous collectors like Philipp Franz von Siebold (1796–1866), and Johannes Tiberius Bodel Nijenhuis (1797–1872) and institutions like the Royal Tropical Institute and the Royal Netherlands Institute for Southeast Asian and Caribbean Studies. With the addition of the map collections of the two institutes mentioned above, Leiden's map collection almost doubled in size in the past years. The incorporation of the libraries of these institutions made Leiden University Libraries one of the leading collections on Asia, and the largest on Indonesia worldwide.

## 1 Leiden University: Origins and Orientalism

Leiden University was founded by William of Orange in 1575, as the first university in the Northern Netherlands. 12 years later, on 31 October 1587, the university library became operational in one of the vaulted rooms of the current Academy Building. In 1595 the library moved to the upper floor of the Beguinage Chapel (Berkvens-Stevelinck 2012: 29). A print of the old, chained, university library was

M. Storms (✉)
Leiden University Libraries, Leiden, Netherlands
e-mail: m.storms@library.leidenuniv.nl

 
M. Altić et al. (eds.), *Dissemination of Cartographic Knowledge*, Lecture Notes in Geoinformation and Cartography, DOI 10.1007/978-3-319-61515-8_20

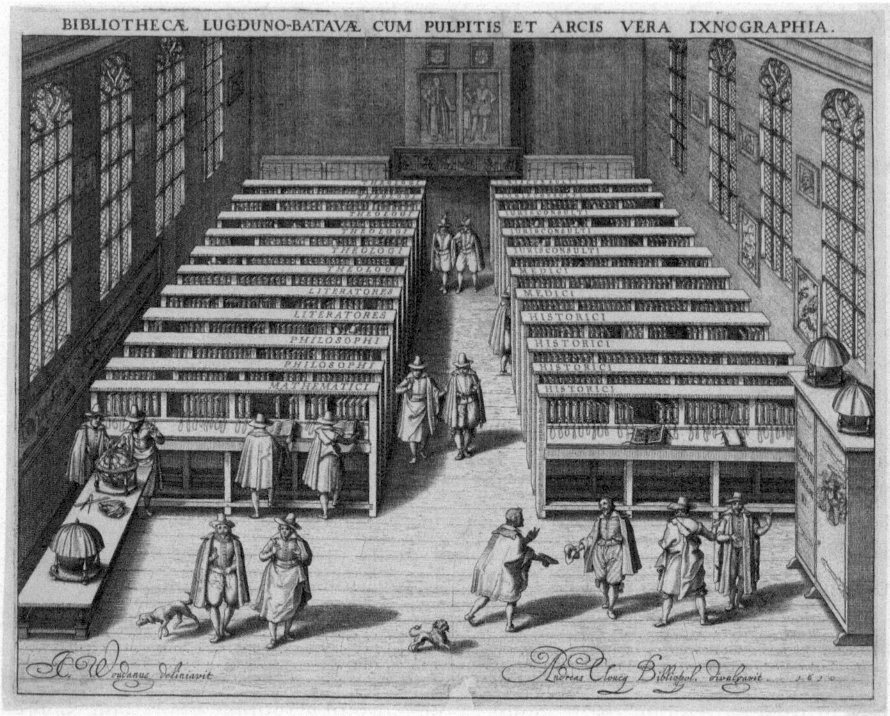


**Fig. 1** J.C. Woudanus, *Bibliothecae Lugduno-Batavae cum Pulpitis et Arcis vera Ixnographia*, 1610 (Leiden University Libraries, COLLBN Port 315-II N 22)

made in 1610. On the northern wall, at the left side of the image, a more than 11-meter-long city profile is hanging (Fig. 1).

## 1.1 Melchior Lorck's Prospect of Constantinople

This is the famous Prospect of Constantinople, drawn by the Danish artist Melchior Lorck around 1560, after he returned from Constantinople where he was assigned to a German embassy to the court of Sultan Suleiman the Magnificent. This prospect was a gift to the library by the university's steward Nicolaas van der Wiele in 1598. This huge drawing symbolized the variety of Leiden University's special collections, of West meeting East, and the dawning of interest in the East that would be so characteristic for Leiden University during the seventeenth century. After the rebuilding of the library the prospect was lost for a long time, until it was found back in 1856 on the attic of the city hall, rolled, damaged and gnawed by mouses.

When we zoom in on two of the central sheets we see that Melchior Lorck added a self-portrait, while working on the drawing. The compass rose tells us that his

**Fig. 2** Melchior Lorck, Prospect of Constantinople (Leiden University Libraries, BPL 1758: 11)

viewpoint is from the north on the other side of the Golden Horn, looking south towards the city centre. An Ottoman with a turban hands him the inkwell (Fig. 2).

## *1.2 Joseph Scaliger (1540–1609)*

The library's collections got a more oriental profile with the legacy of professor Joseph Scaliger in 1609. He bequeathed all his books in Oriental languages to the library (Berkvens-Stevelinck 2012: 61–65). Among this legacy were a few maps and a globe pair. On the 1610 print of the university library we see the two globes on top of the cabinet in which Scaliger books were kept. Unfortunately the globes didn't survive.

## *1.3 Isaac Vossius (1618–1689)*

The first collection with a substantial corpus of maps and geographical works that entered the university library was the library of classical philologist Isaac Vossius.

The cartographic works and maps in Isaac Vossius' collection consist of 188 loose map sheets, 24 printed atlases and 4 manuscript volumes. The loose maps all date from the middle of the sixteenth century and most of them are printed in Venice and Rome, originally bounded together as IATO (Italian, assembled to order) or Lafreri atlases (De Vries 1981: 337–341). This map collection is not specifically devoted to Asia, although it contains some maps in relation to Asia, most of them based on Ptolemaic maps. Here we see three world maps in some of Vossius's manuscripts. The famous map of Henricus Martellus, designed just before America was discovered, a Portuguese work related to the voyages of Ferdinand Magellan and an early sixteenth century copy of Ptolemy's *Geographia*'s maps.

# 2 Japanese Map Collection

## 2.1 *Philipp Franz von Siebold (1796–1866)*

The most exciting story related to a Leiden map collection probably is the so-called 'Siebold Incident'. From 1823, the German physician, botanist and traveler Philipp Franz von Siebold (1796–1866) lived and worked on the small artificial island of Dejima, the Dutch trading post in the bay of Nagasaki. Three years later Japanese officials discovered that he obtained maps of Japan, partly from court astronomer Takahashi Kageyasu. Among the maps were detailed surveys of the disputed northern islands of Hokkaido and Sachalin (Fig. 3). Siebold was accused of high treason and of being a spy for Russia. He was placed under house arrest and finally, in 1829, expelled from Japan. However, he managed to bring most of the Japanese maps he collected home to Holland.

After his exile, Siebold settled in Leiden. There he composed *Nippon*, his illustrated ethnographical and geographical work of Japan. Seven volumes of this *magnum opus* have been published but the work was never definitely completed (Forrer 2000: 25–31). In 1840 Siebold published an overview map of Japan, based on the geographic and cartographic material he had collected during his years in Japan. This map was a major step forward in accuracy of the Western mapping of Japan. At Leiden University Libraries several drafts and annotated proof print fragments of this map are kept (Storms 2008: 127–129).

## 2.2 *Jan Cock Blomhoff (1779–1853)*

The collection of Japanese books, manuscripts and maps is generally referred to as 'Siebold Collection'. However, parts of this collection consists of works acquired by other employees of the Dutch trading post, such as Jan Cock Blomhoff (1779–1853). Cock Blomhoff was store keeper at Dejima from 1809 to 1817,

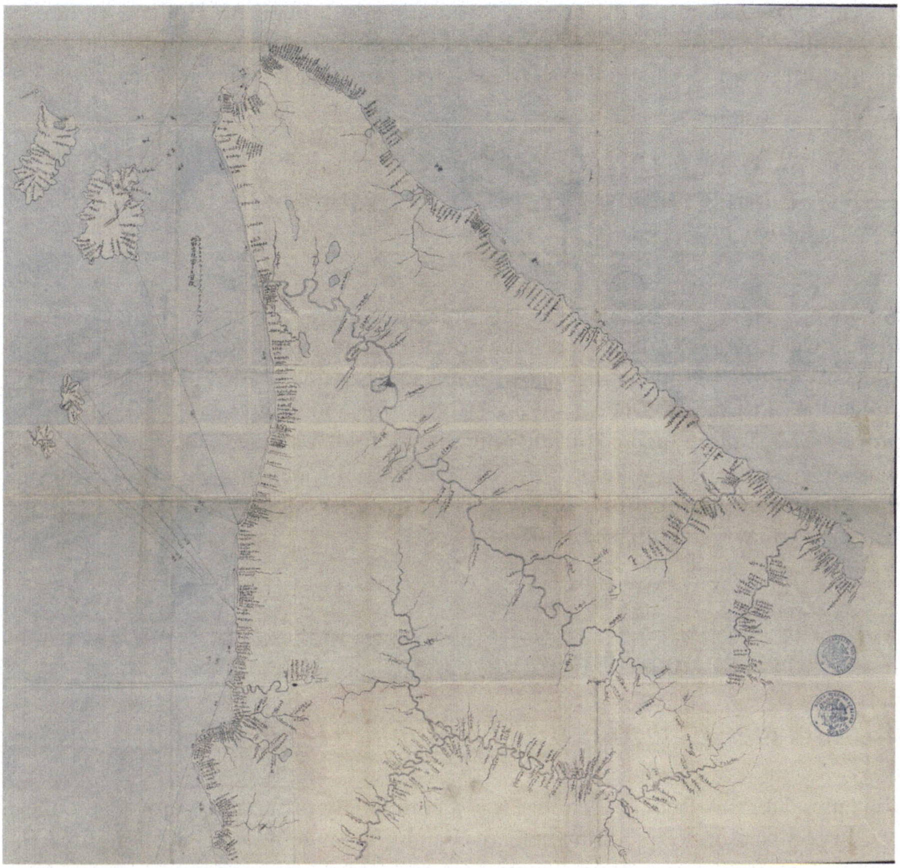

**Fig. 3** One of the three sheets of the manuscript map of Hokkaidô, probably after Mogami Tokunai or Inô Tadataka, c. 1825 (Leiden University Libraries, Ser. 205c)

after which he was appointed director ("opperhoofd") of Dejima, a function he performed until 1824.

Among the maps from the collection of Cock Blomhoff were two almost identical Japanese maps of Korea. The map is designed by Hayashi Shihei (1738–1793), a Japanese military scholar. It is one of the five maps he compiled for his book describing Korea, Okinawa (including Taiwan) and Hokkaido. Shihei's publications arose from his concern about the growing Russian power in East Asia. He advocated an improvement in the Japanese defense. Since his work was published without authorization, Shihei was placed under house arrest by the Edo government in 1792. A woodcut and manuscript copy of the map were send to the Netherlands by Cock Blomhoff, another manuscript copy by Philipp Franz von Siebold. By the way, when Siebold returned to Holland, he probably was unpleasantly surprised. In Leiden he found out that some of the maps, or similar maps, that he had smuggled out of the country, were already brought to Holland earlier by Cock Blomhoff. Siebold must have asked himself whether the risks he had taken, for himself as well as his Japanese friends, had been worth taking (Forrer 2000: 32).

### *2.3 Johannes Gerard Frederik van Overmeer Fisscher (1800–1848)*

Another predecessor of Siebold was Johannes Gerard Frederik van Overmeer Fisscher (1800–1848), first clerk and storekeeper at Nagasaki, between 1817 and 1829. A number of interesting maps from his possessions ended up in the Japanese collections in Leiden. One of the most beautiful and impressive works from his properties is a more than 11-meter-long zigzag folded stroke, or leporello: a route map of the so-called 'Tôkaidô', the eastern coastal route from Edo to Kyoto, the most important thoroughfare in Japan (Fig. 4). This copy is printed in 1772 and colored by hand, embellished with gold leaf.

## 3 Collection Bodel Nijenhuis

### *3.1 Johannes Tiberius Bodel Nijenhuis (1797–1872)*

Johannes Tiberius Bodel Nijenhuis was a completely different type of map collector. In comparison to travelling scientists and officers temporally working in Asia, Bodel Nijenhuis was a typical armchair traveler, barely putting a foot outside Holland. With a enormous collection of 50,000 maps, 300 atlases and over 20,000 topographical prints and drawings he is best characterized as a cartographic omnivore. As far as I could trace he was, based on number of maps, the greatest private map collector worldwide, until recently. Only David Rumsey, who's collection now is kept at Stanford, surpassed him. Basically, Bodel Nijenhuis collected every map he could get. He studied law at Leiden University and as the last generation of Luchtmans (from his mother's side) he was destined to direct the respected Luchtmans publishing house in Leiden. However, he liquidated the family company in 1848 to devote himself to his collections and the role he played in the academic community in Leiden (Storms 2008: 13). Bodel's primary interest were maps in relation to the Netherlands, but in fact his collection covered the whole world. Estimated there are approximately 6000 maps of Asia in his collection. The majority of these maps are related to the Dutch East India Company and the colonial period of the Netherlands East-Indies. Bodel Nijenhuis also was one of the pioneers of map historical scholarship. He is considered as one the first scholars in the long and rich Dutch tradition of cartobibliographical research. As an example, he wrote an article on Johan Nieuhof (Storms 2008: 29).

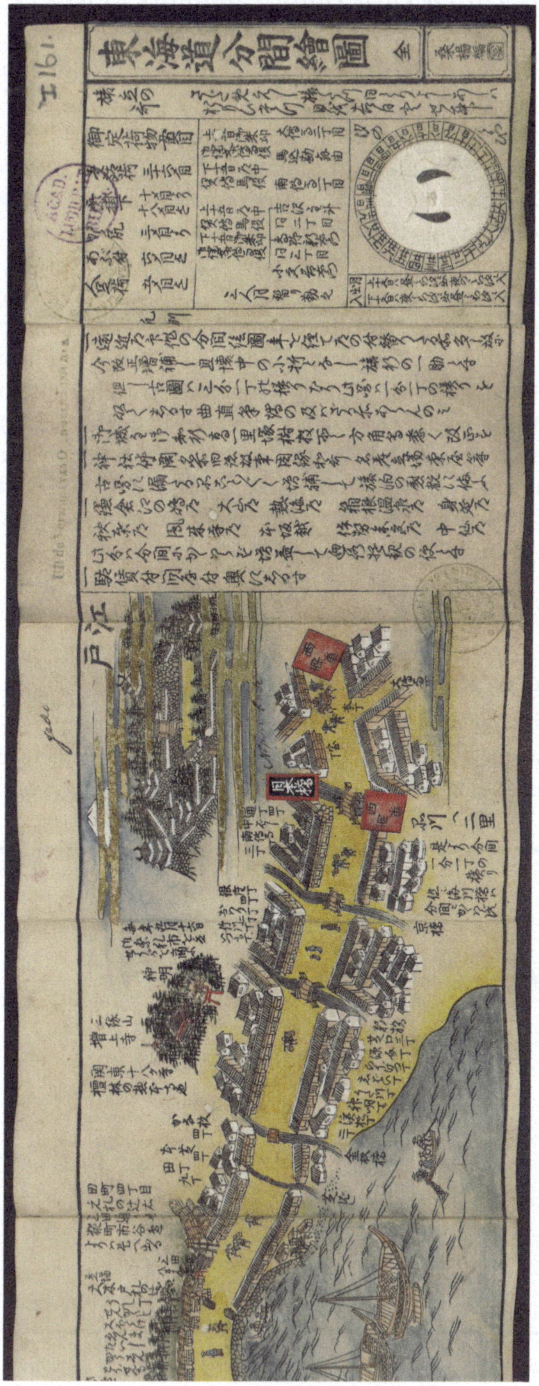


**Fig. 4** Sôyô, Most right part of a route map of the Tôkaidô (sea route from Edo to Kyoto), 1772 (Leiden University Libraries, Ser. 410a)

## 3.2 Maps of the Dutch East India Company (V.O.C.)

Among the most valuable maps in Bodel's collection, are charts of the Dutch East India Company. About 80 manuscript VOC charts are part of his collection of which 15 on vellum. Two of these charts on vellum are signed by Joan II Blaeu (1650–1712), official mapmaker of the VOC: a chart of the Java Sea (1673) and one of the South China Sea (1686) (De Vries 1996: 43, 55). An example of a chart drawn on paper is an anonymous map of Ambon from the first half of the eighteenth century, according to a note of Bodel Nijenhuis drawn by an indigenous hand (Fig. 5).

One of the most appreciated maps, and one of the cutest, in the collection of Bodel Nijenhuis is the so-called 'elephant map' of Ceylon, a manuscript map from the eighteenth century depicting loads of animals, especially elephants, spread over the island. We can even see baby elephants together with their mothers (Fig. 6).

Characteristic for the Bodel Nijenhuis collection is the great number of fortification plans, varying from individual plans to extensive series. Extraordinary is a series of six drawings of Fort Duurstede at Saparua, one of the small islands in the Southern Moluccas, drawn in 1693 by Isaak Hogeboom, just around the time that the VOC fortress was built.

## 3.3 Bodel's Collecting Strategies

Bodel Nijenhuis was the greatest private map collector of the nineteenth century. He purchased antique maps at auctions and received maps, including proof prints, as a gift

**Fig. 5** Manuscript chart of Ambon, c. 1700–1750 (Leiden University Libraries, COLLBN 006–14-011)

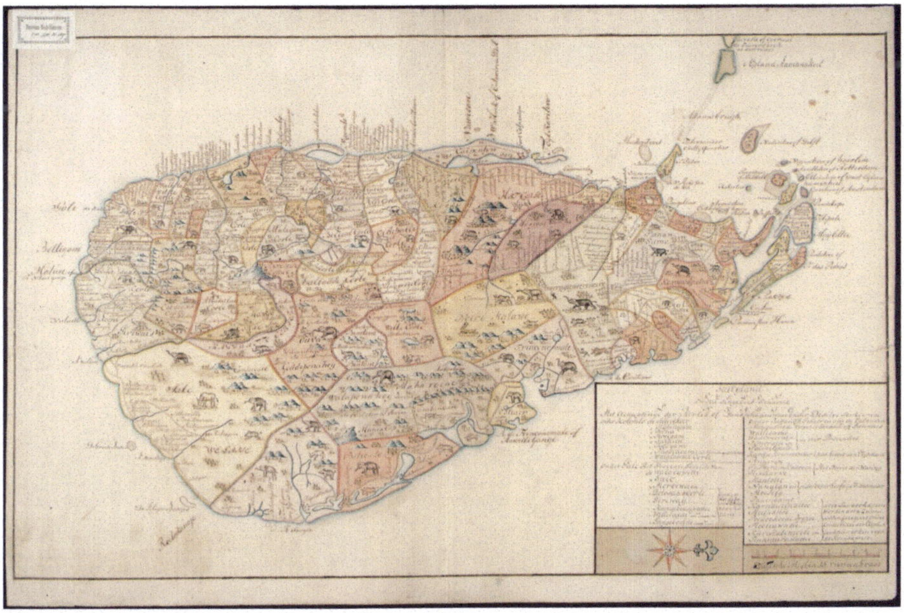

**Fig. 6** *Het eijland Ceijlon in het Singalees genaemt* Lankadiep, c. 1765 (Leiden University Libraries, COLLBN 002–09-031)

from his contacts in the publishing industry. Bodel Nijenhuis is also known, or infamous, for cutting maps and prints from atlases and books. This rather insignificant map shows another collecting strategy of Bodel Nijenhuis. When he didn't had the opportunity to acquire a certain map for his collection, he made sure he got a hand-drawn copy of the map. For years draughtsman Jacobus Cornelis Wendel served as a kind of private copyist for Bodel Nijenhuis, but we also find copies of maps drawn by members of his family and even some sketches made by himself. For example, a map of Great Tibet was kept in the collection of the Royal Zeeland Society of Sciences in Middelburg. Society director Frederik Nagtglas copied the map for Bodel Nijenhuis, together with a transcript from the eighteenth century travel account through India of Samuel van der Putte, to which this sketch map belongs (Fig. 7).

## 4 Institutional Collections

### *4.1 Kern Institute*

The Kern Institute in Leiden is the national center of expertise for South Asia and the Himalayan region. In 1925, the institute was founded by the sanskritist and archaeologist Professor Jean Philippe Vogel. He named the institute after Hendrik Kern who, from 1865, held the first chair of Sanskrit in the Netherlands. Since 2010

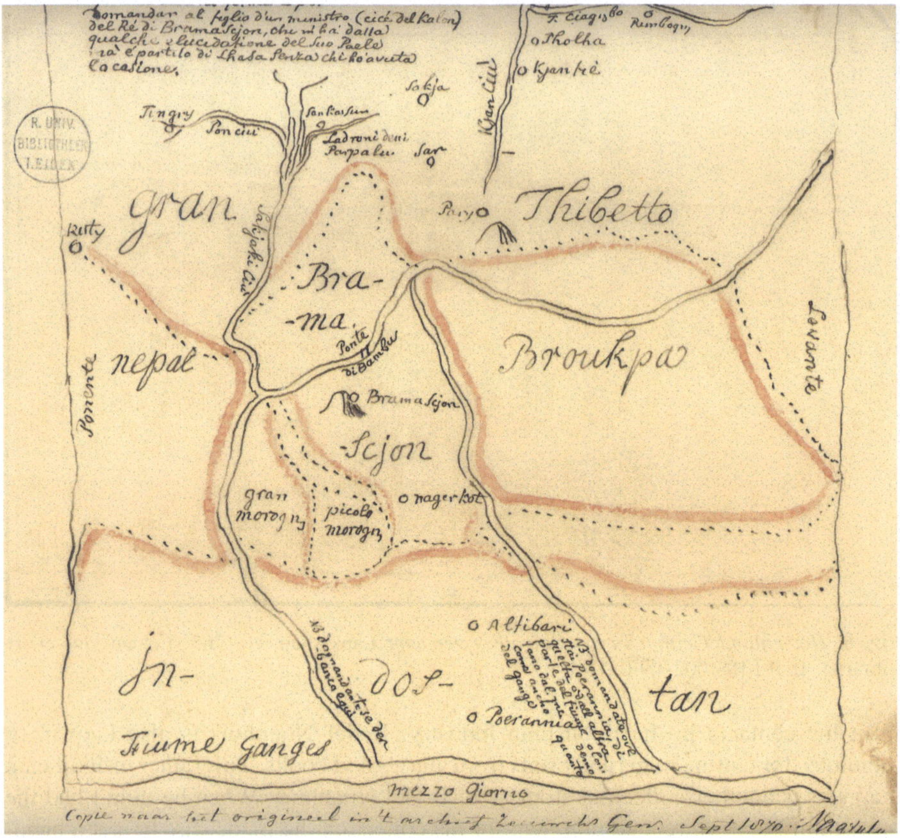


**Fig. 7** Samuel van der Putte, *Map of Great Tibet*, c. 1730, copy by Frederik Nagtglas, 1870 (Leiden University Libraries, COLLBN Port 177 N 71)

the collection of the Kern Institute is included in the university library. The cartographic part of this collections consists of 13 old printed maps and about 600 modern topographical maps of the India Army Map Service, the Survey of India and the National Atlas of India.

## *4.2 Sinology Institute*

Until the summer of 2016 the special collections of the Sinology Institute were located in the East Asian Library of Leiden university. Now these collections have moved to the main library building where more and more Asian materials are gathered under one roof now. The cartographic collection of the Sinology Institute consists of about 50 atlases and 250 maps. Unique is the Route Map for an Imperial Inspection Tour to South China, dated 1784 (Fig. 8).

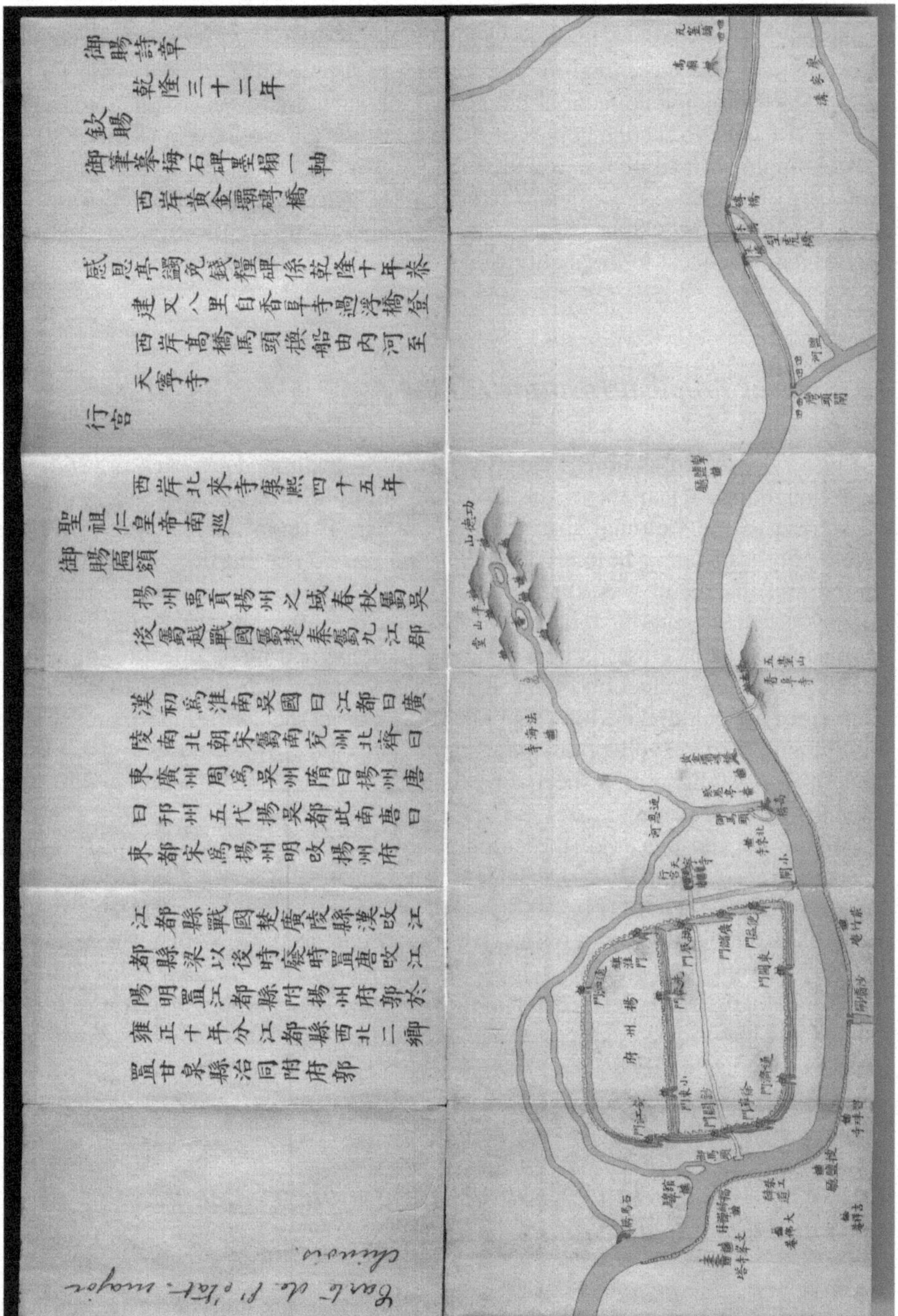

**Fig. 8** Manuscript map and memorial for the Qainlong Emperor's fifth journey to Nanxun in 1784 (Leiden University Libraries, SINOL. VGK 3039.7.6)

The emperors of China made regular inspection tours to South China. They travelled by boat along the Imperial Canal from the capital Beijing to various cities in the south. The Qianlong Emperor made no fewer than six of these journeys. This leporello map was most likely made for him as a preparation for his fifth tour in 1784. The Sinology collection holds two of such maps, each one measuring c. 12 × 208 cm. No second copies of these maps are known to exist (Kuiper 2014).

The Sinology Institute was founded Jan Duyvendak (1889–1954), sinologist and professor of Chinese at Leiden University. He actively acquired many rare and special editions in China. Some of the maps in this collection are part of Duyvendak's bequest to the institute.

## *4.3 Royal Tropical Institute (KIT)*

The history of the collection of the Royal Tropical Institute dates back to 1864 with the foundation in Haarlem of the Colonial Museum. In 1913 the collection was transferred to the Colonial Institute, founded in 1910 in Amsterdam. After the sovereignty transfer to Indonesia in 1949 the mission of the Royal Institute was altered and the name was changed to Royal Tropical Institute. Since then, the library collected material relating to all tropical regions. In 2013 the library of this institute had to close its doors and the heritage and map collection moved to Leiden. The map collection of the Royal Tropical Institute consists of c. 11,000 maps from the colonial period of which 90% is concerned with Indonesia. Next to this colonial heritage collection, the institute collected c. 1000 atlases and c. 15,000 modern topographical map sheets of which c. 6000 map sheets are related to Asia (Levi 2012: 93–94).

Among the top pieces of the colonial collection are the manuscript sheets of Gijsbert Franco baron Von Derfelden van Hinderstein's *Algemeene kaart van Nederlandsch Oostindië* (Fig. 9), as well as the set of nine copperplates of this map. In 2016 the restauration of the set of water damaged copperplates was started. The Royal Tropical Institute developed its own map viewer for their colonial maps. All 11,000 maps are georeferenced and searchable via a comprehensive gazetteer of modern and historical geographical names. The hosting of this map viewer was also took over by our library.

## *4.4 Royal Netherlands Institute for Southeast Asian and Caribbean Studies (KITLV)*

University Library. After a reorganization, the Royal Netherlands Institute for Southeast Asian and Caribbean Studies (KITLV), closed its library. Another 16,000 maps and 500 atlases were added to Leiden University's map collections.

**Fig. 9** Manuscript index sheet of Von Derfelden van Hinsderstein's. *Algemeene kaart van Nederlandsch Oostindië* (Leiden University Libraries, KIT KK 112–05-01)

A key figure for the KITLV collection is Gerret Pieter Rouffaer. He is considered the last explorer of Indonesia and as librarian of the institute he was very important for the cataloguing and growth of the map collection (Ngo 2012: 26). During his journey through the Lesser Sunda Islands in 1910 he made several sketch maps of some of the islands he visited. This map of Adonara (Fig. 10) is one example of that (Okker 2015: 210–211, 218).

With almost 7000 maps of the KITLV collection we just finished a crowdsourcing georeferencing project, named 'Maps in the Crowd'. In 7 months time the maps are linked to Google Maps by the crowd. This project is a first step to make our maps better accessible, findable and usable for education and research. Although biased, because only a selection of maps from one collection are used. The dot map (Fig. 11) gives an indication of the spread of mapping, or the intensity of mapping in Indonesia during the colonial period.

With the addition of the map collections of the mentioned Royal Tropical Institute and KITLV, Leiden's map collection almost doubled in size in the past years. Asia in general and maps of the Dutch East Indies in particular are now more prominent in our profile for collection building (Storms 2014: 37). The library still acquires such maps. The most recent acquisition is a 1,5-meter-long manuscript

**Fig. 10** [Gerret Pieter Rouffaer], *Schetskaart van het eiland Adonara,* [1910] (Leiden University Libraries, KITLV D E 44,5)

map of Southern Sumatra from 1839, the first topographical map of this area at such a large scale.

## 5 The Asian Library of Leiden University

The incorporation of the libraries of these institutions made Leiden University Libraries one of the leading collections on Asia, and the largest on the Dutch East Indies/Indonesia worldwide. Therefore, an extension of the main library building will be opened in 2017: the Asian Library. 2017 will be the Leiden Asia Year. The aim of the Leiden Asia Year is to put the city of Leiden, its museums and the University more firmly on the international map as one of the world's leading knowledge hubs on Asia, covering research, teaching, collections and expertise. In close collaboration with partners in the city, the University will be organizing symposia, conferences, lectures and exhibitions in the city throughout 2017, all on the theme of Asia. Leiden Asia Year which will be enriched with a 2-day symposium on Asian cartography: "Mapping Asia. Cartographic Encounters between East and West", organized in collaboration with the ICA Commission on the History of Cartography.

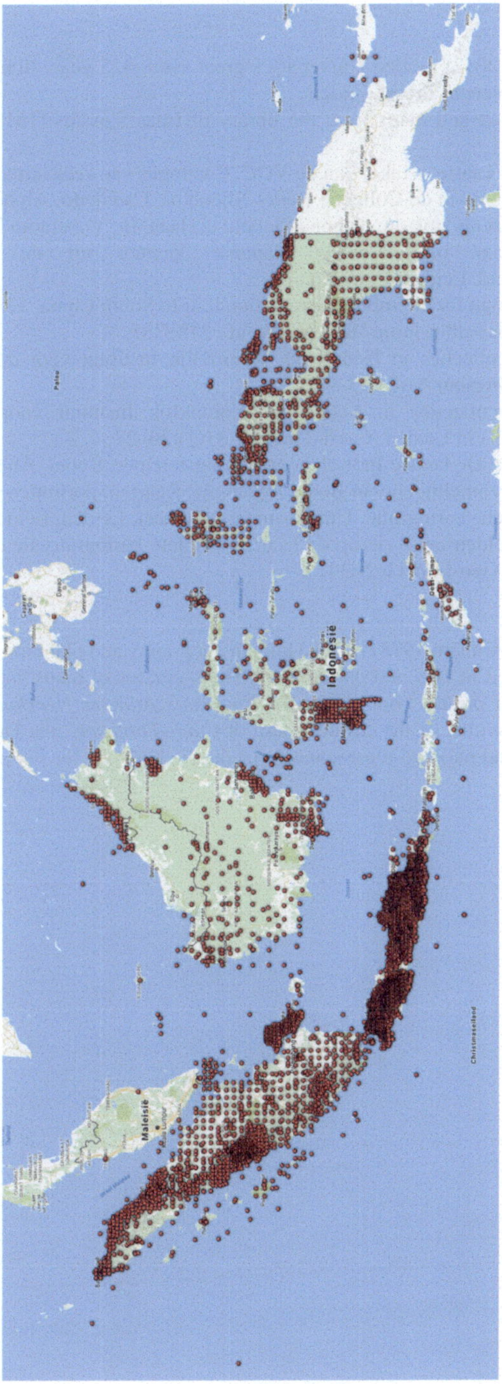

**Fig. 11** Dot map of the central location of the c. 7000 georeferenced maps of the KITLV collection (http://blogs.library.leiden.edu/mapsinthecrowd/)

## References

Berkvens-Stevelinck C (2012) Leiden University's great asset. 425 years library collections and services. Leiden University Press, Leiden

de Vries D (1981) Atlases and maps from the library of Isaac Vossius (1618-1689). Int Yearb Cartogr 21:171–194

de Vries D (1996) Uit de kaartenwinkel van de VOC. Catalogus van zeekaarten van de Verenigde Oostindische Compagnie in de Collectie Bodel Nijenhuis. Canaletto, Alphen aan den Rijn

Forrer M (2000) Een kaart als stok om de hond Siebold te slaan. In: Yonemoto M, Forrer M, Smits I (eds) Staatsgevaar of sierobject. Japanse kaarten uit de Siebold-Collectie. Universiteitsbibliotheek Leiden, Leiden

Kuiper K (2014) Route map for an imperial inspection tour to South China, 1784. https://socrates.leidenuniv.nl/R/-?func=dbin-jump-full&object_id=3188735

Levi P (2012) Kaartencollecties in Nederland. Koninklijk Instituut voor de Tropen (KIT) in Amsterdam. Caert-Thresoor 31(3):93–94

Ngo L (2012) Kaartencollecties in Nederland. Koninklijk Instituut voor Taal-, Land- en Volkenkunde (KITLV) in Leiden. Caert-Thrrsoor 31(1):26–27

Okker F (2015) Rouffaer. De laatste Indische ontdekkingsreiziger. Boom, Amsterdam

Storms M (2008) De verzamelingen van Bodel Nijenhuis. Kaarten, portretten en boeken van een pionier in de historische cartografie. Universiteitsbibliotheek Leiden, Leiden

Storms M (2014) UB Leiden zet Azië op de kaart. Complete kartografische collectie KIT blijft behouden in Leiden. Geo-Info 11(2):34–37

**Martijn Storms** MA (Arnhem 1978) studied human geography and planning at Utrecht University, where he specialised in GIS and cartography. He is curator of maps and atlases at Leiden University Libraries and project coordinator for *Koeman's Atlantes Neerlandici* at Brill publishers. Besides, he is member of the editing board of *Caert-Thresoor*, the Dutch journal on the history of cartography and national representative of the Netherlands for *Imago Mundi*.

# Serbian Cartographers of the Nineteenth Century in the Collection of the National Library of Serbia

**Jelena M. Glišović**

**Abstract** This paper presents the structure and characteristics of the Cartographic Collection of the National Library of Serbia along with a brief history of its establishment and development. The collection is exceptionally diverse and rich in terms of content, purpose and dimensions, as well as production modes and appearance of the cartographic publications. Special emphasis is put on the history of Serbian cartography of the nineteenth century and Serbian cartographers who marked this period of social, material and cultural development of Serbia. The nineteenth century is of crucial importance for Serbian cartography because a change to more favorable historical circumstances finally allowed the completion of the acculturation process and the acceptance of the achievements of the European Enlightenment, including cartography. Along with the establishment of civil society and the development of different cultural and scientific institutions, geographic research and cartographic activities advanced. This paper explains the historical circumstances that accompanied these processes and accelerated the development of cartography in Serbia, as well as the different needs of the citizens and the army which instigated them. Maps from the Cartographic Collection of the National Library of Serbia, created by the most important cartographers of the time, will be presented as the evidence of these processes, which most faithfully represent the period.

## 1 Introduction

As one of the most significant cultural institutions of the country, the National Library of Serbia (NLS), serves to preserve the national identity and cultural, scientific and ethnological uniqueness of the Serbian nation and to present its cultural heritage to domestic and foreign researchers. Besides the more general collections, containing regular printed and periodical productions, special non-book materials are also collected and preserved in the NLS. Since 1947, when it was

J.M. Glišović (✉)
National Library of Serbia, Belgrade, Serbia
e-mail: jelena.glisovic@nb.rs



M. Altić et al. (eds.), *Dissemination of Cartographic Knowledge*, Lecture Notes in Geoinformation and Cartography, DOI 10.1007/978-3-319-61515-8_21

officially formed, the Special Collections Department has amassed all these specific and diverse materials to form, among others, a Cartographic Collection.

This Collection is exceptionally diverse and rich in terms of content, purpose and dimensions, as well as production modes and appearance of the cartographic publications. More than 40,000 sheets of all kinds of maps—from thematic and topographic maps of different scales over settlement plans to navigation and astronomical charts and more than 1200 titles of different types of atlases (general, historical, on specific subjects etc.) are preserved in this collection. The largest part of the Cartographic Collection relates to publications presenting the area of South Slavic countries, South and Southeastern Europe and it testifies to the cultural identity and historical continuity of these countries.

When focusing on the part of the collection which represents the national cartographic production it is important to know that historical, economic and political circumstances were unfavorable for the development of cartography in Serbia. "The territory of what today is Serbia appears on geographical maps at an early date, due to its position in relation to the cores of the ancient civilizations of Greece and Rome. However, for a long time cartographic representations of Serbia remained scant and unequal in quality. Serbia's remoteness from the trends of European cartography meant it was misrepresented, whereby information was copied from old maps, or new maps were drawn based on the descriptions of travel writers. This is why the representation of Serbia on maps and its cartographical heritage must be viewed through two parallel perspectives: that of European cartographic representations of Serbia, and Serbia's own cartography" (Živković 2014: 85).

By the end of the eighteenth century, geographic and cartographic information about Serbian lands were incomplete and partial. The nineteenth century is of crucial importance for Serbian cartography because a change to more favorable historical circumstances finally allowed the completion of the acculturation process and the acceptance of the achievements of the European Enlightenment, including cartography. In the first half of the nineteenth century, Austria, Russia and France were the countries that had a determining influence on the depiction of the Serbian countries on geographic maps. Since Karadjordje's Uprising which happened at the beginning of the nineteenth century, Serbia developed an increasing self-awareness and an awareness of the need to develop its own cartography. However, at that time, it was not yet able to perform such a cartographic task itself, because having only just been freed from Turkish rule, all its resources were directed towards the most urgent state affairs.

Austrian cartographers were the main contributors to the cartography of Serbia. Both in terms of quality and quantity, their maps substantially surpass the maps created by other European powers. The dominant role of the Austrian cartographers is the result of both the great political interest of Austria for the Serbian countries and of the level of development of its cartography. In 1804, at the outbreak of the First Serbian Uprising and the start of fighting of the Serbian nation against Turkey, the book *Novo graždansko zemleopisanije* (New Civil Geography) by Pavle Solarić appeared in Venice. Solarić wrote this book based on the work *Allgemeine*

*Einleitung in die neueste Erdbeschreibung* by Adam Christian Gaspari, as stated by Solarić in his foreword. The book by Solarić was printed in two volumes on 720 pages, and accompanied by an atlas with 37 maps. However, it did not contribute to a better understanding of Serbia, as it does not contain any new, original data. The next year, in 1805, the first Serbian map of the Serbian countries appeared, created by Sava Tekelija and entitled *Zemljoobraženije Srbske, Bosne, Dalmacije, Dubrovne, Crne Gore i pograničnih predel* (Map of Serbia, Bosnia, Dalmatia, Dubrovnik, Montenegro and border areas). The map is a not very reliable compilation and is, to a great extent, based on the Austrian map *Neueste Karte der Koenigreiche Bosnien, Servien, Croatien und Slavonien* by Carl Schütz (Vienna, 1788).

Serbian independence and the territorial expansion of its boundaries were treated by international commissions for delimitation and offered an opportunity to welcome their professionals and topographical officers. Along with the development of cultural, scientific and military institutions, geographic research and cartographic activities were being developed as well. During this period, Serbian countries also continued to be presented within the framework of the Habsburg Monarchy and the Ottoman Empire, but independent maps by Serbian cartographers became increasingly common. On these Serbian maps from the nineteenth century, all the crucial moments of its historical development are displayed, from its patriarchal form at the beginning of the century to a developed society that was formed according to European standards at the end of the century.

## 2 Jovan Bugarski

In the 1840s Serbia was materially and culturally evolving at such a pace that maps became a necessity for state administration, education and defense, and therefore a large number of maps were made during this decade. One of the most important maps of Serbia is the *Karta Knjaževstva Srbije* (Map of the Principality of Serbia) by Jovan Bugarski, Sc. Ing., a former governmental officer (Fig. 1). The map has a scale of 1:345,000, measures 98.5 × 73.5 cm, and was published in 1845 in Belgrade. The top-right corner comprises the title, the author's name and the place and year of publication. In the lower-right corner decorative graphics lithographed by Anastas Jovanović are placed, featuring a group of Serbs gathered around an oak tree with the Serbian coat of arms. The lower-left part of the map comprises a table with labels of districts, their names, the number of villages and the number of inhabitants of each district. Above the table a tribute to Prince Mihailo Obrenović is shown. The top-left corner of the map is a graphical scale ratio given in Turkish, Austrian and Hungarian miles. Near the graphic data there is also another table with the division of districts in the counties. The map has a double black frame with degrees of latitude and longitude.

There is little known about Jovan Bugarski. He is mentioned in the history of Hungarian cartography among the people who measured the Györ county between

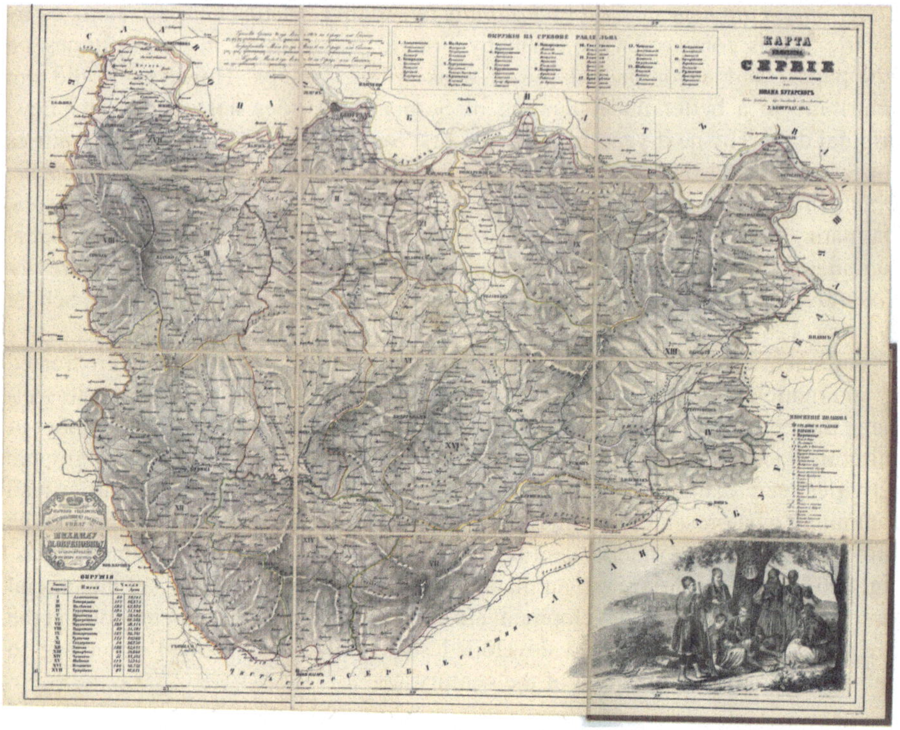

**Fig. 1** Map of the Principality of Serbia, Jovan Bugarski (Kr III 1170, National Library of Serbia, Belgrade, Kr III 1170)

1840 and 1860, after leaving the Serbian civil service, where he served from 1839 to 1844. Based on the available information we know that the Serbian government asked him to draw up a new map of Serbia, but it is not known when and whether he himself measured the terrain and which sources he used. The year of publication indicated on the map is 1845. It was the first time that an original Serbian map achieved general recognition. The map was published in several versions, of which the most common type was the Serbian one. The edition in French entitled *Carte de la Principauté de Serbie, composée d'apres les sources les plus récentes par Jean Bugarsky ingenieur* (Belgrad, 1845) is identical with the Serbian edition (Fig. 2). However, the place names are all transcribed from Serbian Cyrillic into Serbian Roman alphabet, except for the neighboring countries, whose names have been translated. The lower-left part of the French version of the map comprises a tribute to Prince Mihajlo, written on the ancient pole next to the female figure. According to one source, the map was printed by Bugarski and Jovanović in Vienna at their own expense, but the place of publication is not mentioned on the map. Despite the lack of info on Bugarski, his map was used as one of the most important sources by foreign cartographers who wanted to include a representation of Serbia in their maps. Bugarski's map is for example referenced in the subtitle of Heinrich

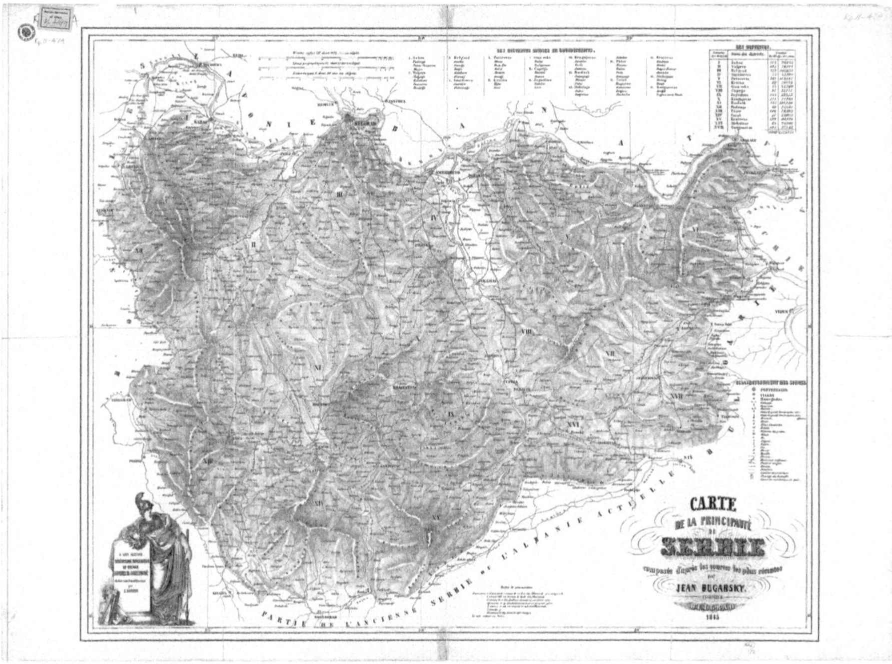


**Fig. 2** Carte de la Principauté de Serbie, Jovan Bugarski (Kr II 474, National Library of Serbia, Belgrade, Kr II 474)

Kiepert's map. The map of Serbia by Bugarski has been well received in the scientific literature as well.

## 3 Jovan Milenković

A map of the *Principality of Serbia* on a scale of 1:300,000, lithographed in the State Lithography on four sheets, was released in 1850 by Jovan Milenković, based on a sketch by the Russian captain Kocebu (Fig. 3). In the top-right part of the map, below the coat of arms in a U-shape, the title *Knjaževstvo Srbije* (Principality of Serbia) is inscribed. The top-left part of the map comprises a graphical scale expressed in geographic miles. The lower-left corner of the map shows a table of the districts with a list of villages and their population. In the lower-right part of the map a list of counties is given. The map has a double black and white frame with degrees of latitude and longitude.

In his work *Source Materials for Cartography and Geography of Serbia*, Ante Aleksić presented information that can help determine the source of the Milenković map. He conveyed a testimony of General F. Zach. Milenković was once designated to follow Captain Kocebu when the Serbo-Turkish-Russian Commission

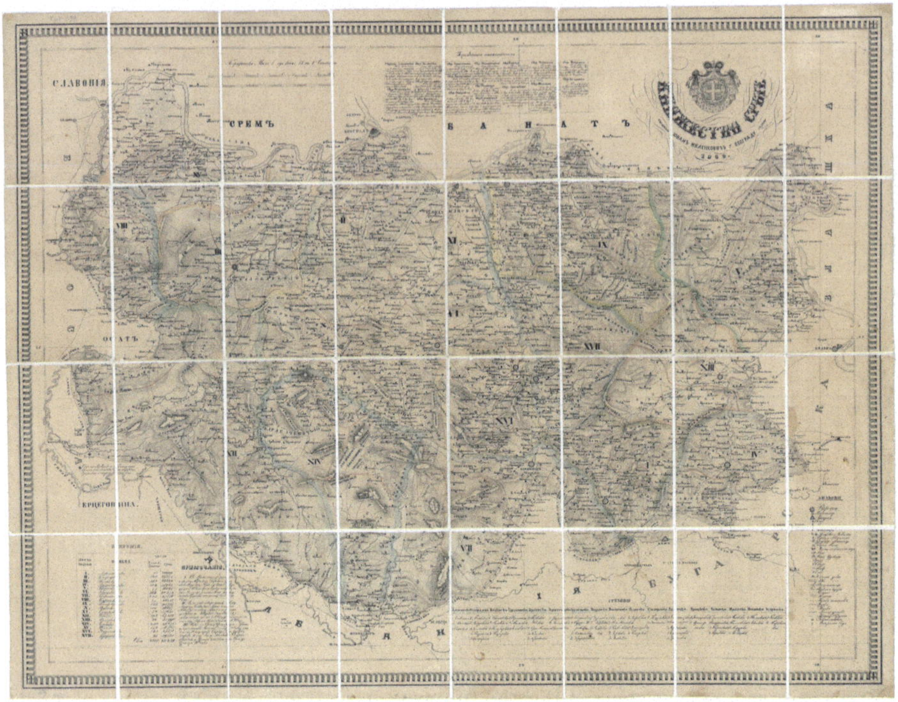


**Fig. 3** Principality of Serbia, Jovan Milenković, 1850 (Kr II 599, National Library of Serbia, Belgrade, Kr II 599)

recorded the borders of Serbia at the time of Prince Miloš. While following the captain, Milenković used the opportunity to quickly copy the sketch of the map at the window pane. In 1881, General Zach donated his library to the Military Academy and on that occasion handed over a draft map of the Principality of Serbia. On the cover of the envelope, General Zach wrote "original document", and the map in the envelope was entitled *Erster Verluch einer Karte von Serbien 1830 von Milenkovitch*. Zach said that Milenković did not have much topographical knowledge, but that this map incorporated many new names of villages, hills, and rivers, which until then were unknown to foreigners. He thus acknowledged the merits of Milenković's work, although overall he assessed the map as bad because the orography was done by a lithographer without much knowledge of the terrain as can be seen from the rivers that go uphill. The plates were done by the stonecutter in the government typography. Milenković was supposed to "allegedly correct or amend" (Aleksić 1883: 261) the work of the lithographer, but he did not have the necessary knowledge to do so, except for knowing the names of villages, hills and rivers. Based on his own observations, Milenković added the rivers and roads of the interior of the country, because Captain Kocebu prepared only a border map with carefully compiled data. The border map of Captain Kocebu was not released to the public so the map of Milenković was an independent publication, without the involvement of the military or interior ministry. The statements by Zach are somewhat unclear. If the map, according to his claims, was made already in 1830, then why was it published only 20 years later?

Aleksić assessed the map of Milenković more positively, because he felt that the map to a great extent explains the geography of Serbia up to the Treaty of Adrianople (1829), as well as the territory that Serbia received as a result of the Treaty. The boundaries of the state and *nahia*, the former principality, as shown on the map of Milenković, were explained by Aleksić like this: "One can see that Serbia was divided into 11 or 12 districts; these were then the nahias. The boundaries of nahias were marked on the map of Milenković, but since they were in certain parts incorrectly and unproportionately drawn, we have drawn the borders of the then nahias based on other contemporary maps, abiding in everything by Milenkovic. The nahias were the following: Šabac, Valjevo, Sokolac, Užice, Rudnik, Požega (Čačak), Kragujevac, Belgrade, Smederevo, Jagodina, Požarevac and Ćuprija or Resava. Milenković did not separate Požega's nahia from Rudnik's nahia, so it looks like they jointly constituted one nahia, but it was not actually like that. . .Milenković did not record on this map our southern border along Uvac but along the mountain range, straight at the distance of Uvac, i.e. from Tornik to Cigota and Murtenica mountain on the original landscape of Vapa so that the very source of this river remains in Serbia. Judging by some points marked on the map, as if there were some variations on the southern border. It must only be considered which of the projected two border lines was better. The question must have been whether a part of Zlatibor with Mokra Gora should belong to Serbia or whether we should establish the boundary over the mere wreath of the mountain. Thus, the Studenica county was bounded in a questionable manner, either belonging to Serbia or to Turkey . . ." (Aleksić 1883: 262–263). In addition to the boundaries shown on the map, Aleksić analysed even the administrative divisions of Serbia. Despite all the deficiencies observed in the topographic accuracy and the mere preparation of the map, Aleksić still called Milenković the first Serbian cartographer.

## 4 Development of Serbian Pedagogical Cartography

The map of Milenković was used for 40 years for education purposes and in all jurisdictions, and served as the only source of knowledge on the territory and the form of the Principality of Serbia. Along with the establishment of cultural and scientific institutions, geographic research and cartographic activities were being developed as well. Civil cartography certainly had a delicate goal—it had to supply the schools and civic bodies with geographic maps and atlases. It developed quite spontaneously and had more of a private character. In the beginning it were for the most part individual geographers and surveyors which were the authors of maps, while the civil government offices only later began their systematic work and produced their first results. Despite all the difficulties, the work of Serbian geographers, surveyors and cartographers points to a continuing desire for progress.

Contributing to the development of Serbian cartography was the establishment of the Lyceum Serbia in Kragujevac in 1838, where, in addition to mathematics and physics, the main subject was practical geometry, i.e. geodesy. The Lyceum was moved from Kragujevac to Belgrade in 1841, and pursuant to Serbian Law, in addition to other courses it introduced a course on physical geography from 1853 onward. Also of great importance for the geographic exploration of Serbia was the establishment of the Society of Serbian Letters in 1842, which published *Glasnik*, containing descriptions of many areas, often with maps attached. All of these were still just attempts of enthusiasts of the still underdeveloped Serbian science, but they paved the way for later, more educated cartographers. In Belgrade in 1853 an atlas was published entitled *Earth maps for church history*, which was lithographed by Kosta Sreplović, and was printed in The State Lithography House in Belgrade. The Atlas contains ten maps measuring 34 × 27 cm. This atlas of Šreplović and the map of Milenković were used in schools and certainly contributed to the development of Serbian cartographic science.

A second period of development was marked by the start of scientific research on European Turkey by Serbian scientists. Numerous works appeared providing geographical descriptions, which is one of the results of the great interest European countries had in the mid-nineteenth century for the Turkish possessions in Europe. Serbia was materially and culturally developed, but the local mapping industry had not yet been able to take on all the work of surveying the country, so it was understandable that some tasks were given to foreigners. In addition to the existing state printing house, the development of Serbian cartography was also aided by the establishment of a private lithographic printing house in Belgrade, especially for the reproduction of town plans and maps.

Numerous maps of Serbia for general, office and school use were published independently or as appendices in professional, scientific or popular geographic works on the geography of Serbia. Most of these maps have no greater importance to the geographic knowledge of Serbia. They were mostly compilations and reworkings of existing maps at a smaller scale, often carried out for commercial reasons, to meet the increased demand for new maps.

## 5 Vladimir Karić

The first map that stood out because of its quality and originality was Karić's *Map of Serbian Country and the Balkan Peninsula*, on a scale of 1: 3,000,000, printed by M. Jorgović in 1884. Jovan Cvijić dedicated his first extensive work (*Geographical research in Kučaj area in the Eastern Serbia*) "to his first teacher in geography, Mr. Vladimir Karić, consul in Skopje". Vladimir Karić produced numerous publications on geography. In his *Geography for High School*, published in Belgrade in 1883, he emphasizes: "There is no class, no science for which geography would not be of use... Geography is not just an assistant to History, but the whole Science of Nature called for help. Without a more extensive geographical knowledge there can be no philologist, no national economist, no politician, no trader, simply no one ...".

After 9 years of work in a school, he published his first textbook. From 1879 until the end of 1893 the students of the third and fourth grade of elementary school and the first to fourth year of high school were studying geography using Karić's textbooks, which were published in several revised and supplemented editions. He also published two school wall maps of the Balkan Peninsula and the Serbian countries. Karić had modern views on the tasks of geography and its separate branches, and can therefore rightly be considered one of the pioneers of modern teaching and one of the first scholarly scientific workers in the field. He also pointed out the significance of maps in geography classes. The extent to which he followed the modern cartographic trends can be best seen on his *Map of the Balkan Peninsula* on a scale of 1:1,000,000, which he compiled according to a then very recent map of Kiepert from 1883 (Fig. 4).

Karić's main work was published in Belgrade in 1887 (Fig. 5). The book entitled *Srbija: opis zemlje, naroda i države* (Serbia: Description of the Land, People and

**Fig. 4** Map of Kingdom of Serbia, Vladimir Karić (Kr II 567, National Library of Serbia, Belgrade Kr II 567)

**Fig. 5** Serbia: Description of the Land, People and the State (front page), Vladimir Karić, Belgrade, 1887 (MAG II 2403, National Library of Serbia, Belgrade MAG II 2403)

the State) was divided into four thematic sections: Earth and its Nature, the People, the State and Neighborhoods and Significant Places. The first part elaborated on the geographical position of the Kingdom of Serbia, its size, borders, mountains, rivers, climate, geology, flora and fauna. The second part described the origin and distribution of the population, its physical characteristics, health status, lifestyle, food, clothing, houses, population of villages and towns, family, traditions, language, religious life, warrior traits, population density and movement of population; then science, literature, art, and scientific and artistic associations; and finally morals, schools, agriculture, mining, industry, trade, currency, measures and traffic. The third part presented the establishment of the Serbian state, the state administration, the state budget, education, church and judiciary. The fourth and final part described the basins of major rivers, fields, towns, historical and cultural sites and others. Karić illustrated his work with four maps and nine cartograms. The unpublished map *Skica geološke karte Kraljevine Srbije* (Sketch of the Geological Map of the Kingdom of Serbia) was relinquished by Jovan Zujović to Vladimir Karić for his book (Fig. 6).

The publication of Karić's book on Serbia was received with great enthusiasm by both local and foreign scholars and experts. The work undoubtedly contributed to the overall geographic knowledge of certain areas of Serbia and is a great source of information on the natural and social reality of Serbia in the late nineteenth century.

Until the end of the nineteenth century the geographical and cartographic publications on Serbia were not subjected to any scientific analysis. In general these works were not compiled by geographers i.e. cartographers, because there were none, but by historians, writers, and officers. However, these men still managed to collect extensive material on the Serbian countries, both in terms of descriptions of nature, the historical development, ethnography and demographic trends, as well as in terms of more accurate and complete representations of the territory on geographical maps. The geographic maps of Serbia were produced and then revised by national authors. The extensive monograph on *Serbia* by Vladimir Karić greatly contributed to the development of a more scientific methodology for conducting geographical research.

## 6 Development of Serbian Military Cartography Within the Geographic Department of the General Staff

Alongside the creation of cultural, scientific and military institutions, both geographical research and Serbian national cartography were being developed. For practical reasons, geographical research in Serbia was primarily in the hands of the military, who acquired the most precious experience with it. The establishment of the Geographical Department of the General Staff in 1876, whose mission it was to prepare a map of Serbia at the largest scale possible, was of a particular importance

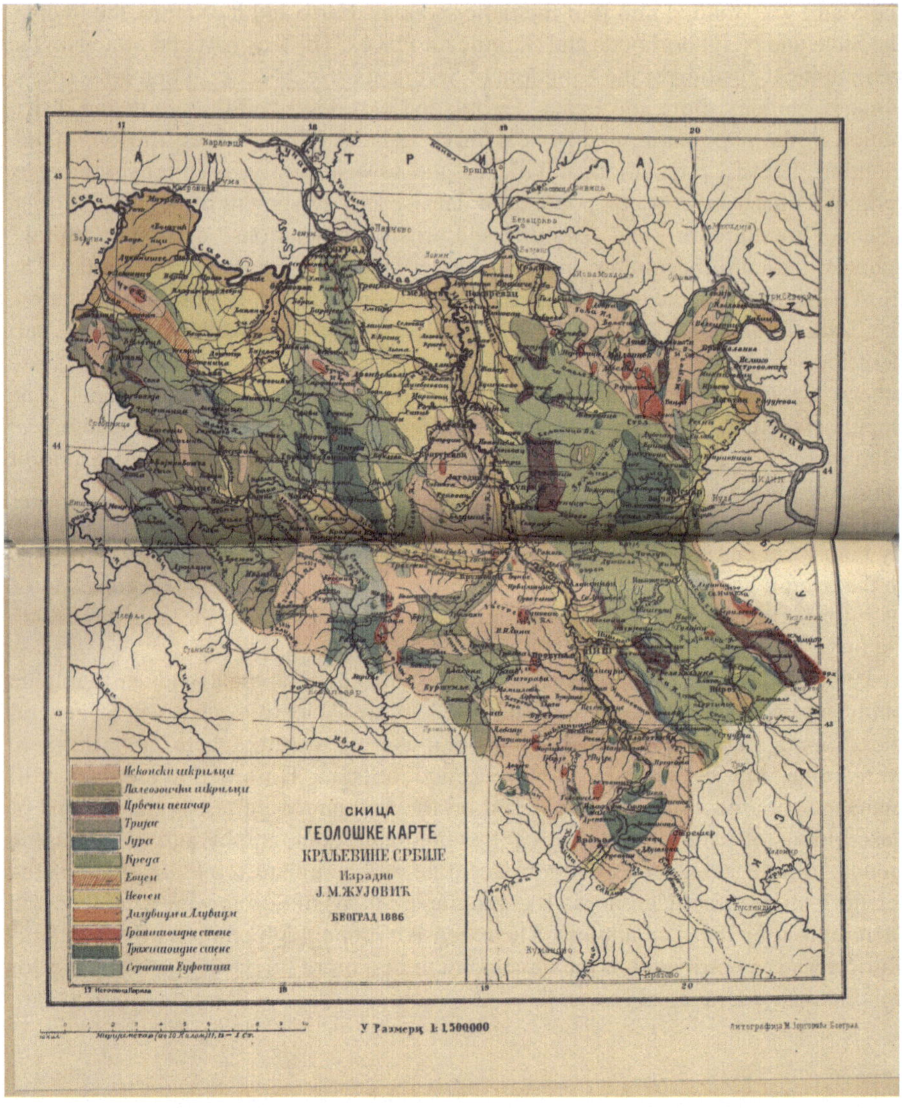


**Fig. 6** Sketch of the Geological Map of the Kingdom of Serbia, Vladimir Karić (National Library of Serbia, Belgrade, MAG II 2403)

not only for the Serbian Army but also for cartography in general, because cartographic activities were now entrusted to a specialized institution. Thus, Serbia followed the process that had already been completed in most European countries, where the government, through military institutions, allocated the funds for the surveying and mapping of its territories. Even with minimal staff (one department head and four officers), but in cooperation with other officers who were being

**Fig. 7** Graphic Sheet Index of General Staff Map of Serbia (National Library of Serbia, Belgrade, Кr II 2)

trained for a general staff profession, the Geographical Department began to, more or less systematically, perform the first topographic surveys of the entire former Kingdom of Serbia. The first exhaustive Serbian topographic map on a scale of 1:50,000 was thus composed and then reproduced in the Geographic Department on a scale of 1:75,000. The map was called *Đeneralštabna karta Kraljevine Srbije* (General Staff Map of Kingdom of Serbia) and had contour lines with a 50 m equidistance (Fig. 7).

Even though it was not exactly made in the most modern manner, it still served its purpose over the course of several decades, giving both soldiers and scientists the opportunity to thoroughly familiarize themselves with pre-war Serbia and in that way solve many practical and scientific problems.

## 7 The End of the Nineteenth Century and the Development of a More Scientifically Based Cartography

At the end of the nineteenth century, Jovan Cvijić played a major role in giving a scientific basis to the further development of Serbian geography, pointing out the direction it needed to take. According to Cvijić the main task of geography was not to offer a description, because this has no scientific value except for areas for which there are no topographical maps. For almost 35 years, Cvijić was consistently striving to conduct his versatile geographical research in a scientific way. Already in 1894, as a professor of the High school, Cvijić founded the Geography Chair, which would become the nursery of scientific geographical thought in Serbia and also more in general a milestone in scientific research work in the country (Petrović 1977: 5).

Serbian cartography of the nineteenth century, which started with a map by Tekelija that was mostly a compilation containing no original data, ended with an exemplary, original, scientifically based work by Jovan Cvijić. Cvijić gathered around him and educated a whole generation of geographers who made great contributions to the field of geography as a science, i.e. in all its disciplines. They created a large number of transparent maps, school wall maps, atlases, and maps with accompanying texts in various geographical publications (V. Karić, J. Dragasević, M. Jorgović, J. Bešlić, M. Ubavčić and others). The development of cartography was especially aided by the establishment of the Geographic Institute at the High School in 1894, because its members became some of the most famous Serbian cartographers.

The completion of the first geodetic—topographic survey and subsequent publication of the General Staff Map of Kingdom of Serbia, followed by the development of academic cartography, enabled the creation of numerous new maps of various content. These were no longer based on unreliable travel accounts or on arbitrary sketches, copies or compilations of previous maps, but rather relied on first-hand knowledge recorded in the field. They were the first true modern geographic maps of Serbia. After the completion of cadastral surveying, all maps were based on the findings of such surveying. They varied in scale and there were technical enhancements as regards printing techniques and use of color, or details were amended, but in general their contents remained the same, while their meaning and intention were changing. Therefore, the end of the nineteenth century is understood as the end of all those influences which substantially changed the image of Serbia on maps. After the cadastral surveying, Serbia's representation on geographic maps was finally given its complete, true and correct form as wished by cartographers over the course of its history.

## References

Aleksić A (1883) Građa za kartografiju i geografiju Srbije [Sources for cartography and geography of Serbia]. Godišnjica Nikole Čupića 5:261–263

Bošković S (1930) Geodetski i kartografski radovi Vojnog geografskog instituta 1878–1930 [Geodetic and cartographic works of The Military Geographical Institute 1878–1930]. Pripremni odbor za III kongres slovenskih geografa i etnografa, Beograd

Petrović D (1977) Srpska geografija XIX veka [Serbian 19th century geography]. Flogiston 3(5):5

Vemić M (2005) Cvijić i kartografija [Cvijić and cartography]. Globus 36(30):159–170

Živković D (2012) Srpska kartografija – od inženjerske do digitalne [Serbian cartography – from engineering to digital]. Glasnik Srpskog Geografskog Društva 92(3):1–10

Živković D (2014) Stanje i perspektiva kartografije u Srbiji [Status and perspective of cartography in Serbia]. In: Grčić M (ed) Geografsko obrazovanje, nauka i praksa: razvoj, stanje i perspektive. Geografski fakultet, Beograd

**Jelena M. Glišović** (1982) works at the National Library of Serbia in the Special Collections Department as the Head of the Cartographic Collection since 2012. She holds a master of LIS degree from the Faculty of Philology, University of Belgrade, and is currently pursuing a PhD degree in cultural studies at the same faculty.

# User-Centered Design of a Collaborative, Object Oriented Historical GI-Platform

**Philippe De Maeyer, Elien Ranson, Kristien Ooms, Karen De Coene, Bart De Wit, Michiel Van den Berghe, Sven Vrielinck, Torsten Wiedemann, Anne Winter, Rink Kruk, and Isabelle Devos**

**Abstract** The project STREAM (Spatio-Temporal Research Infrastructure for Early Modern Flanders and Brabant) aims to create a research infrastructure that will allow spatio-temporal analyses in order to improve our understanding of the demographic, social and economic changes that occurred in Flanders and Brabant (Belgium) between 1550 and 1800. The Carte de Cabinet of count Joseph de Ferraris (1771–1778) offers information on various subjects for that time period and is considered one of the most important products of Belgian cartographic history. Hence this historical map was used as the main source document to develop a vectorial geographical database that constitutes an important step towards the creation of a research infrastructure. To build this geographical database a retrogressive method was used in order to interpret the historical map and its related data in an absolute geographical reference system, which the Carte de Cabinet lacks. Since STREAM results from a collaboration between researchers from different disciplines a specific user-oriented editing platform was developed to support the different actors. This platform allows the digitisation of the historical road network in a geographic reference system based on the current road network by means of a slider, a shift tool and an editing tool. Initial analyses have confirmed the strong geometric distortions of the Carte de Cabinet but also the multiple possibilities for spatio(-temporal) research when combining the information of the Carte de Cabinet with cartographic analyses of other cartographic documents.

---

P. De Maeyer (✉) • E. Ranson • K. Ooms • K. De Coene • B. De Wit • M. Van den Berghe
Department of Geography, Ghent University, Ghent, Belgium
e-mail: Philippe.DeMaeyer@ugent.be

S. Vrielinck • T. Wiedemann • I. Devos
Department of History, Ghent University, Ghent, Belgium

A. Winter
Department of History, Vrije Universiteit Brussel, Brussels, Belgium

R. Kruk
National Geographic Institute, Brussels, Belgium



M. Altić et al. (eds.), *Dissemination of Cartographic Knowledge*, Lecture Notes in Geoinformation and Cartography, DOI 10.1007/978-3-319-61515-8_22

# 1 Objective

Belgium has a rich cartographic heritage going back to the beginning of the Renaissance. One of the most important products of Belgian cartographic history is the so-called *Carte de Cabinet* of count Joseph de Ferraris (1771–1778) (Fig. 1), realised on a scale of about 1:11,520 (De Maeyer and De Coene 2016). This manuscript map of 275 sheets covers more or less the current territory of Belgium, which coincided at that time with the Austrian Netherlands, the Prince-Bishopric of Liège and the Duchy of Bouillon. As such it is the only source that provides a very detailed overview of the delineation and topography of the Belgian localities in the late eighteenth century.

This map contains important information on various subjects (toponymy, physical landscape, administrative areas, road system, waterways, etc.), as was also proven by the map key created by De Coene et al. (2012). However, it lacks a reliable geometrical and geodetic foundation. The map was created by simple plane table surveying, without the support of a national geodetic framework of precisely located points obtained through triangulation (Vervust 2016a). This means that the map cannot simply be transformed to a modern coordinate reference system and it resulted in local errors such as deformations in the orientation of the road system and hydrographic network, incorrect paths, surface distortions, lack of systematic cartographic semantics, etc. A thorough and detailed study of the geometric accuracy of the Carte de Cabinet has been carried out by Soetkin Vervust as part of her doctoral research at Ghent University (Vervust 2016b; Vervust et al. 2017).

**Fig. 1** Detail of Carte de Cabinet sheet 17. Reproduced with permission A3442 of the National Geographic Institute—www.ngi.be

Despite these shortcomings, the Carte de Cabinet remains an important historical geographic source for spatio-temporal analyses (Beyaert et al. 2006). These analyses are only possible if the appropriate tools are available to create, manage and query a vectorial dataset. Therefore an object-oriented vector GIS is needed.

The vectorisation of the Carte de Cabinet is part of the project STREAM which stands for Spatio-Temporal Research Infrastructure for Early Modern Flanders and Brabant (Belgium). STREAM is funded by the Hercules Foundation and consists of a collaboration between the departments of Geography and History of Ghent University, the department of History of the *Vrije Universiteit Brussel* (VUB), the National Geographic Institute (NGI) and the Belgian State Archives.

STREAM was developed by a team of historians, geographers and IT specialists in order to (1) create a series of high-quality datasets relating to the population movements and the social and economic changes that occurred in Flanders and Brabant between 1550 and 1800; (2) to develop a vectorial geographical database to allow spatio(-temporal) analyses of these socio-economic data for these specific areas during that specific period; and (3) to ultimately improve our understanding of regionally and locally differentiated economic, social and demographic developments at the time.

As a first step, important collections of early modern quantitative data (e.g. number of domestic animals, mills, churches, farms, etc.) have been localised, critically assessed, digitised and harmonised after which they were added to information systems with statistical as well as other implementations. The second step was creating a spatio-temporal reference system consisting of vectorised road systems, hydrographic networks and other geographical features with well-known locations, which makes it possible to link the very diverse datasets accumulated in the first step to it, and to analyze and compare them.

Two approaches are possible when vectorising geographical features: (1) the original geometry of the historical map is respected or (2) the historical map and related data are fitted into an absolute geographical reference system. For the STREAM project both approaches were realised in order to answer different research questions. The first approach, referred to as *Ferraris Vectorized*, offers historians a quick tool that is less time consuming to create and respects the source map but has the disadvantage of being spatially incoherent and of therefore not resulting in a true geographical database. In this paper this approach is not discussed further. The second approach, referred to as *Ferraris Georeferenced*, is explained in the following sections. The method is strongly reliant on a correct interpretation of the historical map but when this is achieved, offers better possibilities for spatio (-temporal) analyses because of its geographical accuracy and seamless database.

The development of this vectorial geographical database is a key step towards the creation of a historical spatio-temporal research infrastructure. The information in the database can be used as an offset to localise other data in time and space and to make the data available for spatio-temporal analyses, which offer numerous research possibilities for various disciplines.

## 2 Methodology

To build this vectorial geographical dataset the geometry of the historical Carte de Cabinet (in a raster format) is compared to the accurately positioned vector data of the current topographic map of Belgium Top10-V on a scale of 1:10,000, produced by the Belgian national mapping agency (National Geographic Institute). More specifically it is the current road network which is used as a reference canvas to structure the historical cartographical information in a geographically accurate way, because it is one of the most stable and abundant features on the map. To start the work each map sheet of the Carte de Cabinet was roughly positioned onto the modern map through an affine transformation of its four vertices, after which the vectorised modern roads were used as a reliable pattern for the more detailed comparison. The goal is to ultimately digitise several semantic categories on the Carte de Cabinet, but so far work has been focused solely on digitising the historical road network.

To support the different actors in this project a specific user-oriented editing platform (Fig. 2) was developed to realise the digitisation of the road network. The platform includes a set of tools that are tailor-made so researchers from different fields do not have to immerse themselves in all functions available in a typical GIS-platform, which results in a limited learning curve. Another advantage of the platform is that it makes it easy to compare different source images (historical maps, aerial photographs, soil maps, etc.). It also minimises the risk of data loss through the use of an automatic save function and central back-up-system which registers all actions by the different operators so adjustments can be made later. Thus, the platform allows for all maps to be consulted by multiple people in a practical manner and is the first step towards a spatio-temporal research infrastructure.

**Fig. 2** Overview of the user-oriented editing platform

## 2.1 *Tools*

### 2.1.1 The Slider

Given that the current situation and the one on the Carte de Cabinet have a time gap of 240 years, it is necessary to take other historical cartographic documents (topographic maps from the nineteenth century, early cadastral maps, the present-day soil maps, (old) aerial photographs, etc.) into account when interpreting the geographic features of the Carte de Cabinet. To obtain the best overview, the sources are visualised in multiple windows, if necessary on two separate screens. There is a synchronisation of the cursor in the different windows and within the same window it is possible to compare the Carte de Cabinet with another historical or contemporary map by means of a slider.

A particularly interesting document to consult is the Popp map (1842–1879), named after Philippe Christian Popp, one of its producers. This cadastral map was created in the nineteenth century and its road system resembles the present one in great detail. This map was integrated in the software as a separate layer, which can be consulted by sliding the Carte de Cabinet to the left by using the slider (Fig. 3). Because the Popp map is more recent than the Carte de Cabinet it is safe to assume that current roads missing on the Popp map could also not have existed at the time of the Carte de Cabinet. The Popp map is therefore useful to conduct a first road analysis in which every not-corresponding road is classified as "*delete*" (see 2.1.3).

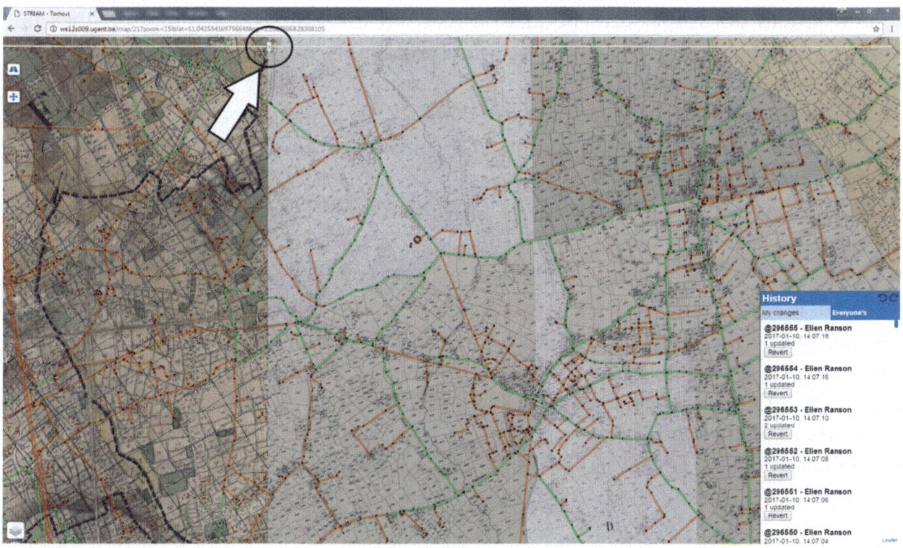


**Fig. 3** Comparison between the Carte de Cabinet and the Popp map by use of the slider

### 2.1.2 The Shift Tool

The goal of STREAM is to obtain accurately positioned vectors that depict the historical road network and can be used for further research and analyses. From this point of view, it would not be useful to conduct a traditional georeferencing task, i.e. to indicate homologous points in order to obtain a static, accurately positioned raster image of the historic map as the end result. Instead, the georeferencing for this project required indicating homologous networks. These cautiously chosen historical network segments can in this way be fitted onto the modern road network.

The selection of the homologous networks can be simplified by using the shift tool. This tool allows the Carte de Cabinet to be moved locally in relation to the current road network (the vectorial Top10-V layer), which is geometrically accurate. Hence, the interpretation of the historical network is much easier (Fig. 4).

### 2.1.3 The Editing Tool

Once a map sheet of the Carte de Cabinet has been roughly positioned on the modern map by georeferencing its vertices, the current road network is visualised by purple[1] vector lines, meaning that these roads are in a "*pending state*". In case of a strong resemblance—based on visual interpretation—between a present-day road segment and one on the Carte de Cabinet, the purple vector line is converted into a "*confirmed state*" and consequently turns green. This conversion happens by clicking the right mouse button and choosing the option "*confirm*". If a current road segment does not correspond to a road on the Carte de Cabinet or does not appear on it, the option "*delete*" is assigned and the purple line turns orange (Fig. 5). Every choice can be revised at any time by simply choosing another option. Once all vector lines have been classified as "confirm" or "delete" a clear picture appears of which current roads were present or absent at the time of the Carte de Cabinet (Ranson et al. 2016) (Fig. 6).

In addition to the options "*confirm*", "*delete*" or "*pending*" it is possible that a historical road does not have a corresponding vector line. In that case the historical road has to be added to the current road network by means of an editing tool. This tool also allows to divide larger road segments into smaller ones. This is necessary when a certain part of a road segment corresponds to a historical road while another part does not. By dividing the road, each new segment can be classified as "*confirm*" or "*delete*". Each segment can be divided as many times as needed. The lines drawn to divide existing roads are not part of the road network and need to be removed. For this purpose an additional option "*destroy*" can be chosen by clicking the right mouse button (Ranson et al. 2016) (Fig. 7).

[1]The colour can be adjusted.

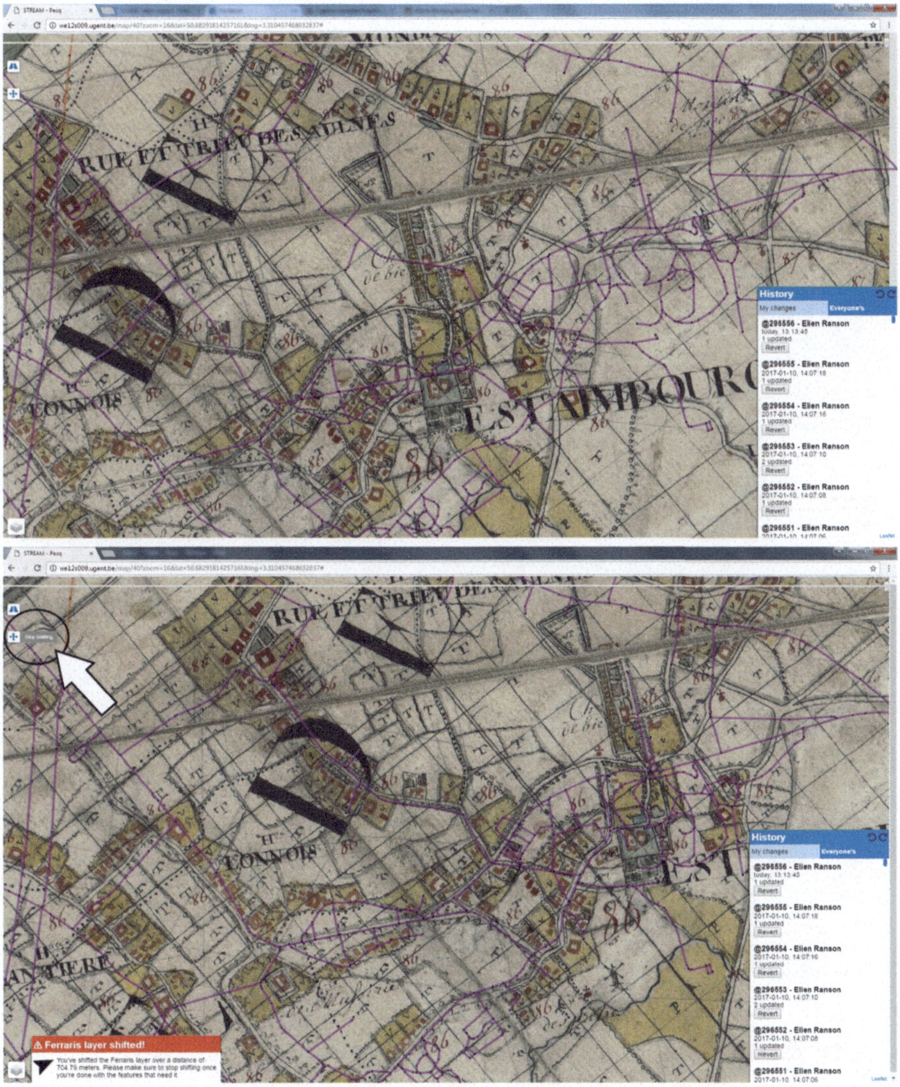


**Fig. 4** Local movement of the Carte de Cabinet in relation to the current road network visualized by *purple lines*

### 2.1.4 Tool to Verify the Integrity of the Data

After completing a series of Carte de Cabinet map sheets, the results need to be verified. Errors can occur due to (1) flaws in the vector data of the current road network, (2) the visual interpretation and (3) the software. To verify the integrity of the data an algorithm has been created to detect roads that are not connected to other roads, thereby creating an *island* that is not attached to the overall road network.

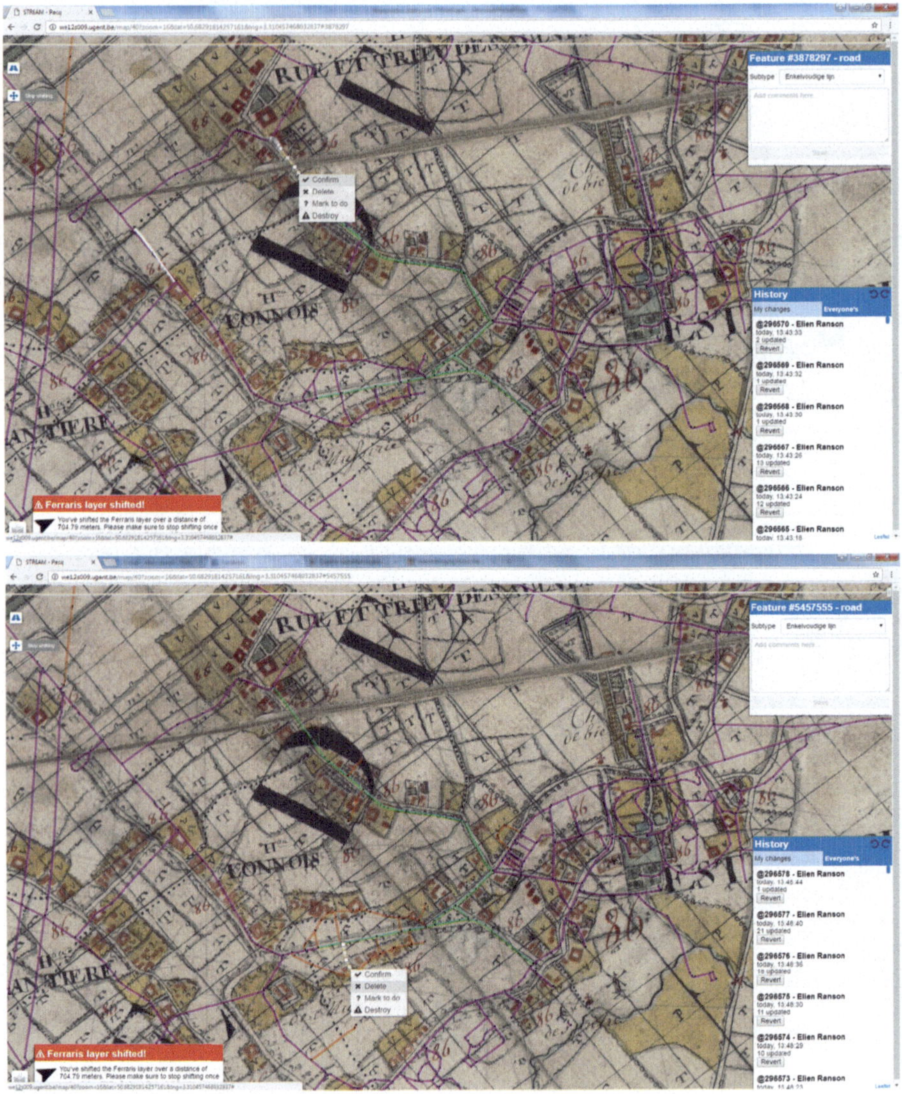


**Fig. 5** Confirming or deleting historical roads

The algorithm also makes it possible to expose undershoots. These errors[2] appear when a digitised line does not connect properly with the neighbouring line thereby creating a gap between roads that should intersect. After the verification of the integrity has been completed, the discovered errors can be resolved. In some cases

[2]Undershoots are in a few cases not actual errors, for instance when dealing with dead-end streets.

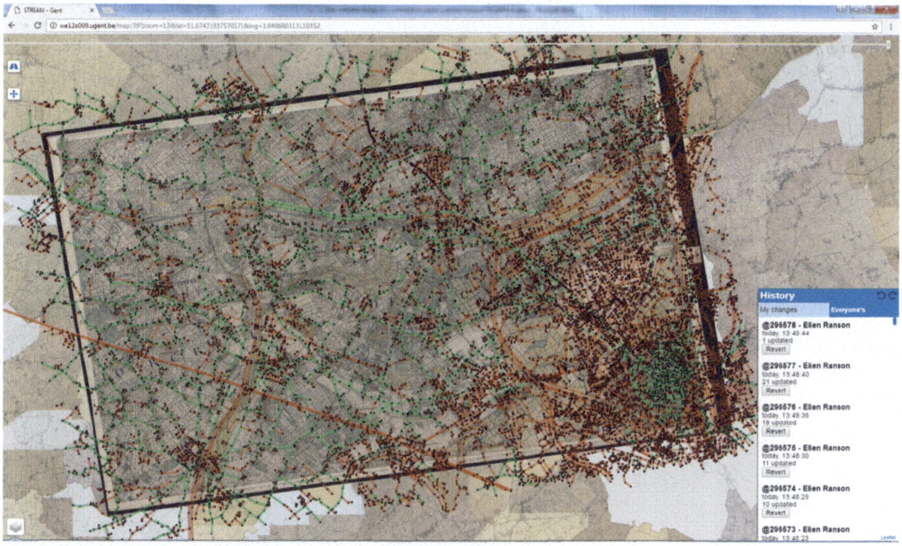


**Fig. 6** View of the historical roads in the vicinity of Ghent

the correction can be done automatically. If not, a manual correction is necessary by inspecting the points that were indicated by the algorithm as possible mistakes.

## 3 Results

The geometric distortions of the Carte de Cabinet can be partly compensated by using the shift tool which allows the Carte de Cabinet to be moved in relation to the current road network. This shift tool could be extended with a rotation tool which would also allow the rotation of the Carte de Cabinet, thereby increasing the chance for a correct interpretation.

Despite the possibilities of the shift tool, the final results are greatly influenced by the geometric inaccuracies of the Carte de Cabinet. A first problem appears when certain historical roads do not perfectly correspond to the current road network but the general shape is present on the historical map. In this case it is useful to consult the Popp map. If the current roads correspond perfectly with those on the Popp map, these roads are classified as "*confirm*" since it can be assumed that the differences on the Carte de Cabinet are due to measuring flaws. If the Popp map shows little resemblance to the current roads, then these roads are classified as "*delete*" since the presence of the general shape on the Carte de Cabinet will then most likely be mere coincidence (Fig. 8).

A second problem is that, despite some preconceived rules, the cartographic and historical analysis remains a process of interpretation. Even when certain roads seem to correspond with one another, it is still possible that this correspondence is

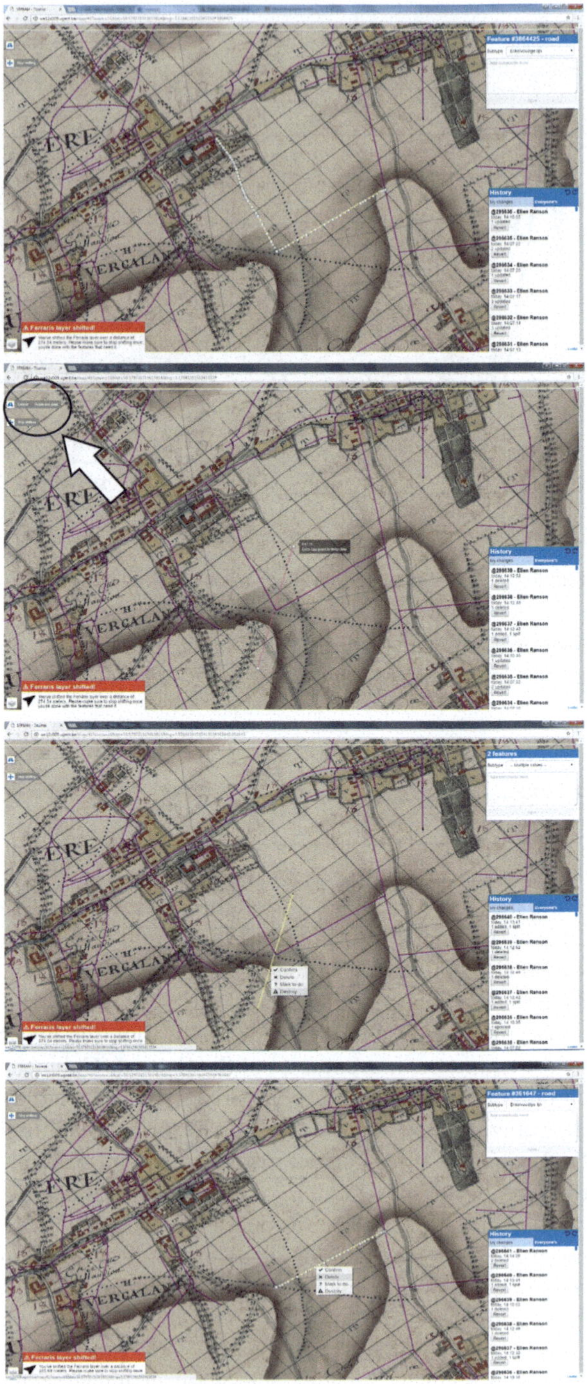


**Fig. 7** Dividing a road segment using the editing tool

**Fig. 8** Interpretation of the historical road network based on the Popp map

an unfortunate coincidence. Decisions based on visual resemblance therefore need further confirmation by combining the visual interpretation with cartographic analyses of other cartographic documents.

# 4 Conclusions

The objective of STREAM is to create a research infrastructure to allow spatio-temporal analyses in order to improve our understanding of the demographic, social and economic changes that occurred in Flanders and Brabant (Belgium) between 1550 and 1800. To reach part of this aim a vectorial geographical database was built which allows the interpretation of the Carte de Cabinet and its related data in an absolute geographical reference system. This approach, also referred to as *Ferraris Georeferenced*, uses a specific user-oriented editing platform that supports researchers from different fields who are not specialised in GIS. This platform includes tailor-made tools that are easy to use and lead to fast results. Source documents can for example be visualised in multiple windows, making it easy to compare different map images which in turn improves the results. Despite some problems that arise from flaws in the vector data of the current road network, the visual interpretation and the software, STREAM offers multiple possibilities for spatio(-temporal) research in various disciplines.

**Funding** The STREAM project is possible thanks to funding from the Hercules Foundation, Ghent University and the *Vrije Universiteit Brussel* (VUB) and the support from the National Geographic Institute and the Belgian State Archives.

## References

Beyaert M, Antrop M, De Maeyer P, Vandermotten C, Billen C, Decroly J, Neuray C, Ongena T, Queriat S, Van den Steen I, Waeyens B (2006) België in kaart. De evolutie van het landschap in drie eeuwen cartografie. Lannoo, Tielt

De Coene K, Ongena T, Stragier F, Vervust S, Bracke W, De Maeyer P (2012) Ferraris, the legend. Cartogr J 49(1):30–42

De Maeyer P, De Coene K (2016) 240 jaar landsdekkende grootschalige basiscartografie in België. Cartogr Visie op de Kaart Jaarboek 15:22–41

Ranson E, Ooms K, De Maeyer P, De Coene K, De Wit B, Van den Berghe M, Thoen E, Vrielinck S, Wiedeman T, Kruk R, Lambrecht T, Winter A, Devos I (2016) STREAM. Een tijdruimtelijke onderzoeksinfrastructuur voor vroegmodern Vlaanderen en Brabant: de vectorisatie van de Kabinetskaart van Ferraris. Cartogr Visie op de Kaart Jaarboek 15:279–285

Vervust S (2016a) Deconstructing the Ferraris maps (1770–1778): a study of the map production process and its implications for geometric accuracy. Ghent University, Ghent

Vervust S (2016b) Count de Ferraris's maps of the Austrian Netherlands (1770s): Cassini de Thury's geodetic contribution. Imago Mundi Int J Hist Cartogr 68(2):164–182

Vervust S, Claeys Boùùaert M, De Baets B, Van de Weghe N, De Maeyer P (2017) A study of the local geometric accuracy of count de Ferraris's *Carte de cabinet* (1770s) using differential distortion analysis. Cartogr J (forthcoming)

**Philippe De Maeyer** is currently senior full professor in cartography and GIS at Ghent University (Belgium). He holds a Master degree in Geography, a Postgraduate degree in Geology and Marine Applications, a PhD in Sciences and a Postgraduate certificate in Business Economy. Since 2008 he is full professor and was appointed head of the Department of Geography. His research focuses on applied GIS, especially the use of GIS and remote sensing in archaeology and risk management, indoor routing (algorithms, landmarks) and the use of GIS in historical (map) studies. Furthermore, he is involved in several projects overseas, especially in Yucatan (Mexico) and Xinjiang (China) and he is the Belgian promotor of the "Sino-Belgian Joint Laboratory for Geo-Information", a partnership between the Xinjiang Institute of Ecology and Geography (Chinese Academy of Sciences) and Ghent University.

The manufacturer's authorised representative in the EU is Springer Nature Customer Service Centre GmbH, Europaplatz 3, 69115 Heidelberg, Germany. If you have any concerns regarding our products, please contact ProductSafety@springernature.com

Printed and bound by CPI Group (UK) Ltd, Croydon, CR0 4YY

25/09/2026

02222882-0002